BIOMEDICINE AND THE HUMAN CONDITION

How to avoid disease, how to breed successfully, and how to live to a reasonable age are questions that have perplexed mankind throughout history. This book explores our progress in understanding these challenges and the risks and rewards involved in our attempts to find solutions. Nutritional experiences and exposure to microbes and alien chemicals may have consequences that are etched into our cells and genomes. Such events have a crucial impact on development *in utero* and in childhood and later affect the way we age, how we respond to infection, and the likelihood of our developing chronic diseases, including cancer. The issues covered include the powerful influence of infectious disease on human society, the burden of our genetic legacy, and the lottery of procreation. The author discusses how prospects for human life might continually improve as biomedicine addresses these problems and also examines the ethical checkpoints encountered.

Michael G. Sargent is a research scientist in developmental biology at the National Institute for Medical Research at Mill Hill, London.

Biomedicine and the Human Condition

Challenges, Risks, and Rewards

Michael G. Sargent

The National Institute for Medical Research
Mill Hill, UK

 **CAMBRIDGE**
UNIVERSITY PRESS

CAMBRIDGE UNIVERSITY PRESS
Cambridge, New York, Melbourne, Madrid, Cape Town, Singapore, São Paulo

Cambridge University Press
40 West 20th Street, New York, NY 10011-4211, USA

www.cambridge.org
Information on this title: www.cambridge.org/9780521833660

© Michael G. Sargent 2005

First published 2005

Printed in the United States of America

A catalog record for this book is available from the British Library.

Library of Congress Cataloging in Publication data

Sargent, Michael G., 1943–
Biomedicine and the human condition : challenges, risks, and rewards /
Michael G. Sargent.
   p. cm.
Includes bibliographical references and index.
ISBN 0-521-83366-3 – ISBN 0-521-54148-4 (pbk.)
1. Health – Philosophy.   2. Medicine – Philosophy.   3. Bioethics – Philosophy.
4. Philosophy of mind.   5. Philosophy, Modern.   6. Philosophical anthropology.
I. Title.
R723.S223   2004 2005
610'.1 – dc22   2004050103

ISBN-13   978-0-521-83366-0 hardback
ISBN-10   0-521-83366-3 hardback

ISBN-13   978-0-521-54148-0 paperback
ISBN-10   0-521-54148-4 paperback

To Jean and Catherine

who brighten the dullest day

Contents

Preface

How to breed successfully, how to avoid disease, and how to live to a decent age are questions that have perplexed our ancestors throughout recorded time. As humans explored new lifestyles and habitats, each new challenge – whether it was agriculture, urban living, colonisation of new territories, domestication of animals, or industrialisation – could have notable rewards but was usually fraught with unpredictable physiological penalties. The history of our species is marked by technical solutions that have made these problems of human biology bearable. Some originated in common sense – the piped water and closed sewers of the nineteenth-century metropolis – and others in the application of science, but even these could sometimes produce less-than-satisfactory outcomes and, in some instances, disaster.

The idea of this book is to examine some of these adventures through the lens of twentieth-century biomedicine and to identify the risks and the rewards involved in each. During the first decade of the last century, crucial developments were afoot; philanthropists and eventually governments were recognising the importance of biomedical science and beginning to devise a financial infrastructure that could support its progress. This process gained an irresistible momentum after the Second World War, when the American government undertook unprecedented investment in the life sciences. Within three decades, profound insights into inheritance and cell biology gave us the powerful tools needed to establish a gene technology industry. At the same time, a better understanding of the glorious mechanics of human biology provided a framework in which the hitherto baffling and intractable mysteries of chronic disease and ageing could be investigated.

I have begun this story of how our species gained such extraordinary control of our biological destiny with the perceptions of an ancient people recorded in the book of Genesis. This may have been unwise, according to a publisher of my acquaintance, because biologists are supposed to

be neurotically hostile to biblical allusions! Their palpitations notwithstanding, I hope my readers are open-minded enough to empathise with
Genesis as a literary artefact that speaks, almost uniquely, of the biological
hopes and fears of ancient times and that gives a certain perspective to
the narrative. The preoccupations of the ancient Hebrews seem not unlike our own, and their oldest tribal myth indicates that they understood
with extraordinary prescience the potential of human intelligence to make
choices that could be hugely beneficial or catastrophically ugly. Their conception of an almighty Deity set the ethical framework for a large swathe
of humanity for more than two millennia and was taken as the explanation
for every cruel twist of fate. Today, secular societies feel no moral obligation
to accept the cruelty of fate and indeed believe it is a humane and proper
impulse to try to overcome hitherto insoluble biological problems, such as
genetic disease and infertility. The acceptability of such innovations is now
considered using ethical frameworks generated by public bodies encom-
passing all strands of society, a process in which the chief issue is whether
any individual is harmed by our actions.

What is "biomedicine"? It is a word that evokes glittering prospects
for clinical science; a notion that, taken literally, would limit discussion
to recent and future developments affecting human health. However, for
lack of a better word, I shall stretch its purview to include every influence
on human biology that made us the long-lived creatures we are today.
In the nineteenth century, this meant the emergence of a political will to
control public health and social welfare, which was informed by scientific
knowledge only latterly. Whatever benefits the developed world may gain
from future clinical successes – promised today with such incontinence –
the vast majority of humanity still struggles to recapitulate the biomedical
progress made in the industrial world in the last century. A wish list for the
Third World would probably place piped water, closed sewers, moderate
population growth, sufficient food, and an adequate national income as
high as vaccines against HIV and malaria.

The issues discussed here are an outline of some core themes of human
biology. Paradoxically, though, students of biology preparing for university entrance in Britain, and probably elsewhere, touch upon many of these
matters only tangentially. Their curriculum is a rigorous preparation for
specialised university courses such as medicine and biochemistry that dwell
on fundamental principles, leaving little space for anthropocentric reflection or frothy ethical debate. One suspects that a substantial proportion of
students without focussed ambitions in science, as well as many laypersons,

would welcome the opportunity to consider personally relevant issues of human biology without undertaking rigorous schooling in the entire subject. A rigorous introduction to biology – fascinating and rewarding though this may be – is not necessary to develop an appreciation of how current developments in biology affect society. In another utilitarian age, Charles Dickens began his novel *Hard Times* with Thomas Gradgrind enunciating his singular testament: "*Now, what I want is, Facts. Teach these boys and girls nothing but Facts. Facts alone are wanted in life. Plant nothing else and root out everything else. You can only form the minds of reasoning animals upon Facts: nothing else will ever be of any service to them.*" One hopes that nothing so grotesque informs education today, but Gradgrind's spectre remains a baleful presence. In Britain, the Nuffield Foundation is developing a curriculum on the social impact of science, aimed at pre-university students who will not become scientists but who want to acquire a rational and well-informed view of scientific developments. My book was written in a similar spirit, for a general reader interested in learning about biomedical innovation in its historical and cultural context without the burden of detail. For the same general reader, I frequently refer to news articles in *Nature* or *Science*; I hope their succinct and accurate reviews of current issues will satisfy requirements for traceability of key facts not cited. Many more primary sources were consulted than are listed here, but the book would not bear the weight of such documentation. In the text, references are given as numbered superscripts keyed to the reference list.

In acknowledging help received in writing this book, I must particularly thank the historian Professor James C. Riley of the University of Indiana, who politely alerted me to the dangers of triumphalism based on clinical science. His own fascinating and original research into the global history of rising human life expectancy reveals the remarkable variety of ways in which this has been achieved (see p. 144). The resources allocated to the paraphernalia of modern medicine, public health management, and local self-help activities such as improved parenting vary widely from country to country. In some places, remarkably high life expectancy has been achieved without much investment in clinical medicine. I also gratefully acknowledge the constructive criticism of several anonymous reviewers of my original proposal to Cambridge University Press, the expert copyediting of Russell Hahn, and my friends Bill and Judy Webster, who read the manuscript in its entirety. I also greatly appreciate the continued support of Katrina Halliday, Kirk Jensen, and the many people at Cambridge University Press who have made this book possible.

1   Challenges, Risks, and Rewards: Learning to Control Our Biological Fate

When our hunter-gatherer ancestors exchanged their timeless wanderings for the momentous opportunities of agriculture, they adopted a lifestyle of which they had had no previous evolutionary experience. This had significant consequences for their nutrition, their reproduction, their susceptibility to infection, and for the development of their children. As they became the dominant life form on the planet, they gained increasing control of the environment, but an understanding of the biology of their own species has eluded them until comparatively recently. In this introductory chapter, we shall explore episodes in our history in which remarkable institutions were established that made possible unprecedented insights into human biology.

Leaving Eden

For more than a hundred millennia our species has roamed the earth multiplying slowly, subsisting on animal and plant food, constantly adapting to changes in climate and to potential prey. We know little of their lives except that the Cro-Magnon people, the immediate ancestors of Europeans, hunted big game with what seems a mystical enthusiasm and buried their dead with some reverence. By the end of the last Ice Age, the herds of deer had dwindled away and human populations were reduced at one point to perhaps just 10,000 pairs, while our Neanderthal cousins became extinct.[1] When the ice receded, the survivors' frugal existence, dependent on small animals and plant foods, began to improve as they learnt to grow plants and to manage domesticated animals.

In the Euphrates valley, a new kind of grass appeared, with seeds that ripened in ears on a stalk. The seeds were not dispersed by the wind; they

1

could be harvested efficiently and used to start another crop in the follow-ing year. Once our ancestors had learnt to sow, harvest, and store grain, the timeless hunter-gatherer lifestyle gradually ended. Families formed permanent settlements, based on organised cultivation of plant foods, and in the good years, when food was plentiful, they would multiply in a way that was impossible for hunter-gatherers. The new lifestyle con-ferred substantial benefits, but it had consequences to which our ances-tors were ill-adapted. Food rich in calories was there in abundance in the good years, but when they abandoned the highly varied foodstuffs of their hunter-gatherer past, their diet became monotonous and contained fewer micronutrients. The archaeological evidence suggests that calcium-sequestering chemicals such as phytic acid – abundant in cereals and usually removed in modern grain processing – may have weakened their bones. Tooth decay and serious damage to tooth enamel also make their ap-pearance in skeletons of the period, reflecting the use of refined dietary carbohydrate and stone-ground cereals. We know, too, from certain dis-tinctive markings of the skeletons that reflect deficiency of protein, vita-min C, and folic acid[2,3] that famine was a constant presence. The eternal battle between humans and epidemic-causing microbes started in those times.

Archaeologists can objectively reconstruct the lives of our ancestors from skeletons and artefacts, but these tell us little about the thoughts of ancient people. To add a vivid and recognisably human dimension to those far-off times, we can look at the best-known fable of the origins of humankind, the book of Genesis. This work, of complicated provenance, derives in part from much older Canaanite and Sumerian texts and even from myths that date from the Stone Age. It conveys in a few terse sentences the situation of the first family after the expulsion from Eden (see Fig. 1.1), desper-ately grubbing a living from the soil, constantly aware of their mortality. However, it is the complicated metaphor of the forbidden fruit, with its strangely prescient conception of the power of knowledge, that is the re-ally curious idea. The serpent promises the first couple an extraordinary revelation if they eat the fruit. *"In the day ye eat thereof then your eyes will be opened and ye shall be as Gods knowing good and evil."* The fruit seems to be a visa to an alternative world beyond Eden, full of extraordi-nary rewards and terrible risks, choices that could be hugely beneficial or catastrophically ugly. Milton, a man profoundly versed in biblical exege-sis, had no doubt, in *Paradise Lost*, about the significance of the Tree: "O

> **The scene** – The Garden of Eden in which Adam and Eve, "naked and unashamed," are destined to live an idyllic and undemanding existence for eternity.
>
> **The crime** – The Almighty warns Adam and Eve that if they eat the forbidden fruit of the "Tree of Knowledge of Good and Evil," they will become mortal and inevitably die. The Serpent entices Eve to ignore the warning with the promise, *"In the day ye eat thereof then your eyes will be opened and ye shall be as Gods knowing good and evil."*
>
> **Retribution** – Famously, the guilty pair become conscious of their nakedness and then God summarily ejects them to ensure that they cannot *"take also of the Tree of Life, and eat, and live for ever."*
>
> **Adam's curse** – *"Cursed is the ground for thy sake; in sorrow shalt thou eat of it all the days of thy life. Thorns also and thistles shall it bring forth to thee; and thou shalt eat of the herb of field. In the sweat of thy face shalt thou eat bread till thou return to the ground"* (see Chapter 6).
>
> **Eve's curse** – *"I will greatly multiply thy sorrow and thy conception; in sorrow thou shalt bring forth children; and thy desire shall be to thy husband, and he shall rule over thee"* (see Chapter 2).

Figure 1.1. "Your eyes will be opened."

sacred, wise and wisdom giving plant, Mother of Science, now I feel thy power."

The first episode after the expulsion – the strange story of Cain and Abel – is more than a tale of a dysfunctional family. It is an allegory that fixes our attention on the clash between a new farming culture and an older nomadic life. Adam and Eve's sons – Abel, the "keeper of sheep," and Cain, the "tiller of the ground" – offer to the Almighty the first fruits of their labour, but only Abel's are "accorded respect." Cain, in a jealous rage, murders his brother and flees. Adam and Eve put aside their disappointment and start again with their third son, Seth, from whom the patriarchs are descended (Genesis 4:2–15). Apprehensive of the agricultural future, the narrator's sympathies are evidently with the herdsman. Similar sentiments inspire ancient Indian texts.

As the stories unfold, we learn how the Hebrews are troubled by fundamental questions about human existence – how to breed sensibly and successfully, how to avoid disease, and how to live to a decent age. With the emergence of Moses as a leader in touch with the Almighty, they get an answer in the form of "the Law." Fiercely and uncompromisingly, he demands of his followers the strictest obedience or otherwise risk the wrath of the Almighty in the form of pestilence, famine, and infertility. The dire warnings of retribution notwithstanding, "the Law" becomes a patently humane set of rules for organising society, even down to sensible strictures about public health.

In the absence of a better idea, the Judeo-Christian tradition has concluded that natural disaster and personal suffering reflect a divine purpose. Indeed, until the time of Charles Darwin the Christian world considered no other explanation of the vicissitudes of human life. Many people reject the idea that "design" plays any part in the origin of life and yet may say inadvertently that "nature never intended us to . . ." or "we were not designed to . . .", a snare I will explicitly avoid. I follow the conventional evolutionist's view that every human physiological function, no matter how miraculous it may seem, has evolved to fit its current role. We will see that these functions are not perfectly "designed" in one important respect; insufficient resources are available to ensure their indefinite maintenance (Chapters 7 and 8).

The Bible has been used frequently as an authority to condemn a variety of human innovations, most notably anaesthetics in childbirth, blood transfusion, heart pacemakers, the use of cadavers for medical research, and heart transplantation. Today, the chief issue in resolving ethical questions is whether harm or benefit might ensue from our actions. Darwin was acutely conscious of how the forces that drove evolution were another kind of explanation of grief and suffering in human life, every bit as dreadful as the arbitrary rage of the Judeo-Christian God. Which view was most comforting was a moot point. In later editions of the *Origin of Species*, he borrowed a phrase from Herbert Spencer, "the survival of the fittest," which was later used by Social Darwinists to glorify the perpetual competition they believed would bring social progress. Darwin had no sympathy with this view and in *The Descent of Man* asserted that human moral sense differentiated us from every other animal and directed us to support the weakest of our species.[4] Although he never developed this position, his philosophy underpins the use of biomedicine to provide solutions to human misfortune.

Challenges, Risks, and Rewards – Introducing a Theme

This book is the story of a long metaphorical journey from Eden to modern times. Each chapter dwells on different challenges that our biology presents, the initiatives we have launched to reduce their menace, and the rewards or otherwise that followed. Chapter 2 addresses a fundamental question, as important today as in the time of Genesis: how to breed successfully? Why should reproduction – the most important enterprise that people ever undertake – be fraught with such difficulties and uncertainties, and what solutions have we found? The next four chapters concern fundamental issues of biology: how life is handed on, cells in sickness or health, how events in embryonic life affect life after birth, and the requirements to keep us healthy. The middle section examines three great curses that afflict human existence: ageing, cancer, and infection. The final section concerns initiatives and innovations meant to overcome the facts of existence that every species faces, but which only we can see are rather gloomy.

The monumental step that put humans on the agricultural treadmill facilitated the appearance of more dynamic civilisations but was perhaps the first great human misadventure that we know about. Humans were ill adapted to the agricultural life, and it had serious consequences for health, susceptibility to infection, and reproductive physiology. While agriculture could feed more people and support bigger families, the ever-present danger of bad harvests and famine created an unstable balance between the birth rate and the food supply. Insanitary conditions inevitably prevailed in crowded settlements, and the communities became highly vulnerable to epidemics once populations approached a critical density. Later, the population growth rate slowed as the high birth rate was balanced by a high death rate[5] – a time when, in Thomas Hobbes' unforgettable phrase, "the lives of men were nasty, brutish and short." Farming families dependent on single staple crops must have suffered from vitamin and protein deficiencies that increased their susceptibility to infection and the severity of disease. We know from contemporary experience that poor nutrition *in utero* seriously damages the immune system and that chronic enteric infections retard the growth of children. The skeletal remains from prehistoric communities in-dicate people often died pitifully starved, deficient in protein, and with an adult height on average fifteen centimetres smaller than Cro-Magnon man. Indeed, estimates of average dietary protein suggest that levels required for people to grow to the same average height as Cro-Magnon man have been reached only in the last two hundred years.[6]

In the great population explosion that started in mid-eighteenth-century Europe, life span increased significantly and infant mortality decreased, but these advances were also the seeds of another misadventure. Industrialisa-tion thrived on a rapidly growing urban proletariat that inhabited cities fatally flawed by their inadequate sewage disposal systems. Many complex factors contributed to a sustained surge in population growth, including a diminishing threat of infectious disease, improved real incomes, improve-ments in nutrition, and a reduction in the age at which women could con-ceive a child. Paradoxically, the most tangible indications of progress were disastrous for the health of the urban poor. In London and other European capitals of the early nineteenth century, unparalleled wealth existed side by side with the most vicious squalor, providing favourable conditions for infectious disease to prosper.

When cholera appeared in Europe for the first time in the 1830s, re-current epidemics decimated the urban poor because drinking water was contaminated by sewage. The epidemic of tuberculosis that would become the major cause of death in nineteenth-century Europe may have begun in the great factories that were the hub of industrial growth. Rickets, the bow-legged disability caused by soft bones, affected most urban children of Britain and northern Europe, probably because the smoke-darkened skies shielded their skin from sunlight sufficiently to prevent vitamin D from being made. In the absence of effective public hygiene, infectious disease dominated the life of European cities throughout the nineteenth century.

No medical treatment of any kind – except smallpox vaccination – affected the course of any infectious disease. The death rate from every single one, including tuberculosis, and most notably the death rate for in-fants, started to fall in industrial countries after 1870. The installation of closed sewers, clean water supplies, and the demolition of foetid slums all played an important part, and the appointment of medical officers of health put sanitary regulation at the heart of public administration. Later, improved care of the sick and infants and rising enthusiasm for hygiene contributed importantly to better life expectancy.

As life expectancy began to improve, the birth rate started to fall mysteriously, to the dismay of some authorities. It seems once couples believed that their children would survive infancy, they managed to control their fertility, even if the methods used were far from satisfactory.[7] As the threat of infectious disease subsided, diseases of a much more complicated character (heart disease and cancer) emerged as the next frontier in our pursuit of greater longevity (see Chapters 6 and 8).

In the remainder of this chapter, we will consider how in the early twentieth century, partly through the perspicacity and generosity of philanthropists, biomedical science began to emerge as the great hope for understanding and improving the human condition.

George Bernard Shaw's Misgivings – Twentieth-Century Optimism

In 1900, children born in Britain (and probably in most of Europe and North America) had an enormously improved life expectancy compared to that of their parents. Their chance of succumbing to an infectious disease was greatly reduced by massive advances in public hygiene and many kinds of social intervention. One of the great intellectuals of the age, George Bernard Shaw, was adamant this had nothing to do with the medical profession and used his art to lampoon them in his play *The Doctor's Dilemma* (1906). The preface to the published version of the play (1911) is a hilarious diatribe against physicians and medical scientists and all the manifestations of their trade. Vaccination, patent medicines, experimentation on animals, lack of statistical proof of efficacy of recommended treatments, pointless surgical operations – all are targets of his withering scorn. The otherwise undisputed triumphs of the age – antiseptic surgery, anaesthetics, and the germ theory of disease – he treats with robust contempt. His only concession was to be glad that cleanliness was ousting godliness in public esteem.

Shaw was invincibly ill informed about biomedical developments, but when he wondered why sick people consulted doctors for a fee when so few complaints could be cured, he was raising an interesting question. Medical historians also wonder what strange psychology gripped desperate nineteenth-century minds and usually conclude patients responded to the mystique of the physician's bedside manner, which worked in much the same way as the placebo effect.[8,9] (Clinicians conducting drug trials often find the control group receiving the placebo – a dummy pill lacking the drug – show unexpected benefits from the trial.)

Shaw's clowning has a certain prescience. In the year the play was published, the Carnegie Foundation's investigation of European medical training criticised British clinical teachers for their failure to inculcate scientific attitudes in medical students.[10] His enthusiasm for the "sunshine and soap" creed of the sanitarians – the early nineteenth-century movement dedicated to cleaning up filth, before the importance of germs was fully appreciated – was not misplaced either. Even so, "the germ" had entered the language as the embodiment of a public enemy that needed eradicating with soap and disinfectants, the use of hand washing and handkerchiefs, and campaigns to prevent spitting.[8,11]

Shaw's polemic came just at the moment when biological science, as distinct from medicine, was making its debut in human affairs. Seminal ideas that would be massively influential later in the century were being launched in the Edwardian twilight, but they were less newsworthy than the arms race or the rising tide of social unrest. In Cambridge, England, Gowland Hopkins was establishing vitamins as essential components of the mammalian diet necessary for a healthy life. In Germany, a thirty-year odyssey initiated by Paul Ehrlich to find a magic bullet that could cure infectious diseases without adversely affecting human health had started. In Austria, Ludwig Haberlandt was discovering a hormone secreted by the ovaries that governed the menstrual cycle, which, he suggested a decade later, could control fertility and be the basis of a contraceptive pill. In New York, Thomas Morgan was laying the foundation of modern genetics, using an obscure little fly, and Alexis Carrel was conducting the first person-to-person blood transfusions. Peyton Rous was identifying a virus that caused cancer in chickens and that would, almost seventy years later, open the way for the discovery of the oncogenes and their crucial role in malignancy. In London, Archibald Garrod was demonstrating how genetic diseases are inherited biochemical deficiencies. By associating particular organisms with infectious diseases, the first generation of bacteriologists was creating a robust basis for the embryonic discipline of public health.

The impetus for such novel ideas came from the great universities of the world, where the problems of nature were acquiring a new focus. There was confidence in the air that humankind no longer had to accept passively the depredations of disease. In anticipation of important developments from the great successes of Pasteur and Koch, medical research institutes sprang up in the major European capitals, funded by direct donations from the public. The Institute Pasteur opened in Paris in 1888, and the Koch Institute in Berlin and the Lister Institute in London opened two years later. The British venture lacked serious public support owing to the virulent anti-vivisectionist propaganda that was prevalent in Britain, in which Shaw played a prominent part. The institute was capitalised with just £64,000, of which half was contributed by a single person. It would have failed quickly had not Lord Iveagh of the Guinness family donated £250,000 a few years later.[12]

Quite soon, the governments of most advanced countries were setting up, on a modest scale, national medical research organisations to tackle important issues of public health and the standardisation of medicine. There was a growing recognition that an understanding of the basis of medical conditions required a deep knowledge of physiology. The first faltering step in this direction was taken, in Britain in 1911, through a footnote to the National Insurance Act. One penny per year for each person paying National Insurance contributions was supposed to generate a fund of £40–50,000 per year to finance medical research through a precursor of the British Medical Research Council (MRC).[13] Though less than a fortune, this was almost the first government money in British history to be spent to the benefit of health. The most important priority of the last Liberal government – and indeed, of Edwardian Britain – is evident in another expenditure proposed in the budget of 1909. This was the 5.3 million pounds to be spent on eight "Dreadnought-type" battleships, necessary to satisfy the public's appetite for nautical hardware.[14]

During the First World War, the MRC's finances deteriorated drastically because military personnel were not required to pay their National Insurance contributions.[13] After 1918, the situation improved, and research was undertaken on a broad front. Rickets, the misshapen legs that disturbingly characterised British town dwellers at that time, attracted particular attention. Quite quickly, the requirement for vitamin D to prevent bendy bones in children was established. Research quickly gained momentum in many directions, but in an atmosphere of confrontation with the medical establishment as enshrined in the Royal Colleges of Medicine and Surgery.[15] Shaw, in *The Doctor's Dilemma*, captures brilliantly the foibles of a bunch of contemporary medical luminaries. He presents the Royal Physician, full of oily charm and ignorance, and the fashionable surgeon whose one-fits-all special operation is supposed to cure any kind of ill health, for a fee. Other roles include a grumpy old cynic who believes there is nothing he could usefully learn, and a successful general practitioner whose success is attributed to promising his patients that "cures were guaranteed." The crux of the play concerns whether any of them can save from tuberculosis a supremely talented but morally despicable artist. Shaw extracts the maximum of fun from the medical profession, knowing perfectly well that no treatment on offer, anywhere at that time, could cure tuberculosis. One of his protagonists, Sir Colenzo Ridgeon, is based on his "friend" Almroth Wright, the celebrated Edwardian bacteriologist and inventor of the first anti-typhoid vaccine. However, it is Wright's now-forgotten and fallacious invention, the Opsonic Index – a measure of resistance to disease – that Shaw (usually thought of as a scientific ignoramus) chooses to mock with such daring prescience.

Wright contributed one more footnote to medical history (that Shaw might have appreciated). The spectacular outcome of Florey and Chain's clinical trial of penicillin in Oxford in 1942 was announced in *The Times* – rather oddly, without mention of the authors. Wright saw fit to rectify this, according to his perceptions, by a letter to the newspaper. Using the secret language of great men of those days, he suggested that "the laurel wreath should be decreed," "*palmum qui meruit ferat*," to Alexander Fleming.[16]

Walter Fletcher, the first secretary of the MRC, contending with characters not unlike Shaw's creations, constantly challenged the suitability of the Royal Colleges to oversee scientific research. With outlooks fashioned in the previous century, they were suspicious of research and saw it as a threat to their professional position. Their last victory was celebrated in the late 1930s, when in the technicalities arising from a cancer bill introduced to Parliament, they persuaded the government to leave cancer research to the voluntary sector.[15]

The American National Institutes of Health (NIH) emerged rather late in the day (1930) from a small federal bacteriology laboratory founded in 1866. Until then, American biomedicine had relied on its many excellent medical schools and philanthropic patrons. Once in Bethesda, near Washington, it grew in forty years to become a substantial city and the greatest concentration of medical science the world has ever seen.

Fine universities existed in every industrialised country during the nineteenth century, but the system created in the German states was the one most universally admired. Research, both scientific and clinical, was closely associated with teaching. Training for professional researchers was conducted on a scale unheard of elsewhere for the ascendant German chemical industry, which attracted young men of many nations to extend their education in medicine and science. The German system was also the model for other university medical schools, such as Johns Hopkins in the United States. Formal disciplines concerned with every aspect of the human body took shape during this period and would become the foundations for the epic advances of twentieth-century biology.

By the 1920s, medical research was institutionalised throughout the developed world, with a number of well-supported research institutes and a growing number of research-oriented universities supported tenuously by endowments, charities, and medical fees. In industry, pharmacological research became the most active discipline, with many companies, especially in the United States, trying to use their chemical skills to bring new versions of old folk medicines to market. In Britain, Boroughs-Wellcome, a successful pharmaceutical company founded by Americans, also undertook highly regarded fundamental research.

The Great Philanthropists

In the United States, some of the wealthiest people who had ever lived were starting to finance philanthropic works on an unprecedented scale. Andrew Carnegie sold his steel interests in 1900 to spend the rest of his life distributing almost his entire fortune (said to be 400 million dollars), based on his famous aphorism "a man who dies rich, dies disgraced." He endowed important public facilities such as the Carnegie Institution of Washington, the Carnegie Corporation of New York, and the Scottish universities. A report commissioned by the Carnegie Foundation for the Advancement of Teaching (1910) had the most profound consequences. This report proposed a wholesale restructuring of American and Canadian medical schools in order to provide high-quality medical education and research based on the model of the Johns Hopkins University. The reformed medical schools rapidly became world leaders in medical research. The Carnegie Institute for the "study of human evolution" was set up, but this had less commendable outcomes when it became deeply involved in the now discredited notion of eugenics. A second report on medical education in Europe, completed in 1912, criticised clinical teachers for not engaging in research or inculcating scientific attitudes toward medicine and played a significant part in reforming British medical education.[10]

Carnegie's disapproval of inherited wealth attracted enormous interest and inspired John D. Rockefeller to donate a large part of his fortune (said to be 530 million dollars) to philanthropic causes. This extraordinary fortune was amassed in building the Standard Oil Company into a huge nationwide monopoly during the tumultuous expansion of the American economy after the Civil War. Carnegie's dictum went right to the heart of this austere Baptist just as he felt the first intimations of mortality. His astute associate, the Reverend Frederick Gates, who knew the realities of contemporary medicine from his pastoral experience, advised Rockefeller that medicine needed properly paid full-time researchers if anything was to improve medical science. Rockefeller was dismayed at the powerlessness of medicine at that time, and under Gates's influence he came to regard health as the world's greatest social issue. The final straw for Rockefeller was the realisation that nothing could prevent the death from scarlet fever of his beloved grandson. The last forty years of Rockefeller's long life (and the entire life of his son) were spent in dispensing this incredible wealth to organisations that they hoped would rectify the situation. In the 1890s, Rockefeller founded the University of Chicago and an organisation to finance medical schools. Then, in 1901, he set up the Rockefeller Institute for Medical Research in New York (now Rockefeller University) with an endowment of sixty million dollars, organised in the style of the Institute Pasteur. This in turn became the prototype for other institutes. His most revolutionary idea, the Rockefeller Foundation, established in 1913, was to be a vehicle for his philanthropic mission "to promote the well being of mankind throughout the world." In time, the foundation tackled intractable problems ranging from tropical diseases to population policy, the genetic improvement of staple crops, and deficiencies in the public health infrastructure that prevented existing health systems from making progress.

The Rockefeller Foundation exerted immense influence in the years between the world wars, when profound economic and political difficulties meant that investment in medical research was not a high priority for governments. Major grants were made to improve the quality of medical education and public health both in the United States and abroad, even in relatively rich countries such as Britain and France. Fundamental biological research underpinning the objectives of medical research was also financed throughout the world with grants to laboratories for research, new facilities, and fellowships for scholars to train at the world's leading universities.[10] When the Nazi nightmare descended on Germany in 1933, the foundation did not hesitate to assist Jewish scientists to make their homes in friendly countries.

The foundation focussed particular largesse on London, which it saw as a strategic centre for promoting health amongst the underprivileged throughout the British Empire by training personnel. The London School of Hygiene and Tropical Medicine was created for postgraduate education in preventative medicine. The foundation contributed almost half a million pounds for buildings; the British government agreed to pay for recurrent expenditures.[10] A five-million-dollar donation to University College, London, was used to endow a number of chairs and construct a great medical centre. The plan was to make research, teaching, and medical care an integrated unit and a model for other medical schools in Britain and throughout the empire. Similar endowments to the University of Wales, Cambridge, and Edinburgh followed.[10]

The Rockefeller Foundation supported specific research projects, too, with grants or scholarships that were, with hindsight, a roll call of much of the most important European science of the interwar years. Grants were made to Paul Ehrlich for work on antimicrobial chemicals, to Charles Harrington in London for his work on the thyroid hormone, and to Leonard Colebrook of the Hammersmith Hospital in London for the first clinical trial of sulphonamides. William Astbury received a grant for X-ray crystallography, Ernst Chain in Oxford for penicillin, and Hans Krebs for his work on metabolism in Cambridge, among many others. In all, the foundation has supported the work of 137 eventual Nobel Prize winners. Research into parasitic and viral diseases that were so devastating in the tropics was pursued with special fervour. A thirty-year campaign against yellow fever waged by the foundation culminated in a successful vaccine and a Nobel Prize for Max Theiler in 1951.

The foundation also embraced genetics with missionary zeal during the 1930s, starting with the work of Thomas Morgan and George Beadle, both of whom would win Nobel Prizes for their seminal contributions. By the 1950s, the foundation was supporting genetics in organisations such as the California Institute of Technology, at a time when grants from NIH were not forthcoming and the relevance of genetics to medicine was not very clear. The development of the physical techniques, such as X-ray crystallography, that were ultimately to underpin molecular biology was another farsighted activity.[17,18]

New charitable foundations, especially in the United States, soon matched the magnificent philanthropy of Carnegie and Rockefeller. Katherine McCormick financed the research of Gregory Pincus on the development of the contraceptive pill from 1951 into the 1960s, at a time when NIH was forbidden to support any research with an explicit connection to birth control (Chapter 2). The vast Ford Foundation supported

research on eradication of parasite diseases, population policy, and reproduction research and supported Robert Edwards's work on *in vitro* fertilisation during the 1970s.[19]

Conspicuously absent from the ranks of the great philanthropists were
the ancient land-owning European aristocrats with vast fortunes, who saw
no virtue in returning any of their loot to the community that had supported
them for so long. By contrast, men who made their money in commerce
such as Sir William Dunn, Lord Nuffield, Lord Iveagh, and Jesse Boot
endowed institutions dedicated to improving the human condition.[10]

The old established foundations' support for biomedical research was
eventually overtaken by government, but they diverted their resources to
important projects that were not government funded. The Rockefeller
Foundation's singular vision supported the development of dwarf wheat
varieties that grow well in Mexico to help improve the food supply. More
recently, it has financed development of a genetically engineered variety of
rice that makes a building block of vitamin A, intended to combat a vitamin deficiency that afflicts about 400 million people around the world.
Today, even wealthier foundations – the Howard Hughes Medical Institute,
the Wellcome Trust in Britain, and the Gates Foundation – are playing a
leading role as patrons of biomedical research throughout the world.

Investment in the New Biology

The episode that utterly changed the way the study of biology would be
used to improve the human condition was the extraordinary investment
made after 1945 by the American government to foster academic research
in health science. Influenced, possibly, by the risk of a recession if public
spending stopped abruptly at the end of the war, government planners redi-
rected wartime resources toward the National Institutes of Health (NIH)
and the National Cancer Institute (NCI). During the war years, the American public learnt to appreciate the benefits of research and was not disappointed when antibiotics, cortisone, and the polio vaccine provided a
spectacular stream of advances.

Biomedical research was the chief beneficiary of a confrontation between President Truman and the American Medical Association (AMA).
Truman, a man steeped in the New Deal tradition, hoped to introduce a national health insurance scheme, but this was anathema to the AMA. In the
horse trading that followed, medical research came to be seen as the "best

insurance." Almost immediately, a vast organisation appeared, committed to biomedical research on the most radical scale ever conceived. Starting with a budget of eight million dollars in 1947, the NIH was receiving more than one billion dollars a year by 1966.

Revolutionary ideas underpinned the new arrangements. Biological science was to be supported for its own sake, at the taxpayers' expense, so that issues of medical importance could be tackled from a fundamental viewpoint. A sophisticated and democratic method for distributing funds was created. Established research scientists were invited to propose projects that would be reviewed by a committee of their peers. If accepted, grants would provide the salaries and expenses of laboratories dedicated to these projects. They also participated by reviewing the proposals of others, thereby shaping a huge creative collaboration. More than half of the budget was to be spent in university-based research that would create many training opportunities for graduate students and postdoctoral fellows and massively expand the universities. The public, and probably most congressmen, believed that this massive investment was devoted to the elimination of particular diseases, but what emerged was unprecedented and comprehensive support for every aspect of biology. The objective was evidently an understanding of the basis of life in many kinds of organism, from humble bacteria to man.

The new dispensation had far-reaching consequences. The membership of professional societies dedicated to biomedicine increased rapidly. In the United States, the membership of the Federation of American Societies for Experimental Biology increased from 469 in 1920, to about 3,000 in 1950, to more than 50,000 in 1999. A similar trend developed in Europe.[20] The effect of American largesse was felt around the world, because the new scholarships could be awarded to non-Americans to study and work in the United States. Many would return to their own countries, but a significant proportion of every generation of graduates from some countries would be lost forever to the United States.

Politicians were sometimes sceptical of the value of this far-reaching scheme. In 1965, President Lyndon Johnson, thinking perhaps of funding military involvement in Southeast Asia, began to look at the NIH budget. "We must make sure no life giving discovery is locked up in the laboratory," he drawled, but after listening to the head of the NIH present the case for supporting fundamental research, he backed off.[21,22]

The NCI benefited in the same way, although its brief was nominally to solve the cancer problem. The decision was taken that cancer research would achieve nothing while it was narrowly focussed on patients and chemotherapy; progress would be made only when the growth and the replication of cells and viruses were better understood. The perspicacity of this judgement is illustrated by one outcome of a program started in the late 1940s that awarded fellowships for work on viruses of bacteria – agents that could never be directly relevant to the cancer problem. One recipient was the young James Watson, who within a few years had published with Francis Crick a proposed structure of DNA. This would transform the landscape of biology as a discipline and was probably the most scientifically influential thousand words of the twentieth century. Meanwhile, the study of bacterial viruses became the bedrock of the infant science of molecular biology.

Outside of the United States, few countries contemplated expansion of scientific research in the aftermath of war. Britain, with a large wartime scientific establishment and close involvement with penicillin, was in a strong position to "beat swords into plowshares." However, Britain chose a different course and spent more than ten percent of its GNP in the early 1950s on the manufacture of its own atom bomb and the legendary V bombers, which would never fire a shot in anger. American commentators watched with incredulity as Britain's opportunity slipped away. Some years later, the eminent British developmental biologist Conrad Waddington presented a critique, with others, of American science policy in which he lavishly praised the American model. The system for formulating policy, he noted, was based on wider consultation than was ever considered appropriate in Europe. Investment in higher education as a fraction of GNP was unmatched, and the sheer effectiveness of mass education in American universities impressed him enormously. Financial support for research was naturally much more generous than could be considered in Europe. He was also impressed by the accessibility of the grant system to young American academics, unfettered by the administrative heads of their own institutions who might block the ambitions of young researchers for obscure reasons. Waddington's only anxiety was that increased competition might inhibit creativity.[23] Fascinatingly, an editorial in *Science* thirty years later reiterated much the same thoughts.[24]

The level of investment in science education and research in the United States was, of course, astonishing and established a commanding position for the United States once the biotechnology industry was launched. By 1995, American biomedical research was receiving about 15.8 billion dollars of government money and about 18.7 billion from industry and charities. Comparable figures for the United Kingdom were 1.7 and 0.9 billion.[20] However, the many awards of Nobel Prizes for medicine and chemistry to non-American scientists indicate how good ideas in science can emerge from less lavish financial support systems.

Questions from a Sceptical Public

In the 1950s and 1960s, the success of clinical medicine and the rapid improvement in the quality of life on many other fronts established confidence and trust, but as the public became more demanding and more critical, the mood changed. Legislative demands for proof of the efficacy and safety of drugs, and then demands through advocacy groups for action on cancer and AIDS, were crucial early episodes in the shaping of the public's attitude toward biomedicine. Later, with the emergence of a biotechnology industry and reproductive technology, many controversies arose to which we will return in later chapters.

With the thalidomide affair in mind, and wishing to outlaw the peddling of fake cures to desperate victims of cancer, the American Congress enacted the Kefauver–Harris amendment to the Food, Drug and Cosmetic Act in 1962. This demanded that all drugs sold to the public be tested to ensure that they are safe and efficacious. The new law had far-reaching effects. Every manufacturer, anywhere in the world, who wanted its products to be acceptable in the United States had to submit to a three-phase drug assessment process overseen by the Food and Drug Administration (FDA). This process has gained the trust of the public and remains the only objective criterion by which physicians, patients, and investors in the biotech industry can judge the quality of new products.

Objective evaluation of drugs has been attempted for several centuries, but a universally acceptable scheme that unequivocally eliminates unconscious bias on the part of the clinician did not emerge until randomised double-blind trials were introduced in 1948. The British statistician Austin Bradford Hill, who devised this test to assess the efficacy of streptomycin, gave the subjects either the test drug or a placebo – a dummy pill lacking the drug – randomly allocated according to a code.[26] The legislation now requires manufacturers to meet this stringent test of efficacy. The placebo (a Latin word meaning "I will please"), however, is not entirely neutral, as many well-controlled trials indicate that three or four of every ten patients receiving a placebo report some measure of pain relief.[27]

In spite of this far-sighted legislation, which provided the best assurance available of the quality of medicines, enthusiasm for so-called alternative medicine has continued to grow. Alternative medicine is most accurately described as drugs that are not evaluated by a test that permits no bias, and manufacturers of alternative drugs are rarely prepared to submit their products to this challenge. Incredibly, Americans today spend two dollars on alternative medicine for every ten dollars spent on prescription drugs, and the situation is no different in Europe.[25] Why this should be is not apparent; perhaps such remedies are beneficial through the placebo effect; and if so, who should deny their value? Recognising this implied rebuke by the public, NIH has set up an organisation that will evaluate alternative medicines and establish whether they are beneficial or hazardous. The first report shows an extract of St John's wort, widely promoted as an alternative antidepressant, has some benefits in trials but seems to reduce the efficacy of a number of other medications, including contraceptive pills.

Families of the generation that reached middle age fifty years ago were sometimes faced with cancer for the first time in their collective memory. The death rate from cancer seemed to be increasing steadily, and the cause was not obvious. Powerlessness against this terrible illness created a militant fund-raising movement that by the 1970s was achieving enormous influence on policy decisions in cancer research (Chapter 8). For a time, an all-out frontal assault on the problem was all that the movement would consider, asking the question, "If they can spend billions of dollars sending a man to the moon, why not spend it on cancer research?" The essential breakthroughs, however, did not come from the so-called war on cancer but from subtle molecular biological studies of cancer cells that indicated where the next cancer-fighting drugs would be found. Twenty-five years later, the problem is better understood, and prospects for effective treatments are increasing steadily (Chapter 8).

The AIDS crisis of the 1980s precipitated another episode of research targeted at an urgent problem, but when no effective therapy was forthcoming, the effort was quickly diverted into fundamental immunology research.

Biomedicine Comes of Age

An industry based on molecular biology was a logical development once the means of making proteins in cellular factories became available, although few scientists, let alone their political paymasters, would have predicted this as late as 1970.[28] Six years later, though, the first gene technology company was founded; it would revolutionise the pharmaceutical industry and start making human proteins that could solve health problems. Another ten years elapsed before the first product of gene technology was actually sold. President Johnson did not live long enough to see the fruits of the investment in biological research he had questioned in 1965, but he would no doubt have congratulated Congress on its foresight. Within a few years, the disastrous spread of hepatitis, AIDS, and prion disease through the use of human blood proteins illustrated just how valuable products made by gene technology would be in manufacturing definitively safe products. New biotechnology companies have appeared at a phenomenal rate since the 1980s, providing a way to finance research initiated in academia that promises to bring to market many remarkable and precious products. A commercial agenda quite different from the one that drove biomedicine in the last century is likely to shape the future (see Chapters 12 and 15).

GENERAL SOURCES: 8, 29, 30, 31, 32.

2 Learning to Breed Successfully

Man's earliest writings speak of the hope invested in offspring and in the generations to come, but "it is a truth, universally acknowledged" that reproduction, the most important enterprise people ever undertake, is fraught with difficulties. Women can have more than twenty babies in a lifetime, but populations have rarely grown at a rate that remotely reflects this possibility. On the contrary, the high mortality rate of infants and of their mothers was a painful truth at the heart of the history of the family. At the same time, unwanted pregnancy brought misery to people barely able to support a family and to their offspring. Unsatisfactory efforts to limit family size began in the late nineteenth century, marking the start of an era in which the birth control movement would gather momentum, culminating in the 1950s in the discovery of safe, effective hormonal contraception. In the midst of all this fecundity, the poignant regret of those who cannot have children could be heard, but daring schemes to assist reproduction have now made child bearing possible, at least for some of them. Control over reproduction has given us the means to breed successfully and to create possibilities for individuals that never before existed, but challenges remain.

"In Sorrow Thou Shalt Bring Forth Children"

Those who think that motherhood in prehistoric times was any easier than it is today should consider Eve's curse (Fig. 2.1). "*I will greatly multiply thy sorrow and thy conception; in sorrow thou shalt bring forth children,*" writes J, the plausibly female author of the oldest part of Genesis.[32] To her, childbirth was clearly inevitable, painful, unsafe, and too

Hope for the future is invested transparently in offspring in the book
of Genesis. To Abraham, struggling to secure a foothold in Canaan,
it is "to make a great nation." With old age in mind, it is the Fifth
Commandment, *"Honour thy father and mother"*; and to the redoubtable matriarchs, it is the unconditional love of tiny babies.
Desperation for offspring drives most tales from Genesis. Abraham's
wife, Sarah, permits her maid to be a surrogate for perpetuating the
patriarchal line but is mad with jealousy when the maid conceives
(Genesis 18:12). Her merriment when her pregnancy is prophesied is
perhaps the first recorded laughter of happiness. Rachel, desperate to
have a child, dabbles in magic (the notorious mandrakes) to give her
Joseph, the golden boy (Genesis 30:14). Lot, the nephew of Abraham,
lost in a desolate place with his two daughters after the destruction
of Sodom, is tricked into inseminating them because they believe no
other man is left on earth (Genesis 19:31–36).

Fertility cults were commonplace in the prehistoric Near East. One
anthropologist sees in the placing of genitalia of male animals on the
top of sacrificial pyres in the tabernacle evidence that the sacrifices
were also a plea for fertility. The fierce demands for disciplined sexual
conduct that rage through Leviticus and Deuteronomy suggest that
the Israelite leadership connected loss of fertility with loss of sexual
health. The Law reserved its sternest anathemas for the Phoenician
cult of Moloch (Leviticus 18:21) that involved (it is believed) the
sacrifice by fire of human babies.

Figure 2.1. The reproductive imperative in Genesis.

frequent. Indeed, Rachel dies giving birth to Benjamin, her second child
(Genesis 35:19).

Throughout the insanitary Dark Ages, the greatest risk to women in the
prime of life was puerperal fever, an opportunistic bacterial infection that
struck in childbirth, frequently introduced by persons assisting with a birth.
Famously, Ignaz Semmelweis at his clinic in Vienna in 1850 proved that
if doctors attending a birth washed their hands with chlorinated lime – a
fierce antiseptic – the mortality rate dropped dramatically.[33] Ten percent of
women died in childbirth almost everywhere before the nineteenth century,

and childbirth remained a substantial risk until the 1930s, except in the Netherlands and Scandinavia, where fastidious hygienic midwifery was a tradition. In Britain, the maternal mortality rate reached a twentieth-century peak of 0.44 percent in 1934, of which about 40 percent could be attributed to puerperal fever, little different from the figure for sixty years earlier. This appalling figure quickly plummeted with the introduction of sulphonamides in 1937 and continued descending when penicillin and improved methods of caesarean section made their debut. By 1960, the maternal mortality rate in Britain was 0.039 percent; it was less than 0.01 percent in 1990.[34]

All primates except humans give birth alone and unaided, with little risk of anything resembling puerperal fever. The evolutionary steps that created big-brained, bipedal *Homo sapiens* required novel anatomical developments not easily accommodated by the primate skeleton. The large head could reach full size only if a substantial part of its growth occurred after birth; otherwise the birth passage would be too small. In evolution, the pelvis could not enlarge usefully without impairing mobility. The consequence is that the head of a human baby is almost too big for the birth passage and may distress the mother at birth. In less hygienic times, damaged tissues would have been easily infected, making childbirth exceedingly dangerous.[35] About one in thirty births are not head-first (breech births). Today, the potential hazard to the child posed by this eventuality is met either by "turning" the baby in the womb to ensure that the child is delivered head first or by a caesarean section.

Human offspring normally enter the world head first, but unlike other primates they face away from the mother, making it almost impossible for the mother to deliver her own baby safely. The earliest human literary artefacts (Genesis 35:17, 39:28; Exodus 1:15–21) indicate the existence of midwives and the need of women for companionship at birth. Anthropologists argue that historically, the menopause protected women from death in childbirth and thereby conferred an important evolutionary advantage on the family unit by facilitating her survival as grandmother. She strengthened the family unit in this role by assisting with the birth of her grandchildren and with their feeding, once they were weaned.[35] Recent investigations of pre-twentieth-century family histories indicate that children whose grandmothers assisted in their upbringing had more descendants than those without.

God's curse on Eve became a cause célèbre in the English-speaking world of the late 1840s when James Simpson discovered that chloroform could reduce the pain of childbirth. Religious leaders and conservative doctors of the time were passionately convinced that to remove pain from childbirth would contravene God's will and would be an affront to the "natural order." Simpson tackled this public relations problem head on with ingenious sophistry, persuading his Scottish countrymen that the "sorrow" of the text related only to the effort of childbirth and not to the pain, thereby making it acceptable to mitigate pain. Resistance remained fierce for some years in England and North America, until the news broke in 1853 that Queen Victoria had given birth to her sixth child with the help of chloroform. From that moment, nobody could refuse English women chloroform because (it was said) Victoria, the head of the Anglican Church, would never act against the will of God.[33,36]

Reactionaries of another kind surfaced in the 1930s when Grantley Dick-Read formulated his ideas of natural childbirth. The pain of normal childbirth, he argued, was caused by fear reflexes that withheld oxygenated blood from the uterus and prevented the muscles of the uterus from exerting their full force. This could be avoided if the mother undertook a programme of breathing exercises to encourage relaxation and to permit inherent natural muscular contractions to drive childbirth. The value of his method is still recognised in Britain in the classes of the National Child Birth Trust, and similar techniques are practised throughout the world. Dick-Read's views were regarded as a "needless glorification of the primitive" by his peers, who believed the way forward was more and better drugs.[37]

Today, the caesarean section is used increasingly. This procedure, once fraught with danger, is now no more hazardous than a normal delivery, and in 2000 it was used for about 20 percent of births in London and other major cities. Probably half of these are used for breech births, but many doctors are surprised that so many are necessary. They suggest that this could be a consequence of hospital work schedules; others suspect more controversial cosmetic reasons. In Brazil, the majority of affluent women now have babies by caesarean section, in the absence of any identifiable biological risk, apparently in the belief that a traditional delivery is an inherently plebeian procedure.[38]

Changing Reproductive Habits

In the 1960s, anthropologists believed that important clues to understanding how people adapted to agricultural life in prehistoric times

could be gleaned from comparing contemporary hunter-gatherer populations to their kinsmen who had ceased nomadic life and started farming. At this time, about five percent of the !Kung of the Kalahari Desert
were still following a prehistoric hunter-gatherer lifestyle (the ! indicates the characteristic click used in the languages of southern Africa).
They seemed to be living in a harmonious equilibrium with their harsh
environment – healthy, long-lived, and by no means struggling to find
enough food, but reproducing very slowly. Their babies were breast-fed
for more than three years, with intervals between births of more than
four years; each mother had only four or five children in a lifetime. Reproductive physiologists believe that the women did not conceive while
they were breast-feeding because ovulation was suppressed by a hormonal
response to suckling.[39] Nomadic !Kung girls did not start menstruating
until their fifteenth year, and their first child was not born before their
nineteenth, providing a brake on the population growth rate. Overall,
the nomadic !Kung could double in number only once every three hundred years, while the !Kung who had gravitated to farming reproduced
faster. Breast-feeding stopped earlier as mothers weaned their babies onto
cow's milk and grain, and once the inhibition of ovulation caused by
lactation had stopped, the interval between births decreased. Girls men-
struated earlier than their nomadic kin and had their first babies
earlier.[40]

Whether the agricultural transition of groups like the !Kung mirrors the
momentous start of the first agricultural revolution will probably never be
resolved. The fertility and population growth of nomadic people may have
been lower for reasons we cannot know, but infanticide, sexually transmitted diseases, and limitations of diet could all have played a part. On
balance, though, women would probably have had more babies after the
agricultural transition, and infants would probably have survived longer,
with an overall increase in population growth even if mortality from infection increased. The interbirth interval can be very short, and humans since
the earliest days of agriculture have probably managed to feed a family
of small, very dependent children of different ages using weaning foods.
This facility permits humans to breed very rapidly, unlike the great apes,
who breast-feed their offspring until they are nutritionally independent.
In certain times and places, human populations have grown at the highest rates that are theoretically possible, but during most of human history,
births have exceeded deaths by only the slimmest of margins.

What we know about the reproductive physiology of the agricultural transition is largely conjectural, but we are on safer ground in interpreting recent changes in the birth rate. The age of first menstruation (menarche) has decreased steadily from an average of 17.5 years in 1840 to about thirteen today.[41] Although only a small change, this has a potentially big effect on the birth rate of a population that has no artificial birth control.[39] Physiologists used to believe that menarche began when the body weight reached fifty kilograms, with ovulation starting a few years later. Today, scientists believe that the timing of menarche is established *in utero* or soon after birth. This reflects the nutritional status of the foetus *in utero*, although body mass and timing of menarche are usually correlated.

Sexual maturity arriving far in advance of intellectual maturity is usually seen as the key element in the emergence of the distinctive teen culture that makes late twentieth-century society so different from previous cultures. In the 1970s, one reproductive physiologist was urging society – belatedly, perhaps – to recognise that this profound biological change means premature sexuality is virtually inevitable in modern society. Rather than regarding the sexuality of teenagers as the "stigmata of a decadent and permissive society," he suggested that a better response would be to offer contraceptive advice.[39]

BURDENS IMPOSED BY THE MENSTRUAL CYCLE: Changes in reproductive physiology are perhaps the most obvious consequences of adaptation to the post–hunter-gatherer lifestyle. The !Kung women probably have only four years of menstrual cycles during their entire lifetimes.[39] By contrast, women of the industrial world now have, on average, only two pregnancies in their reproductive life but may undergo menstrual cycles for thirty-five years. This may make modern women more susceptible to malignancies of their reproductive organs. During every menstrual cycle, the concentration of steroid hormones in the body surges up and down, causing bursts of cell division during which cancer-initiating events occur with a very small but significant probability. The longer the time interval between menarche and first birth, the greater the risk of breast cancer; but the risk of ovarian cancer is diminished in women who reach the menopause early. About one in eight women in the industrial world will be touched by breast cancer, a problem on a scale greater than any recent infectious epidemic (see Chapter 8). There is also speculation that conditions *in utero* in the twentieth century may predispose girl babies towards developing malignancies to which a less nutritionally privileged generation may not have been vulnerable.

Enormous resources are deployed in diagnosis and treatment. Natural dietary phyto-oestrogens or drugs such as tamoxifen that compete with steroids for receptors might be a simpler way to reduce the risk of cancer of the reproductive organs. Possibly, there will come a time when steroid-based pills will be available to reduce the frequency of menstruation, both to minimise the risk of hormonally induced cancer and to prevent the unpleasant side effects of menstruation.

BREAST-FEEDING: Chimpanzees and other great apes breast-feed their progeny for a quarter of the mother's life, and the nomadic !Kung breast-feed their babies for more than three years. Creative thinking about alternatives to this commitment began in prehistoric times, when humans started using the milk of domesticated animals to feed human offspring. The question was debated actively in ancient times: we know that Galen, the doyen of Graeco-Roman medicine, forcefully advocated breast-feeding for at least three years. Plainly, babies whose mothers lacked milk would be doomed without a substitute food, an eventuality that mothers seem to have addressed, at least since the time of Genesis, by taking a wet nurse (Genesis 35:8). The custom died out only in the early twentieth century.[42]

Industrialisation was catastrophic for infant feeding. Women working in factories abandoned breast-feeding on a massive scale and resorted to disastrously filthy preparations of cow's milk and unhealthy proprietary products. The infant mortality statistics of British working-class districts in the nineteenth century are truly appalling – 40 to 55 percent in 1858.[29,43] Gastrointestinal disease and the intestinal form of tuberculosis caused by *Mycobacterium bovis* typically originated in contaminated milk and were major causes of death among nineteenth-century children. Disgracefully, tuberculosis from *M. bovis* persisted in Britain until the 1940s because farmers were hostile to pasteurisation and tuberculin testing of cattle herds.[44]

Breast-feeding was abandoned steadily through the early twentieth century, falling to 60 percent in the 1920s and to 24 percent in the 1970s in Britain; similar figures exist for other countries.[45] Commercial dried milk for babies, based on full cream cow's milk supplemented with sucrose (cane sugar) that could be reconstituted with water, was used with little circumspection. The new baby foods were sterile and disease-free – unlike

Victorian baby foods – but the idea that cow's milk was equivalent to human milk was a naïve and untested assumption.[43] The milk of every mammal is a uniquely adapted cocktail containing many energy sources and nutrients (but not sucrose), plus maternal antibodies against many kinds of microbe. It includes vitamins and a vast range of proteins – some comparable to hormones that stimulate the growth and function of particular types of cells with antimicrobial or anti-inflammatory properties. Cow's milk, for example, supports an offspring that grows at four times the rate of a human baby.

In the 1950s, when the more blatant evils of artificial baby milk had been eliminated, surveys seemed to suggest that breast-feeding was not particularly beneficial, but twenty years later authorities definitely knew they were wrong. Many baby foods were harmful because concentrations of salts and energy sources were too high. Some parents were making feeds more concentrated than the manufacturer's recommendation. Public health authorities now strongly endorse breast-feeding for the entire first year of life, and public enthusiasm has revived substantially, although better artificial baby foods are available.

The alternative to breast-feeding spells potential catastrophe for the developing world. Apart from the obvious benefits – that mother's milk is free, sterile, and nutritious – breast-feeding protects against infection through the maternal antibodies present in milk. These are retained in the gut of the infant, protecting against gastroenteritis, the major cause of infant mortality in the world today.[42] Surveys prove overwhelmingly that infant mortality in these settings is less common when babies are breast-fed. International health authorities are often outraged by baby food manufacturers, such as Nestlé, who persistently pressure credulous Third World mothers into abandoning breast-feeding for the convenience of prepared food and the ever-present risk of gastroenteritis from contaminated water.

Baby foods are made to reflect the composition of human milk, but no absolute standard exists, as women vary enormously and even individual women may fluctuate. The possibility that hitherto unrecognised minor constituents may be of great importance is still credible. The role of the essential fatty acid linolenic acid in brain development was unknown until the 1990s.[45] Breast-fed babies generally put on weight faster for two or three months and then grow more slowly than bottle-fed babies, although growth measured by length or head circumference is essentially the same in the two groups. Babies fed on human or cow's milk acquire a very different microbial flora soon after birth; their exact significance is unknown but may relate to the ability of breast milk to protect against food allergies. Long chain polyunsaturated fatty acids supplied in breast milk may also be involved in establishing optimal control of blood pressure during infancy, as artificial baby food supplemented with them seems to confer lower blood pressure in later life.[46]

Surviving Childhood

Demographers reckon that between A.D. 1 and 1750, the doubling time of the world's population was only about 1,200 years. However, by the end of this period births were exceeding deaths by a substantial margin in the great population centres of the world, and the growth rate was set to accelerate. Improved nutrition and sanitation were significantly reducing the spread of infectious disease, but probably the key element affecting life expectancy was the diminishing impact of epidemic diseases such as plague and smallpox. Formerly, a sudden ferocious outbreak would cull a substantial proportion of a population, fatally dislocating society and restraining population growth. Well-authenticated epidemics fatally wounded the Greek and Roman civilisations in the fifth century B.C. and fifth century A.D., respectively; plague brought catastrophic devastation to medieval Europe and China, arresting population growth for several centuries. During the seventeenth century, just as Europe was beginning to industrialise, the Amerindian civilisations of Central and South America were utterly annihilated by their fateful exposure to smallpox and a succession of European diseases (see Chapters 9–10).[47]

The sudden acceleration of population growth in eighteenth-century Europe began when previously epidemic diseases became endemic. During the previous few centuries, sea and overland travel had brought the big population centres of the world into contact, and diseases such as plague spread along these routes. At first, these contacts ignited ferocious epidemics amongst the previously unexposed populations, but over time they became less virulent and increasingly confined to children who had no immunity. The burden they placed on children remained great, but births began to exceed deaths by a significant margin. In mid-eighteenth-century Europe, China, and the North American seaboard, explosive population growth had begun.[47] By the start of the twentieth century, child mortality rates were falling sharply in the industrial world. Hygiene was improving steadily as sanitary regulation entered the heart of public administration.

We will see in Chapter 9 how improvements in sanitary engineering and a host of official interventions began to limit the harm done to the survival of children by poor hygiene. By 1905 in Britain, a new profession of health visitors was created, whose task was to introduce to socially disadvantaged families instruction in the skills of hygienic mothercraft. They won the trust of their clients and contributed hugely to the improved survival and health of children during the early years of the twentieth century. Appalling rates of infant mortality still exist in the Third World. Amongst the poorest nations, 10–20 percent of infants die in the first year of life because of enteric infections, measles, and malaria, a horrifying statistic that is unlikely to change until dramatic advances occur in primary care for children.[48]

Although birth rates in most industrial countries fell during the twentieth century, population growth continued for much of the century. Many developing countries experienced population explosions in the twentieth century, once endemic diseases came under better control and food production was assured. The population of Bangladesh, for example, has increased fourfold in the last fifty years. The world population has increased sixfold in the last two hundred years; indeed, it doubled in just forty years during the last half of the twentieth century.

MALTHUS'S GLOOMY PROPHECY: By the nineteenth century, Europe had a vast urban proletariat that was multiplying steadily, restrained only by the appalling mortality of infants and mothers (see Fig. 2.2 for British statistics). Lives were longer than in previous centuries, but recurrent epidemics, malnutrition, overwork, and alcohol abuse made existence a living hell for many. Relentless reproduction by people barely able to support their families brought another dimension to the misery caused by infection and malnutrition in Victorian society. Contemporary records indicate that countless infants succumbed to infanticide or death by neglect in "baby farms" – homes to which unwanted infants were given for a fee.[49] Former "hospitals for foundlings" still exist in many European cities, bearing poignant witness to the scale of unwanted pregnancy. Thomas Malthus, writing in 1798, prophesied catastrophe because the poor were reproducing relentlessly while failing to provide a decent life for their families. The outcome, he predicted, would be famine, pestilence, and war unless population growth could be contained. Today, we would say the situation he dramatised so effectively was the worst kind of unsuccessful breeding. As a clergyman, he saw this dismal scenario as God's way of rewarding laziness and unchasteness and seems never to have suspected that improved

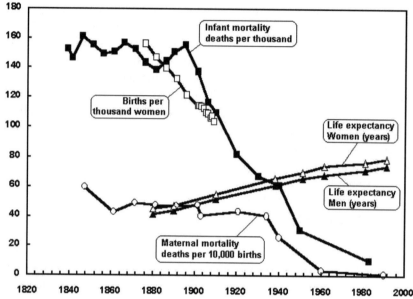

Figure 2.2. Life expectancy, birth rate, and mortality rate of mothers and infants in England and Wales. *Sources:* life expectancy: 50, 51; birth rate: 29, 51; infant mortality: 50, 51, 52; maternal mortality: 34, 52.

agriculture, sanitation, and science could undermine the Almighty's gloomy design. Later in the century, the risks he predicted were acutely realised in the devastating famines that hit Ireland and India. In Ireland, a period of rapid population growth following the introduction of the potato as a staple crop ended in extreme tragedy when the crops succumbed to potato blight. In India, repeated failure of the monsoon was the final straw for a populace already on the brink of starvation. The famines provoked no urgent desire by the political authorities to solve the problem; indeed, they were accepted, complacently, as evidence of the correctness of Malthus's ideas. His unconstructive outlook notwithstanding, Malthus demonstrated with crystal clarity that humankind's most pressing problem was population growth and unwanted pregnancy.

By the end of the nineteenth century, human reproductive behaviour was again adapting to new circumstances. Starting in France during the years of revolution and then in the rest of Europe and North America, the birth rate began to fall. A trend that began amongst the highest strata of society spread steadily to working-class families. Remarkably, little evidence exists of exactly what was practised, or indeed why, but some sort of birth control – perhaps *coitus interruptus* or abstention – was practised.[7,29] Demographers suppose that the decline in fertility was precipitated by parents' growing optimism that their children would survive.

Analysis of population trends in different countries over the last 150 years suggests that a decline in infant mortality is followed by a decline in the birth rate ten or twenty years later. In Victorian England, it appears, couples who believed that their children were likely to survive infancy limited the size of their families, and the statistics show that children were more likely to survive in smaller families. An economic motive to limit the size of families also appeared once child labour was prohibited and schooling made compulsory, as the children could no longer contribute to a family's earning power.

In Victorian Britain, abortion was commonly procured illegally by inducing miscarriage, using poisons from native plants or physical methods whose outcome was often septicaemia and death.[49] In working-class districts, there was no shame attached to inducing a miscarriage until the foetus "quickened" at three months.[7] Desperation sometimes inspired the use of extraordinary poisons to cause a miscarriage. The merits of the lead-containing antiseptic Diachylon – eventually registered as a poison in 1917 – circulated by word of mouth in the British Midlands in the 1890s.[7]

The establishment view of birth control was guided by a strange tale from Genesis. The father of Onan (one of Jacob's grandsons) instructs his son to impregnate the wife of the son's dead brother according to the ancient Hebrew custom and to *"raise up seed to thy brother."* Onan refuses, for reasons unknown, and famously *"spilt his seed upon the ground"* (Genesis 38:8–9). This unmistakable act of *coitus interruptus* apparently displeased the Lord, and the unfortunate Onan was slain without further ado. When the church, judiciary, and medical profession noted the first recorded decrease in the birth rate in Britain in the 1870s, they concluded that "conjugal onanism" was being practised on a large scale. The august columns of the *Lancet* deplored such a wilful reduction in the nation's growth rate. Interestingly, the statistical evidence of reduced family size was first seen in the families of doctors.[7] The prosecution, in 1877, of a republican member of the British Parliament, Charles Bradlaugh, and his friend, the ardent freethinker and labour activist Annie Besant, indicates the strength of feeling in establishment circles. Their crime was the publication of a British edition of *The Fruits of Philosophy*, written forty years earlier by the American physician Charles Knowlton, in which the author daringly advocated birth control. Notoriously, the pair were sentenced to six months in gaol plus a substantial fine, for "depraving public morals" – although they were released on appeal. The extraordinary publicity undoubtedly contributed to the gathering momentum of the case for birth control.

In 1916, the year of the Battle of the Somme, the birth rate in Britain was 30 percent below its 1871 peak. This statistic prompted the Anglican National Birth Rate Commission to denounce all forms of birth control other than abstinence. The Bishop of Southwark reiterated Saint Augustine's timeless silliness with a warning that sexual intercourse was intended only for procreation. Fourteen years later, the Anglicans heard a different message and no longer opposed birth control; they were followed in later years by every other denomination except the unmarried Fathers of Rome.[53]

Birth Control

Although contraception during the early twentieth century was unsatisfactory, there was little overt interest in finding better methods, and in most countries, reproduction was actively encouraged by governments for economic and military reasons. Meanwhile, physiologists and chemists were taking an interest in some newly discovered secretions whose identity was yet to be established but that evidently played a key role in reproductive physiology. The Austrian endocrinologist Ludwig Haberlandt, whose pioneering work established the basis of the menstrual cycle, demonstrated in 1918 that a secretion from the ovary could prevent pregnancy in animals and therefore had potential for birth control. Some years later, this hormone was identified as the steroid progesterone, but it was active only if administered by injection and therefore had no realistic prospect as a contraceptive. The Rockefeller Foundation quickly perceived the importance of steroids and supported a number of laboratories.

A serious search for a contraceptive compound that could be taken orally became possible when Katherine McCormick, under the influence of Margaret Sanger – a pioneer activist in the American birth control movement – proposed financing the necessary research. Gregory Pincus, already a major figure in reproductive physiology, proposed testing Haberlandt's idea by searching for steroids that could prevent ovulation when taken orally. Pincus and the Catholic physician John Rock started trials in Boston, in 1953. Very quickly, two steroids emerged as extremely potent inhibitors of ovulation when taken orally.

Trials of both drugs on large groups of women in Puerto Rico and Mexico were extremely successful, but almost immediately the business side of the companies involved got cold feet. Hints emerged from the Vatican suggesting that all products of the company might be boycotted by the faithful, but Pincus convinced them that the commercial potential of the product far outweighed the risk. However, the FDA was diplomatically asked to license the products as drugs to regulate menstrual disorders and not as contraceptives in order to circumvent any further difficulties with the church. Full-scale trials of the Pill, carried out in 1957 in Puerto Rico, Mexico, and Los Angeles, endorsed emphatically its safety and efficacy and showed conclusively that fertility was restored when the treatment stopped. Within a few years, millions of women were taking the Pill, with few reports of serious adverse consequences. Later, when the number of subjects was very large, a small but significant risk of serious blood clots was apparent. This was greater for smokers, but still substantially less than the risks of childbirth. In order to improve safety, the oestrogen element of the Pill was lowered as a way of minimising the problem.[54]

The development of the Pill was never financed by the American government. Before 1959, NIH was expressly forbidden to support birth control research, probably because of archaic state and federal laws. The Comstock Act of 1873, in particular, prohibited birth control by married couples and the sale of contraceptive devices.[55] In an important ruling in the case of *Griswold v. Connecticut* in 1965, the U.S. Supreme Court declared such laws unconstitutional; it later ruled on other laws relating to unmarried couples. Meanwhile, as the world birth rate accelerated to a doubling time of just forty years, organisations such as the World Health Organisation (WHO) were increasingly convinced that every aspect of population policy needed research. Later, the U.S. government became a major sponsor of reproductive research, but during the Reagan and Bush presidencies all contributions to the United Nations family planning organisation were cancelled.[56] Some poor developing countries, recognising the importance of the problem, dedicate a large proportion of their research budget to the goal of preventing unwanted births.

Well-known and far-reaching social changes followed these events. With greatly reduced risk of pregnancy, women were free to pursue careers and to manage relationships in a way that revolutionised every aspect of society. The birth rate is now so low and family size so small in Western Europe that population growth of the indigenous people has almost ceased. Surveys suggest that married women, particularly in formerly staunchly Catholic southern Europe, are refusing to have more than one child unless they can be certain their partner will contribute equally to the upbringing of children. The impact on people of the developing world was inevitably less direct, as the cost of the Pill was far beyond their resources. WHO campaigns to introduce contraceptive implants and the raising of consciousness about other forms of contraception are making some impact. Nonetheless, according to the CIA, which apparently finds no difficulty in collecting this kind of data, contraception is practised very infrequently in the world's poorest nations.[48]

The enthusiastic acceptance of the Pill by millions of women was an important watershed in the evolution of attitudes toward ethics and moral authorities. The case against birth control was that the life of a new individual begins at conception, so that prevention of implantation of a fertilised egg or termination of the life of a *conceptus* was morally equivalent to a murder. In the opposite corner, biologists and philosophers argued that no violation of personhood was involved in contraception or early-stage abortion because an embryo acquires the moral status of a person only when human attributes are developed.[57] The precise moment when personhood was established was open to some debate. Most thought this was certainly after the time when twin formation (i.e., division of the conceptus into two separate individuals) was still possible. However, it was arguable that the embryo acquired personhood after it was fully formed in miniature, or indeed, just as plausibly, at the moment of birth. Moreover, the idea that life becomes "sacred" at conception ignores the status of the eighty percent of embryos that spontaneously abort in normal reproductive life and the sperms and eggs that are indubitably "alive" before fertilisation. We will see certain people argue that every termination of pregnancy, or indeed any failed implantation, is equivalent to the loss of "a human right" normally associated with a living human being. As the fate of most fertilised eggs is oblivion, this is clearly sentimental rather than rational. In spite of the unyielding stance of the Vatican, probably most Catholics ignore its authority by preventing unwanted pregnancies.

MEDICALLY INDUCED TERMINATION OF PREGNANCY: The central ethical issue of when the embryo acquires the moral inviolability of a person was decided in British law in 1967, not by metaphysical jousting but in the rough-and-tumble of Parliament. The law, the first of its kind in Western countries, was drafted on the advice of the British Medical Association (BMA). This proposed legalising abortion for women whose health was threatened by a pregnancy or who were carrying babies with a serious abnormality, up to twenty-eight weeks after conception. During the debates, the legislation was amended, against the wishes of the BMA,[58] to permit abortion ostensibly in situations where the physical or mental health of other siblings was at risk. The legislation eventually became the legal basis for abortion for "social reasons," or in other words to solve the problem of unwanted pregnancy. Within a few years, the number of abortions per year reached 100,000 (one for every two hundred women between fifteen and forty-five years old), and 98 percent of all abortions were performed for "social reasons."[59] Other countries eventually drafted laws that explicitly permitted abortion on demand during the first trimester. Abortion was legalised in the United States in 1973 following the case of *Roe v. Wade* in the Supreme Court. The Court ruled that a state could not penalise a woman for having an abortion approved by a doctor in the first three months of pregnancy, and until six months if the mother's health was at risk. Effective hormonal contraception has been available for forty years, but planned parenthood still remains elusive. In 2000, one abortion occurred for every three live births in Britain – and very similar statistics exist for many other countries.[48] Thoughtful commentators suspect that secular society treats the issue too casually and glosses over the anguish of women who feel compelled to agree to an abortion, particularly after twelve weeks of pregnancy. In 2003, one in eight abortions in Britain were for foetuses over twelve weeks old.

The British law initially allowed abortion up to twenty-eight weeks after conception, but as babies born at twenty-four weeks survived increasingly well, there was a compelling ethical reason to revise the law. Moreover, two years before the British law was enacted, *Life* magazine published pictures of a foetus *in utero* with the most explicit evidence that it was largely formed at sixteen weeks (see p. 91). As the law stands, the use of abortion is legal and entirely a matter for individual conscience, although the pro-life movement remains implacably hostile to abortion if the mother's life is not threatened. Debate continues in Britain on what is an acceptable cause for abortion after twenty-four weeks. The case for abortion is absolutely clear if the mother's life is at risk or the foetus is seriously malformed and unlikely to survive after birth. But what about a foetus that may develop a cleft palate or club foot? The pro-life community believes that such an abortion is a crime and "wrongs" a person who could have had a life as worthwhile as people currently living with the same disabilities. They argue that the corrective surgery required is relatively trivial.

With such enormous demand for abortion, a search for "morning after" treatments or medically induced termination was inevitable. The first
FDA-approved treatment, which appeared in the early 1970s, was a Pill
containing progesterone and oestrogen derivatives that stopped implantation of the fertilised egg within a few days. The method was safe and more
than 75 percent effective, but the adverse effects of nausea and vomiting led
to the development of novel steroids, such as levonorestrol, that prevent
implantation without side effects.[60]

A radical new method of inducing termination of more advanced pregnancies without surgical intervention was invented in France in the 1980s.
This was based on another steroid, called mifepristone, that antagonised
the progesterone receptor, stopped menstruation, and caused ejection of
the embryo. Unlike patients undergoing surgical termination, the subject
could be treated up to seven weeks after the last period and needed no
hospitalisation, although a drug (prostaglandin) to promote expulsion
of the embryo is needed within two days. Promising clinical trials inevitably made mifepristone a flash point in the conflict with the antiabortion
lobby.[61]

As soon as market authorisation was granted in 1989, the WHO and the
Population Council (a New York–based, nongovernmental reproductive
rights group) wanted to supply the drug for trials in developing countries.
At a moment when most companies would be feeling some satisfaction at
the success of their project, the management of Roussel suddenly withdrew
the drug, fearing a boycott of their other products. The Chinese government, with powerful reasons to control the population explosion and none
for respecting papal influence, promptly developed their own version. To
outflank the opposition at Roussel, the director of the project persuaded
the French government, who collaborated in the project, to declare the
drug to be "the moral property of women." The product went on sale at
authorised pharmacies in France beginning in April 1990, to general satisfaction and minimal protests from general practitioners. Within ten years,
it was in use throughout Europe, but the company still vetoed foreign
licenses by its subsidiary companies. Roussel's moral scruples eventually
crumbled slightly when they assigned the European rights to a company
called Exelgyn and the American rights to the Population Council, but the
company still refused to supply the raw material for the U.S. market. Intervention by President Clinton and a declaration of safety by the FDA in 1996
were not enough to get mifepristone a license for sale in the United States.
However, this changed when the financiers George Soros and Warren Buffet

took responsibility for its manufacture by a small company in September 2000.

Assisting Reproduction

Humans are notably successful at reproduction and yet paradoxically are not particularly fertile compared to, say, baboons or cattle. The highest chance of fertilisation in each month may be only about one in four for a twenty-five-year-old woman in good health at the peak of fertility.[39] About eight out of every ten fertilised eggs undergo spontaneous abortion during the first ten weeks, the conceptus disappearing unnoticed by the mother. Fortunately for the species, couples persist until they are successful, but nonetheless a small proportion of healthy young people are unlucky. Fertility may be affected by a hormone deficiency, blocked fallopian tubes, incomplete development of the reproductive system, defective sperm, or apparently symptomless sexually transmitted diseases such as *Chlamydia*. Some of these problems are soluble by treatments such as artificial insemination, use of drugs that induce ovulation, and surgical correction of anatomical problems. When all of these options are exhausted, *in vitro* fertilisation (IVF) is a possibility – the venerable scientific Latin means "in glass" or outside of the body. The feasibility of IVF was first demonstrated in 1937 using rabbit eggs; it became widely used for animal breeding during the 1960s and 1970s, with no apparent deleterious effects on the offspring. American authorities, seeing the way research was moving, imposed a moratorium on IVF of human eggs in 1975 in laboratories supported by federal funds.

THE FIRST TEST TUBE BABY: Louise Brown, born on 25 July 1978 in Oldham, Lancashire, was the first baby conceived by IVF. This event was the culmination of collaboration between the gynaecologist Patrick Steptoe and the Cambridge University physiologist Robert Edwards and had originated in work started by Edwards in the 1960s.[19] The idea of fertilisation outside of the body was not particularly daunting to a zoologist who believed that he could recreate conditions comparable to those in the uterus and thereby facilitate fertilisation. The fertilised egg would then be transferred to the prospective mother's womb to implant on the wall of the uterus. Edwards was convinced that birth defects were unlikely, because when early cell divisions went wrong the embryo aborted spontaneously, and the early embryos of other mammals were surprisingly robust at the ball-of-cells stage (the blastocyst). Indeed, if a few cells are damaged, the outcome of fertilisation is usually unaffected. Obviously, significant birth defects attributable to the treatment would make the procedure totally unacceptable.

Obtaining eggs from the prospective mother at the optimal time is still the critical step in the procedure. In the clinic, an injection of a mixture of hormones – chorionic gonadotrophin – initiates the release of eggs, and about thirty-four hours afterwards, ripe eggs are recovered with great care. This is done under general anaesthetic using a technique called laparascopy, which permits the physician to view the ovary and eggs, using fibre optics, through a small incision in the egg donor's abdomen. Each egg is placed in a droplet of embryo culture medium along with a sample of sperm, diluted to the right concentration, and enclosed in an inert oil. Over the next few hours, a sperm enters the egg and the egg starts to divide. Two days later there should be eight cells, and four days later a ball of about one hundred cells, now ready for implantation. Normally, the uterus is prepared for a pregnancy in each menstrual cycle by the hormones oestrogen and progesterone, which are secreted by the ovary when the egg is released. Edwards and Steptoe reasoned that the uterus of the waiting mother must be prepared in the same way by exposure to these hormones in order to ensure successful implantation.

At various landmark stages in the project, progress was reported in scientific papers to *The Lancet* and *Nature*, signalling unambiguously the intention of eventually performing IVF. Typically, these papers provoked a barrage of comment. The publication of 14 February 1969 was condemned as "morally wrong" by Dr George Beck, the Roman Catholic bishop of Liverpool, but welcomed by the veteran labour politician Shirley Summerskill, who could not see the ethical problem. Indeed, she remembered it was St Valentine's day and declared that "no better day could be found to celebrate the meeting of sperm and ovum. It is a matter of a woman who wants to be a mother and who is unable to obtain fertilisation in any other way." An IVF pregnancy was still ten years away, but the newspaper leaders made a conceptual leap to a contemporary report of the "cloning" of frog embryos and the possibility that IVF would be used for "selective breeding." In spite of the widespread clamour, the Oldham Hospital ethics committee, composed of lay people and clergy, approved the plans of Steptoe and Edwards. The quest for an IVF pregnancy was sustained by a succession of infertile volunteers who knew they were unlikely to benefit individually.

The Medical Research Council of Great Britain was not inclined to support the work, but the Ford Foundation had no hesitation.

Ten years elapsed before the first successful IVF was performed on a woman whose fallopian tubes were completely blocked. The pregnancy progressed normally, and a baby was delivered by caesarean section. The report of Louise Brown's birth appeared in the *Daily Mail* and not in a scientific journal, which *Science* – a high-profile representative of the genre – suggested was a deliberate snub to the scientific community.[62] Sceptics sniffily implied that success was just a flash in the pan and would not be repeatable, but very soon it was abundantly clear that this was a great golden nugget. Within a few years, the procedure was in use routinely throughout the world, blessing many couples with the child they thought they could never have.[63] By 1983, Edwards and Steptoe could report success for one in three implantations. The outcome of the treatment was most favourable when more than one embryo was placed in the uterus and when the mother was under forty years old.[64] The accumulated statistics of their efforts have vindicated their faith completely, apparently showing no greater risk of birth defects after IVF than in normal birth.

"REPRODUCTIVE TECHNOLOGY": Within a decade of Louise Brown, IVF was widely available through national health services and private clinics and was even covered by insurance in the United States. Later, fertility clinics were often privately funded and sometimes encouraged desperate couples to participate in speculative or risky schemes. Very early embryos of many mammals can be frozen and stored at extremely low temperatures until the time is ripe for implantation into a foster mother. The idea of freezing human fertilised eggs in preparation for implantation was an attractive strategy for IVF clinics and indeed has become standard practise. Because they can be used one at a time until a successful pregnancy ensues, the cycles of donor and recipient need not be synchronous, and the resulting foetus is not harmed. In recent years, evidence has emerged that potentially dangerous ectopic pregnancies are three times more likely to arise after implantation of frozen embryos.

A solution to male infertility caused by low sperm numbers, nonmotile sperm, or missing sperm ducts is now available using intracytoplasmic sperm injection (ICSI). This is done by introducing a single sperm, using a fine glass needle under a microscope, into the cytoplasm of the egg cell – the matter that surrounds the nucleus. As in normal fertilisation, the two nuclei fuse; the embryo then develops *in vitro* initially before being transferred to the womb of the prospective mother. Following successful trials with rabbits and cows, the procedure was used for the first time on humans in Belgium in 1992, and four healthy babies were born to couples with infertility caused by severely impaired sperm. Desperate couples around the world were soon demanding the treatment, and the method was modified to use immature sperm from a testicular biopsy.[65]

No manipulation of human fertility can be contemplated without some anxiety about the possibility of birth defects, but after more than 20,000 births worldwide, they appear to be no more common using IVF than in natural conceptions. Clearly, boys conceived by this method are likely to become infertile adults if their father's infertility was genetically determined by the Y chromosome. Although this does not threaten the genetic health of the species, the method is clearly flagrantly bypassing the normal rules of natural selection. Very influential figures in the field are clearly sceptical of whether the method is adequately understood.[66]

Reproductive technology can provide humane solutions to hitherto insoluble personal problems but may provoke extreme consternation.[67] Women without ovaries but with an otherwise normal uterus can nourish an IVF egg from another source and can become mothers after oestrogen treatment. Women who fear they carry a hereditary disease can avoid the risk of bearing a genetically damaged child by becoming pregnant with someone else's eggs. Women suffering from cancer can have their fertilised eggs frozen before they undergo chemotherapy and then have them implanted once chemotherapy is complete. American fertility clinics will pay donors for eggs – an illegal practice in Britain – and furthermore will put a premium on eggs from beautiful and intelligent women. Even more challenging to tradition is the notion, frequently floated in the media, that women with prospects of a glittering career might want to freeze their fertilised eggs until a golden twilight when they will devote themselves to child rearing. That ethical checkpoint is some distance away – and indeed, most clinics would think it pointless to accept clients for IVF who could have their own children – but the commercial imperative might change that. Young women who have undergone chemotherapy (e.g., for leukaemia) present a more urgent problem, because they are probably not ready to contemplate storing fertilised eggs. However, a solution may be at hand, as unfertilised eggs have been frozen successfully and fertilised by ICSI after thawing, although little is known about the usefulness or safety of the procedure.[68] Early in 2004, it was reported that a frozen ovary implanted in the peritoneal cavity could develop mature eggs, which could be removed and fertilised by IVF. The immediate application is for women who must have chemotherapy, to be able to preserve their eggs, to be fertilised later. The press quickly realised that any woman could in principle delay having her babies until she was postmenopausal.

The first great shock to traditional attitudes that followed the invention of IVF came in 1985, when Kim Cotton was famously paid £6,500 to be a surrogate mother for an infertile couple whom she had never met. Ms Cotton believed that her role in the affair was a supreme gift that could never be adequately rewarded monetarily, but the press affected horror and the British government promptly outlawed surrogacy. In 1990, they changed their minds in instances where surrogacy is not for profit (i.e., expenses only). By contrast, surrogacy has become a commercial enterprise in California, where the cost of a baby is reckoned to be forty-five to sixty thousand dollars. Illuminated in the cold light of ethical thinking, this practice is often called "commodification of reproduction," because a service is purchased and the absolute historic right of a mother to her child is sold. The precise degree of odium that should be attached to the transaction is hotly debated, but the practice is arguably less reprehensible than business relationships with the "oldest profession," another example of commodification of a sacred part of reproduction.

The womb is still capable of nurturing a baby after the menopause, so inevitably the moment came when a postmenopausal woman would become a surrogate mother. In one instance in 1996, a fifty-two-year-old woman, five years past the menopause, became the host for her twenty-two-year-old daughter's fertilised eggs. The daughter, born with ovaries but without a womb, would never be able to have her own child. Once again, a laudable and humane solution to a distressing personal problem created a precedent and an ethical checkpoint. How should one view an IVF pregnancy of a woman in her sixties? The child could be harmed by a setback late in pregnancy or by the mother's lack of fortitude in her eighth decade, as the child reaches adolescence – or alternatively, Jeremiah's notwithstanding, the story could end happily. The press usually regards such a development as an exhibitionist challenge to the status quo, somewhat less admirable in a grandmother than bungee jumping, and in France the practice is now forbidden by law. The possibility that aged fathers may not survive the childhood of their offspring is not considered relevant.

To improve the success rate of IVF for women who have been unlucky, another novel and questionable technique has been used. This involves injecting a small amount of cytoplasm from a healthy donor egg into the subject's eggs in the hope that mitochondria – the source of chemical energy in the cell – will somehow kick-start development of the conceptus. Although healthy children have been born using this procedure, the FDA, alarmed at reports of an increased risk of chromosome abnormalities, has insisted on a full clinical trial. Others believe that the indeterminate genetic status of the progeny caused by the mitochondrial DNA is unsatisfactory.[66,69]

Concern about another aspect of IVF came to a head in the British House of Commons in 1985. A bill was proposed to make any "experimental handling" of fertilised human eggs illegal if the purpose was other than to implant them in a named woman with the aim of bringing them to term.[70] The intention of the bill, masterminded by the archconservative Enoch Powell, was to prevent the use of embryos in research. Embryologists successfully petitioned the government to prevent the bill from becoming law, arguing that without research, progress in improving fertility and the diagnosis of genetic diseases would cease.[70] Most governments have regulated embryo research since 1991, and in Britain, ethical guidelines formulated by a committee stipulate that frozen embryos must be destroyed after five years. Investigations of fertilised eggs can be made only on the "pre-embryo" in the two weeks after fertilisation during which the embryo and the placenta are indistinguishable. Edwards argues that fertilised embryos in excess of the needs of an IVF patient – what he calls "spare embryos" – are ethically acceptable material for research within the two-week period.

More than two decades after the birth of Louise Brown, no more than 50 percent of IVF embryos implant, and only 20 percent result in a live birth, but more disquieting news has appeared in recent surveys.[66,71] These indicate a significantly higher incidence of major birth defects (such as cerebral palsy) in IVF and ICSI babies. IVF babies are also more likely to be underweight, which carries with it an increased risk of impaired health in middle age. Some of these problems originate when two or more embryos are implanted, but even when only a single egg implants, the development of the embryo may be impaired by the conditions used to induce ovulation or culture the embryos. Some reproductive physiologists advocate – perhaps belatedly – more careful monitoring of the outcomes of the new technology.[66]

Challenges, Risks, and Rewards – a Summary

Reproduction is arguably the most important thing that humans do. Its imperative dominated the ancient world – nowhere more powerfully expressed than in the Book of Genesis (see Fig. 2.1) – and continues to this day. In every era, the challenge of breeding successfully has perplexed people from every social stratum, from monarchs looking for heirs to the very poor faced with raising numerous offspring. Slow, steady population growth, with high rates of infant survival and high life expectancy, emerged in the twentieth century as the almost universal ideal of successful breeding, although it still eludes many parts of the developing world. By contrast, within living memory, in the developed world, some women have had more than twenty babies, although such a prospect now seems unthinkable. Such fecundity probably became possible only when our ancestors discovered that babies could be fed with milk from domesticated animals or that babies could be weaned onto solid food after one year of life. In ancient times, these discoveries permitted high birth rates, which were usually restrained by high rates of infant mortality. As hygiene improved, our species was propelled rapidly into becoming the most numerous vertebrate species in existence. Today, with a global population of about six billion – perhaps reaching nine billion in 2050 – the environment of the planet is seriously threatened by the sheer mass of humanity. In the Third World, infant mortality remains high, at least in part because of the shameless pressure from Western baby food manufacturers to override the biological common sense of breast-feeding.

The discovery of safe, effective hormonal contraception in the 1950s signalled unequivocally that unwanted pregnancy was avoidable, at a time when the world population was doubling in forty years. During the same era, many countries legalised abortion. Unwanted pregnancy and the personal distress it entails to mothers and children is avoidable, but planned parenthood still seems elusive. British statistics indicate one abortion for every three live births, and similar figures exist for most developed countries.[48] Average population growth is slowing encouragingly, indeed below replacement level in Western Europe, creating a new challenge, an unprecedentedly aged population.

Now that infant survival is almost assured, the reproductive imperative that dominated the world of Abraham is more subdued, but couples who would otherwise be childless are still driven to undertake "assisted reproduction." This, too, was at first a risky enterprise, pioneered in a climate of official disapproval. Nonetheless, such was the demand that many developments of the IVF technique have been explored, and many seemingly insoluble difficulties in procreation can now be circumvented. As reproductive technology advances into new areas, the view is sometimes expressed that certain practices (e.g., "spare parts babies" – see Chapter 14) are not the "purpose of procreation." However, in a world where the majority of children are conceived unintentionally this seems an irrelevance, as babies generally provoke the unqualified love of their parents irrespective of whether a purpose inspired their conception.

GENERAL SOURCES: 7, 39, 47, 54, 67, 72.

3 How Life Is Handed On

The twentieth century will be remembered as the era when humankind first glimpsed an answer to life's most thought-provoking question: How is life handed on? The key, unknown until fifty years ago, lies in the singular chemical character of DNA (or, in a few cases, RNA), from which every inheritable feature of every organism originates. DNA can replicate to allow genetic information to be passed on to each succeeding generation and is also the primary template for making the protein products in almost every living organism. Most of our genetic information originates in chromosomes selected from the parental sets, but specific maternal and paternal contributions also exist.

DNA Transmits Genetic Information

Once scientists started to understand how the characteristics of animals and plants were inherited, they inevitably wondered what part of a cell carried genetic information. The answer to this profoundly important question came from an unexpected quarter – a lowly pneumonia-causing microbe called *Pneumococcus* – and the penetrating insight of a clinical pathologist, Fred Griffith, in London in 1929. In order to understand his esoteric discovery, one needs to know that disease-causing *Pneumococci*, when cultured in a bacteriologist's Petri dish, grow as smooth slimy-looking colonies; colonies of harmless strains have a rough appearance. Griffith's interest was aroused by the curious outcome of an experiment. Mice had been injected with the "rough" harmless variety of *Pneumococcus* together with sterilised cells of the "smooth," disease-causing version of the same bug. Mysteriously, he could recover cells of the disease-causing form from the mice, from which Griffith made a perceptive and bold deduction. His idea

45

was that a substance from the "smooth" bugs got into the living "rough" ones and conferred on them not just "smoothness" but also the capacity to cause pneumonia. Moreover, this was permanent and inheritable. The study of the genetic traits of animals and plants was well established in 1928, but the idea of treating bacteria in the same way was a novel one that may have seemed obscure to biologists in other fields. A thought experiment conveys something of the oddness and importance of this observation. If a mystery substance in food caused a change in the coat colour of a litter of mice, which was inherited by their offspring, most thoughtful persons would conclude that this was an extraordinary occurrence. As things were, Griffith's experiment caused just the tiniest ripple in the scientific pond during his lifetime.

Fortunately, the ripple did not pass unnoticed. Oswald Avery of Rockefeller University, who was also interested in pneumonia, saw the published account of the experiments and realised the immense importance of the observation, if correct. Avery and colleagues repeated the experiment, describing the process as a *transformation* of the genetic constitution of the recipient microbe. However, they took it one step further by performing the transformation in the culture medium used to grow the bacteria without any help from the mice. With extraordinary persistence, over fifteen years, they purified what they called the "transforming principle" (i.e., the material that caused the transformation), overcoming the difficulties imposed by its peculiar character. Their reward, eventually, was to find that it was identical to deoxyribonucleic acid (DNA) – a chemical curiosity, with no known function, discovered in many organisms in the nineteenth century by Friedrich Meischer. This important conclusion was corroborated by the discovery that highly purified enzymes (biological catalysts) that could destroy authentic DNA specifically could also destroy the transforming activity. Conversely, enzymes that could destroy proteins had no effect. The conclusion seemed clear beyond a shadow of a doubt. DNA must contain genetic information, though how this chemical information was expressed was unimaginable at that time.[73]

Few people were able to appreciate the significance of this work when it was published towards the end of the Second World War, when speedy communication was impossible and the attention of normally receptive people was directed elsewhere. The observations were extended to another bacterium called *Escherichia coli* in France, and later to another genetic characteristic, the capacity to survive the killing action of penicillin. The news made little impact on most biologists at the time, but with hindsight, scientists recognise it as a discovery of monumental importance. The conventional wisdom of the 1930s, based on misconceptions and an ill-conceived prejudice against DNA, was that a protein of some kind controlled heredity, but this was completely overturned by the Avery experiment. The niggling idea that minute amounts of contaminating protein, and not DNA, might be the genetic material remained problematic for some time, although the level of proof offered by the authors was, even by today's standards, exceptionally high. In particular, the powerful testimony of the DNA-degrading enzyme should have silenced the critics. The singular experimental situation of the Avery experiment impeded quick acceptance of its implications, although in reality the circumstances were uniquely favourable. The appearance of the bacterial strains on a Petri dish was very distinctive, and they differed in disease-causing potential. Nonetheless, the reluctance of thoughtful scientists to accept such a far-reaching conclusion at face value seems entirely understandable.

Almost as revolutionary as the idea that DNA was a repository of genetic information was the notion that naked DNA could change the genetic constitution of a microbe. Twenty years later, systems of transformation of much greater efficiency would be in use for making genetically modified organisms. The word "transformation" would persist to describe every instance in which the genetic constitution of a cell is modified by DNA. By then, the idea that the genetic constitution of bacteria was constantly undergoing genetic modification in the wild was well established. Today, the idea that DNA is the source of all genetic information is one of those special pieces of general knowledge of which few people are unaware. Indeed, this knowledge is so robust, we can exploit many of the components of viruses and cells in diverse biomedical applications to make proteins.

NEW PERCEPTIONS FROM A CHEMICAL MODEL: By the 1950s, as chemists were beginning to eye-up bigger molecules on which to test their analytical skills, DNA inevitably attracted attention. In an extensively dissected collaboration of the early 1950s, James Watson and Francis Crick deduced the structure of DNA by a process of model building.[74] One of the many fascinations of the story is their conviction – unique to them, it seems – that once they knew how DNA was constructed, biology's most profound mystery would be in their hands, which indeed it was. The last line of their momentous letter to *Nature*, the esteemed journal, written in arch tones reminiscent of Conan Doyle's famous detective, is hardly explicit or indeed persuasive. It ends as follows: "*It has not escaped our notice* *that the specific pairing we have postulated immediately suggests a possible copying mechanism for the genetic material.*" What was meant was: "Because of the capability of base pairing, each strand is a template on which a complementary copy can be assembled to make a pair of double strands that are replicas of the parent molecule." A new word, replication, rapidly entered the lexicon of biology to signify this copying of existing strands of DNA.

Publication of the letter had little immediate effect on published work about DNA during the 1950s, least of all amongst those most active in the field.[75] Forty years later, Crick remembered Arthur Kornberg – the discoverer of DNA polymerase – had shrugged it off as "mere speculation."[76] Nonetheless, an important audience did respond, because within a few years the inferred scheme for copying had been substantially proven. Importantly, it gave a physical basis for understanding how life was handed on. It also gave a physical meaning to the idea – first discussed by August Weissman in the late nineteenth century – of the difference between lineages of cells that continue into the next generation and cells that die with our bodies. Weissman called the immortal lineage – the one that makes our sex cells – the *germ line*; and he called the remainder the *somatic* lineage (from *soma*, the Greek word for body). DNA, as a self-replicating molecule, can be part of this eternal cycle of fertilisation and reproduction or can be replicated with each division of each of our mortal cells (Fig. 3.1).

If DNA contained genetic information, it raised immediate questions. How could it be read? Ever since 1940, the most widely discussed idea in genetics had been the notion of "one gene, one enzyme" that originated in the experiments with flies and fungi of the Californian school of genetics. Crick and friends quickly grasped that the building blocks in DNA must be written in a code that specifies the order of amino acids in a protein. The simplest scheme that fitted the available facts seemed to be that each successive set of three bases must represent a single amino acid; and of the three possible reading frames, only one would make sense. The chemical basis of the code was cracked in the 1960s – perhaps the greatest intellectual achievement of that decade.

In conveying how DNA could deliver information, linguistic ideas were used to describe the process, and these quickly acquired a precise physical meaning. People began to talk of the *transcription* of DNA into what became known as messenger RNA (mRNA) and the *translation* of mRNA into protein. Proteins were clearly the end point of the process, because

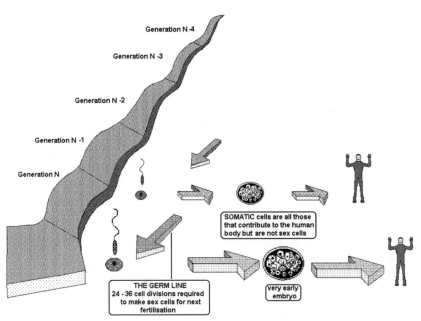

Figure 3.1. The distinction between the germ line and the somatic line. The germ line includes sex cells and the cell lineage in an embryo that makes the next generation of sex cells. Mutations can occur in either lineage, but only those in the germ line would be inherited by progeny. Somatic mutations can create malignant cells that are potentially immortal.

they formed much of the structure of the cell and were the enzymes – the vital elements that catalysed every cellular chemical reaction. By 1966, a plausible concept of the gene had emerged. Ten years later, the technology was born for manufacturing products from nature on an industrial scale in living factories (Chapter 12), demonstrating beyond the slightest doubt on what robust foundations the subject had been built.

Crick liked to call this scheme – the one-way flow of information – the Central Dogma. Apparently, he mistakenly believed that a dogma was an idea for which there was no reasonable evidence rather than a proposition that no true believer could doubt (as the Vatican would have it).[74] The idea was that genetic information flows in one direction, from DNA to protein, and that proteins never insert information into DNA. For Crick, this seemed logically impossible, because if it were true, the Lamarckian idea of evolution by inheritance of acquired characteristics became possible. Although retroviruses can convert RNA information into DNA-based information, only in very exceptional circumstances can proteins modify the information content of DNA.

Once the idea was established that the sequence of bases in DNA conveys information that gives meaning to genetic characteristics, it was a small step to connect changes of base sequence with the idea of *mutation*. Very early in the development of genetics, investigators were aware that particular genes could have several forms (controlling, say, eye colour), created by mutation.

Nucleic Acids and Heredity

By the 1950s, there was already abundant evidence that DNA was associated with chromosomes in every cellular organism, and there was every reason to suppose that DNA could be the basis of heredity in these organisms. Since then, we have come to talk of all the genetic information in any organism as its *genome*. The Avery experiment demonstrated that DNA contains important genetic information but not how it works; Watson and Crick's model amounted to a suggestion of where to look for the complete answer. The challenge was to find experimental systems in which to demonstrate the relationship between the structure of DNA and its manifestations in the character of an organism. Opportunities to test the effect of well-characterised changes in DNA sequence on particular proteins soon occurred. Instances examined ranged from the effect of the human sickle-cell anaemia mutation on haemoglobin structure, to proteins made in the test tube programmed with specific sequences of ribonucleic acid. A vast body of evidence now exists that genetic information is transmitted from DNA to protein in every organism, using a universal genetic code.

The question of how life is handed on was very pertinent to the study of viruses. The need for vaccines against viruses was of great medical importance, but the basis of their life was almost unknown, and an alternative system of inheritance was a theoretical possibility. As the subject developed, a variety of different genetic organisations was uncovered, involving many novel biochemical steps that would find use in the developing gene technology industry. Later, when a serious epidemic emerged caused by the previously unknown human immunodeficiency virus (HIV), the value of this investment was very obvious.

Many viruses contain DNA in the two-stranded form described by the Watson–Crick model, but other viruses have substantially different designs.

Certain small bacterial viruses (bacteriophages) contain just a single strand of DNA that replicates immediately after infection to create a complementary strand; the double-stranded DNA then acts as a template in the usual way. Bacterial viruses propagate in two distinct ways. Some infect a cell, replicate themselves, and then destroy the infected cell – virologists call it the host – disseminating the virus for another round of infection. Others can burst a cell in the same dramatic fashion but also have the option of integrating permanently into the chromosome of the infected cell. Once there, they are passed on from generation to generation, with the host chromosome doing no apparent harm, but occasionally they may be activated by a random stimulus from the environment.

A number of classes of virus contain only RNA. The ability of RNA to play a role in heredity exactly equivalent to that of DNA became apparent in 1956 from investigations of the tobacco mosaic virus, a plant pest famous since the 1930s. The purified RNA of this virus is able, alone, to infect a tobacco plant and create a generation of new virus particles. Several years later, a bacteriophage was discovered that uses RNA in a different way. A single strand of RNA is present in each viral particle, which becomes a template to make a complementary strand once injected into bacteria; this is then copied repeatedly to make RNA identical to the infective form of the virus. RNA viruses of medical importance with other schemes of replication are known. The influenza virus, for example, contains eight separate strands of RNA, each encoding proteins necessary to secure an infection. In addition, an enzyme exists that makes these molecules double-stranded, so that they become a template for making messenger RNA. Polio is caused by a virus that contains a single strand of RNA that is the messenger used as a template for protein synthesis.

RETROVIRUSES: Every family of viruses has taught us important lessons, but none has provided more diverse or far-reaching biological insights than the tumour viruses. In 1911, Peyton Rous showed that an extract of a chicken tumour could induce an identical tumour when injected into another healthy chicken. This venerable observation attracted interest again during the 1950s, when the capacity of the viruses to infect cultured chicken cells became known. The genetic material of these viruses was RNA, but the virus could be propagated through many generations as though it were integrated into the chromosome in a passive state (called a provirus) in the manner of the bacteriophages described earlier. In the absence of a biochemical route by which this could happen, Howard Temin made the bold suggestion that RNA might be copied into a double-stranded DNA version that was inserted into the chromosome. His ideas were not well received at first, but the sceptics were soon confounded.

In 1970, Temin and David Baltimore independently found that these virus particles contained a novel enzyme called *reverse transcriptase* that could copy RNA into DNA, exactly as Temin had proposed. This unusual enzyme immediately assumed enormous practical importance as a means of making a DNA copy of a messenger RNA, and would contribute importantly to the emerging science of gene technology (see Chapter 12). In recognition of this remarkable discovery, Temin and Baltimore were awarded the Nobel Prize for Medicine in 1975.

The AIDS virus, HIV, and many viruses that can cause malignancies belong to the class of RNA tumour viruses that have become known as retroviruses because of the reversed flow of information. The DNA copy of the infecting RNA is copied a second time to make a double-stranded DNA molecule that inserts into the host chromosome at a random site by breaking and rejoining the chromosomal DNA strand. The provirus is stably integrated into the chromosome and is passed on to successive generations in each daughter chromosome. The host enzymes also transcribe the genes of the integrated DNA to make a new virus (i.e., its RNA, its coat, and more reverse transcriptase), from which more virus can be packaged and possibly released from the cell. This is an example of a highly successful parasitism that, in the case of HIV, is exceedingly dangerous, because it permits a long latent period in which a victim can spread the disease unawares. Meanwhile, the virus is gradually inactivating cells that are vital for making an immune response to infectious agents (the T-lymphocytes).

We will see later in this chapter how retroviruses or their degenerate descendants have biological consequences that are almost wholly bad. However, they provide one priceless gift, reverse transcriptase, without which gene technology would be almost impossible. They are also being considered as vectors for gene therapy.

PRIONS: Every virus investigated in the 1950s and 1960s contained either DNA or RNA, but one extraordinary infectious agent seemed to challenge this generalisation. A strange particle that causes scrapie, a disease of sheep, seemed to contain only protein and no nucleic acid. Later, it would acquire great notoriety as BSE, the bovine prototype of transmissible spongiform encephalothapies. Ten years after Watson and Crick, a self-replicating entity that contained no nucleic acid seemed impossibly unorthodox, and who could be sure a tiny amount of nucleic acid was not present to account for its hereditability? However, another kind of evidence was suggesting the absence of nucleic acid and the possibility of quite a different kind of organisation. By the 1950s, the capacity of ultraviolet radiation to kill living things had been attributed to chemical damage to nucleic acid. Although different kinds of cells varied in their sensitivity to ultraviolet light, it seemed likely that every kind of organism would be affected. Scrapie particles were unusual in that they were completely unaffected by doses of ultraviolet radiation that were lethal to all other viruses and bacteria. A British mathematician, J. S. Griffith, then came to the rescue with an ingenious and prophetic scheme. Griffith's idea supposed that the particles were composed solely of protein and that cells contained another version of the scrapie agent folded into an alternative shape that could be forced into the pathology-inducing form by infectious scrapie protein.[77] Later, when prions had become a pressing medical problem, Stanley Prusiner extended this theory and found conclusive experimental evidence in its favour. Today, we accept that the infectious agent is indeed a *protein* – known rather obscurely as PrP^{Sc} – that differs from another protein found in normal cells (called PrP^c) only in shape but not in primary structure.[78] The infectious form mysteriously converts the normal cellular form of the prion into PrP^{Sc} when it enters a cell. This essentially amplifies the total amount of infectious material in the cell, giving the impression that it propagates itself, just like an organism (see Figure 3.2). However, the prions are not outside of the Watson–Crick scheme for heredity, because PrP^c is inherited in the chromosomal DNA.

Genetic Contributions to a Fertilised Egg

Most of the genetic information in a newly fertilised egg is contained in the pairs of homologous chromosomes, one of which is derived from each parent. In addition, there are three more kinds of genetic information that come only from one parent. These are mitochondrial DNA, the Y chromosome, and certain genes whose ability to function is modified so that only one parent's contribution is active in the growing embryo. Not all of the DNA in animal cells is located in nuclear chromosomes. The mitochondria, the structure whose special function is to be a kind of power station for the cell, contains about one thousand identical small circular pieces of

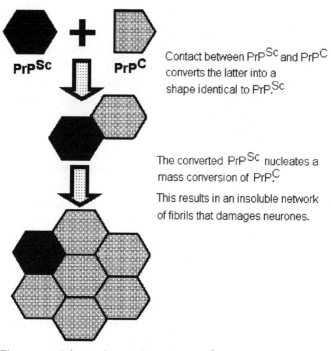

Contact between PrP^{Sc} and PrP^{C} converts the latter into a shape identical to PrP^{Sc}.

The converted PrP^{Sc} nucleates a mass conversion of PrP^{C}.

This results in an insoluble network of fibrils that damages neurones.

Figure 3.2. Scheme that explains how PrP^{Sc} is amplified by prion infection.

DNA that are derived only from the female parent. This DNA encodes only thirteen genes that make proteins important in mitochondrial function. At fertilisation, male mitochondrial DNA enters the egg but is quickly and completely eliminated from the embryo. Owing to its proximity to the energy-generating machinery of the cell, mitochondrial DNA is particularly susceptible to dangerous by-products of the energy-generating mechanism called free radicals; indeed, it has a notably high rate of mutation.[79] Moreover, a number of rare and serious genetic diseases are inherited only from the mother on mitochondrial DNA.[80] More than one percent of men have genetic deletions of mitochondrial DNA that have no perceptible effect on their health but that seriously affect the motility of their sperm because the mechanism for generating energy is impaired.[81]

A disease that seemed to be a mitochondrial genetic disease, one that causes a progressive paralysis of the eye muscles, had puzzled geneticists for a long time but has recently been explained. We know now that it is caused by a chromosomal mutation in a gene for DNA synthesis; this fault causes an accumulation of mitochondrial DNA deletions that progressively damage mitochondrial function and then muscle function. The deletions are inherited from the mother, but the root cause is the chromosomal mutation.[82]

The human Y chromosome, which is derived only from the male parent, famously contains the key gene that determines the sex of the offspring. Mutations in this gene can prevent individuals from developing into typical males and will make them develop female characteristics from early in embryo development. Females have two copies of the X chromosome, but males have two different chromosomes, the X and the Y. This has potentially serious repercussions for males, because if a gene on either chromosome is damaged by mutation, an unaffected version that can compensate for any unpleasant consequences that might ensue is not available. Examples include Duchenne muscular dystrophy (a wasting disease of muscle) and haemophilia (the notorious bleeding disease), which are found on the Y and X chromosomes, respectively. The Y chromosome is much less variable genetically than other chromosomes, probably because genetic changes affecting sperm viability are lost from the gene pool.[81]

The third kind of unequal genetic contribution to the fertilised egg is represented by two sets of genes that are present physically in both parents but their functional contributions to the conceptus are not equivalent. In order for the embryo to develop properly, one set from the mother must be active and the paternal equivalent inactive; another set from the father must be active and the maternal equivalent inactive. One class of bases (cytosine) in the DNA of either parent can be chemically modified in such a way that the gene is no longer functional. The people who discovered this phenomenon call it "imprinting," to convey the idea that the mother or father imposed a modification to the basic information encoded.

We know that products of imprinted genes from both parents are required for the development of the embryo because serious congenital diseases follow if mutations disrupt any of these or if the products are derived from the wrong parent. These genes have subtle but poorly understood physiological roles, but we know that the father's contribution favours growth in size of offspring, while the mother's contribution seems to restrict the size of the baby. Many genes are involved in this process and seem to have far-reaching effects on the development of the foetus that extend even to language and behavioural attributes.[83]

We might conclude from the forgoing that the contribution of mother and father to the DNA of any individual should be totally unambiguous, and that DNA not present in the parents would indicate that a child's parentage is uncertain. A recent report involving an incredibly rare exception to this rule illuminates an extraordinary embryological phenomenon. Two fertilised eggs present in a woman's uterus may, instead of developing as fraternal twins, fuse and develop into what biologists call a chimera (in honour of the mythical beasts made up of different animals). Any genetic test that distinguishes between the parents, by showing if organs had either maternal or paternal characteristics, would detect chimerism. If there were no pressing need to investigate, a chimera would normally be unrecognisable unless the individual had different coloured eyes or ambiguous genitalia. Recently, a woman undergoing tissue-matching tests to see if one of her children would be a suitable kidney donor was told that tests indicated that she could not be their biological mother.[84] Subsequent investigations showed she was chimeric; some of her organs had the genotype of her children, but her blood cells, on which the first tests were done, were unlike those of her children. The blood cells were, however, similar to her brother's and therefore to one of her parents.

What Is a Gene?

We know about the existence of genes at two levels. Most obviously, we see their influence in the complete organism in the inherited characteristics of people and animals, such as eye and hair colour, and in genetic diseases such as haemophilia. However, molecular biologists think of the gene as a specific stretch of DNA that is *transcribed* into RNA that is in turn *translated* into a protein.

In transcription, information is selected from the DNA by means of a DNA sequence just adjacent to a gene (called the *promoter*) that initiates transcription. Our 30,000 genes generate particular messenger RNAs under circumstances that are controlled by protein factors that bind to the promoter and that are activated by signals from elsewhere in the cell. These signals may exist in every cell or only in certain types of cell. The cells that become red blood cells, for example, are the only cells in the body that make globin, the protein part of red pigment haemoglobin, and are the only cells that receive the right signal. This signal lets loose an enzyme called RNA polymerase that makes a faithful copy of one strand of DNA. The transcript is made in the cell nucleus, edited by a process we need not consider here, and finally passes out of the nucleus into the cytoplasm, ready for translation as an RNA sequence that codes for one protein.

Translation literally means rendering the language of messenger RNA into the language of proteins. The building blocks of protein – the amino acids – assemble into beads on a string in a biochemical machine called a ribosome. Three successive bases in the RNA correspond to one amino acid. Translation starts and stops at definite signals. The start signal is almost always at the first occurrence of a base sequence that reads ATG and codes for an amino acid called methionine; translation stops at one of three triplets. Sixty-four possible sets of triplets exist, some of which encode single amino acids (such as methionine); others (in some cases, six different triplets) encode the same amino acids.

If we want to know what a gene does in the human body or in any other organism, we can infer its role from situations where the gene is defective, as in a genetic disease. In fact, the study of human genetic diseases during the first decade of the twentieth century produced the first real insight into how a mutation affects an organism. The British physician Archibald Garrod reached these important conclusions from studying a condition now known as alkaptonuria, in which the patient develops arthritis and produces urine that becomes dark on standing. Garrod inferred that the disease occurred only in individuals who inherited a defective gene from both parents. He suggested that they had a nonfunctional gene that blocked an essential metabolic process – "an error of metabolism" – that made the mysterious telltale dark substance that accumulated in their urine.

In time, the idea caught on that every inheritable defect maps to a particular place on a linear map of human chromosomes where the fully functional version of the gene would normally be found. Physically, this could represent a single change in the DNA sequence, altering a single amino acid in a crucial region of a protein or one that creates a "stop signal" that prematurely truncates the protein.

Dark Secrets of DNA

A "tone-deaf conservatory where the noise is preserved along with the music." We owe this singular image of the genome to Jacques Monod, Nobel laureate and pioneer of molecular biology, writing in 1972, when the real turbulence of its inner life was barely apparent. Monod, perhaps more than any of his peers, liked to celebrate the decisive role of molecular biology in establishing a secure foundation to the mechanism of Darwinism. "Man has to understand that he is a mere accident. Not only is he not the centre of creation, he is not even the heir of a predetermined evolution."[85] Monod's view that "pure chance, absolutely free but blind" was the root of evolution was too bleak and existentialist for some commentators.

Monod's cryptic reference to "the noise" alludes to the evidence of evolution that litters the genomes of all living things. Later we will see how there is overpowering evidence for a massive accumulation of variations in the human DNA sequence, caused by mutation and constant rearrangement of the chromosomes. The Human Genome Project tells us that DNA in different individuals differs by a single base at millions of sites, creating the extraordinary variation among people both in appearance and in the biological outcomes of their lives.

MUTATIONS: Mutations are changes in the chemical structure of DNA. They are caused by so-called free radicals generated in cells by mitochondria and peroxisomes and by the ultraviolet component of sunlight, radioactivity, and certain chemicals that may occur in the environment. Some kinds of damage are repaired with no harm done, but others persist to disrupt replication, creating a catastrophe for the individual cell or a modified sequence that will be passed on to future generations as a mutation. A mutated sequence may be transcribed into RNA and translated into a protein in which the usual amino acid is replaced by an incorrect one. Mutations in sex cells are passed on to progeny, where they may cause genetic disease. Mutations in somatic cells (typical body cells) occur spontaneously, at low frequency, and may eventually contribute to cancer, chronic diseases of late middle age, and even the ageing process itself. Cells contain systems to remove these chemical mistakes, but they are not sufficiently efficient to remove every error. Nature is like a parsimonious administrator that always chooses the most cost-effective safety device. It settles for a system that allows an organism to pass its reproductive years with just a small chance of breakdown, but thereafter problems are likely to develop, such as malignancies and indeed all the attributes of ageing.

Mutations occur more frequently in the sex cells of males than of females and with increasing frequency as males get older. This reflects the number of cell divisions required to make sex cells, and also a marked excess of metabolic activity in male cells.[81] Between the conception of a female and the development of her eggs, probably only twenty-four divisions occur, and thereafter the eggs are stored for the rest of her reproductive life. The number of divisions required to make sperm is thirty-six by puberty and thereafter increases by twenty-three every year.[86]

Several rare genetic diseases tell us just how valuable our imperfect repair process is. Patients suffering from *xeroderma pigmentosum* are unable to repair damage caused by the ultraviolet light in bright sunshine and develop skin cancer readily after exposure to moderate doses.[87] The rare genetic disease *ataxia telangiectasia* is caused by a mutation in a repair mechanism that mends double strand breaks. This kind of break is potentially catastrophic, as chromosomes could be completely severed. Patients with defects in this gene suffer from a range of symptoms including sensitivity to radiation, a high risk of cancer, immune deficiency, partial brain degeneration, and sterility.[88] Mutations occur in every living thing, but when they occur in infectious agents, their effects are profound. The rate of mutation of RNA viruses, such as the influenza and HIV, is substantially greater than in DNA-based inheritance and increases the chance of quickly developing resistance to antiviral drugs or antibodies.

We are usually protected from the consequences of mutation because chromosomes exist in homologous pairs containing two complete sets of genes, so that in general a good copy of a gene can compensate for a mutant gene. Mutant genes are described as *recessive* if a good member of the pair compensates for the malfunction of the other. However, if both parents have a mutation in the same gene, the offspring may develop the disease. Mutations are said to be *dominant* to the normal gene if they cause disease when only one copy of the mutation is present. We will see in Chapter 14 how this is an important consideration in understanding genetic disease. Geneticists use another pair of words to describe the complement of copies of a single gene. If they are identical, they are *homozygous*, and if they are different, they are *heterozygous*. All somatic cells are *diploid* (i.e., have a double set of genes); sex cells (i.e., sperm and eggs) have a single set and are known as *haploid*. Lastly in all this barrage of nomenclature, we should mention the word geneticists use for a fusion of two haploid cells, the *zygote*, which will start a new individual.

ERRORS OF CHROMOSOME SEPARATION: The process that allocates chromosomes to the daughter cells makes mistakes at a low rate, which often have serious consequences. The most common problems originate in potential egg cells. These are laid down *in utero* with the chromosomes arrested part way through the maturation process. During maturation, errors may occur in the allocation of sets of chromosomes, and a child may be conceived with an extra chromosome or with one of a pair missing. The probability of this happening is very small in a young woman but increases with age and is probably related to the length of time for which the maturation process was suspended.[89] The consequences of inheriting an extra copy of most chromosomes are so serious that the foetus usually aborts spontaneously. However, children are occasionally born with an extra copy of chromosome 13, 18, or 21. Their physical and cognitive development are always adversely affected, the most common condition being Down's syndrome (chromosome 21). Science has found no way of preventing the misallocation of chromosomes or overcoming the consequences, but the conditions can be diagnosed early in pregnancy in order to permit a termination if that is the wish of the mother. Exactly why an extra copy of a chromosome is harmful is not known, but excess production of certain gene products probably disturbs delicate balances between opposing processes in the foetus. In the case of Down's syndrome, the primary defect is a too-narrow pulmonary artery that leads to reduced delivery of oxygen to the brain that is eventually damaging to brain development. Mistakes in meiosis that result in one too many or one too few X or Y chromosomes seem to be more frequent than for other chromosomes and are less threatening to the foetus, as they affect only sexual development and fertility.

In somatic cells, failure to allocate chromosomes correctly to daughter cells usually activates programmed cell death, thereby eliminating further problems. However, a failure of this kind could propel the cell towards malignancy.

"JUMPING GENES" AND CHROMOSOMAL REARRANGEMENTS: The genome of every organism is constantly vulnerable to disquieting disruptions in which pieces of DNA rearrange at random, mostly with bad consequences. When the key protagonists in these events, racily known as jumping genes, enter new sites, they create the conditions in which a malignancy may start or in which a sex cell could transmit a genetic disease. In the 1940s, as geneticists were beginning to think of chromosomes as strings of genes in precise positions, one voice was trying to say that in the long term, genomes were astonishingly fluid. This was Barbara McClintock, whose meticulous and imaginative analysis of maize genetics suggested that sections of chromosome changed position freely from time to time under the influence of particular entities – the jumping genes – that cause this instability. For many years her ideas were misunderstood and ignored, until independent evidence for jumping genes was found in microbes. Thereafter, her efforts were recognised and eventually honoured by the Nobel Prize for Medicine in 1983. The influential role of jumping genes – or transposable elements, in formal scientific language – is now well established in the recent evolution of mammals and higher plants as a cause of genetic disease and of cancer. The key effect of a jumping gene is to join hitherto separate pieces of coding sequence, which may result in an organisation that makes weird hybrid proteins or hybrid control sequences that turn on genes at inappropriate times. These sequences are often called "selfish DNA," in honour of their incorrigible and thoughtless behaviour and particularly because they engineer uncontrolled replication of themselves.

Jumping genes invariably bring bad news in the short term. Typically, damage inflicted by a transposon (or a transposition event) affects both the regions surrounding a new insertion and the region from which it moved. An element known as *alu*, present in almost one million copies in human DNA, is a notorious and typical genetic mischief maker. Their great abundance allows them to join-up noncontiguous parts of the genome, causing deletion or duplication of sequences that initiate malignancies or germ-line genetic disease. About 0.3 percent of human genetic diseases are caused by the machinations of *alu* fragments,[90] and insertion events occur in about one in seventeen births.[91] Transposable element DNA is now reckoned to be 45 percent of human DNA,[92] but it is less active in humans than in mice or fruit flies.

The inappropriate and deadly overproduction of hormone-like molecules that may be initiated by transposition events can make cells antisocial and potentially malignant. The infamous Philadelphia chromosome – named after the city in which its association with a certain type of leukaemia was first discovered – exemplifies the joining of chromosomes 9 and 22. This places a growth factor gene from one chromosome under the control of a powerful promoter on another. This drives secretion of a hormonelike substance at a disastrously high rate that stimulates cell division of the pathological cell type. Such transposition events occur in single cells – for example, of the developing blood – even before birth. Their progeny divide increasingly rapidly, fail to take up their proper fate as blood cells, and evolve into a childhood leukaemia.[93]

Transposons have contributed so powerfully to the evolution of human genes that at least five hundred human genes contain a transposable element.[94] Some of these episodes of transposition happened billions of years ago and created important features of mammalian chromosomes, such as the antibody-forming system and the telomeres (the ends of chromosomes).[95]

Retroviruses, like the transposons, have the ability to roam through the genome, creating mutations at the sites they vacate and where they insert. The viruses can also transfer pieces of adjoining DNA to new sites, where they may unleash a powerful promoter that can drive strong and unregulated transcription of a growth factor or hormone that causes uncontrolled cell division. Such events are infrequent, but in a human lifetime they could start a malignancy. About 15 percent of the human genome is retrovirus-like sequences that must have caused extensive rearrangements in our ancestors and that are still actively at work rearranging chromosomes.[96]

Other, more subtly destructive forces that we have not begun to comprehend are at work in the human genome. A set of genetic diseases in which the pathology is attributable to a triplet of nucleotides mysteriously reiterated many times first came to light when the gene for Huntington's disease was identified. This devastating disease starts in middle age with strange uncontrolled jerky movements of the limbs followed by an inexorable decline in mental faculties and a long, slow descent into deep depression and inability to manage life activities. Typically, affected individuals have inherited a section of DNA in which a triplet of nucleotides (CAG) that codes for the amino acid glutamic acid is repeated many times rather than the normal six to thirty-five times found in unaffected people. The disease can arise spontaneously if more copies of CAG are generated by mysterious genetic changes in the germ line. The fragile-X syndrome, the most common cause of congenital mental defects (about one in every thousand boy babies), is another disease of this kind, characterised by an insertion of a piece of DNA containing repeated triplets. The condition causes an unusual apparent fragility of the X chromosome, and, as with Huntington's disease, in each succeeding generation the condition gets worse through amplification of the repeated triplets. The explanation of these weird repetitions remains an unfathomable mystery.

SEX AND THE JUMPING GENE: Low fidelity replication of chromosomes sounds like a recipe for disaster. If mutations accumulated in every generation, would not genes become a hopeless muddle, to the detriment of the species? Having two sets of chromosomes (i.e., diploidy) partly solves this problem, because it provides a back-up copy that works most of the time. Sexual reproduction becomes a critical safeguard, because fertilisation of an egg provides an opportunity for homologous pairs of chromosomes from two different individuals to find their way into the fertilised egg. This provides an important evolutionary opportunity for an individual combining beneficial traits from both parents to emerge, although the possibility of combining two negative elements is also present. The key step that prepares sex cells for their role in fertilisation is the so-called reduction division or *meiosis*. Two successive cell divisions separate a pair of homologous chromosomes and package them into prospective sex cells. Mutations in the germ line can appear in a new individual after conception, but they have no serious consequences if they are recessive to the gene expressed on the other member of the pair. In evolution, this set of "accumulated errors" sometimes proves to be a winning combination of great advantage to an individual.

JUMPING GENES IN THE MICROBIAL WORLD: The unseen fluidity of the human genome is subtler than the hectic dramas enacted in microbial genomes. We saw how Fred Griffith's observation of transformation from rough to smooth *Pneumococci* was the first revelation of the restless world of microbial genetic interactions. In the late 1940s, a form of genetic exchange we now call *conjugation* was demonstrated in *Escherichia coli*, in which a large part of the DNA in one strain of microbe was transferred to another. The "male" microbe contained an infectious agent that seemed to facilitate the transfer of an entire chromosome of the donor cell into the recipient cell. The agent was a small self-replicating circle of DNA, now called a plasmid.

Plasmids and conjugation seemed rare phenomena at first, but perceptions soon changed, as examples of small self-replicating circles of DNA became known under circumstances of great importance for public health. During the late 1940s, when antibiotics were introduced to control bacterial diseases, varieties of bacteria that were resistant to antibiotic therapy were encountered in greater frequency. In Japan between 1956 and 1964, and later in London and elsewhere, half of all dysentery cases were caused by organisms that carried antibiotic resistance genes, and some were simultaneously resistant to four antibiotics. Importantly and surprisingly, all four were found on a single plasmid that could be transferred en bloc to other strains and even to other species. This had immensely serious consequences for individual patients, who then required a new antibiotic, and was a warning that new antibiotics might be swiftly compromised. Today, bacteria carrying as many as eight separate antibiotic resistances are ubiquitous.

The explanation of this extraordinary and unexpected phenomenon was that bacterial transposons comparable to the maize jumping genes described earlier, could capture sequences from anywhere and move them onto alternative plasmids that could also infect other species. Antibiotic therapy created a new dimension to the problem because it provided a powerful selection pressure that favoured any self-replicating entity that conferred antibiotic resistance. Later, it became apparent that plasmids and genes for resistance to antibiotics had been extant before the discovery of antibiotics by humans; they were present in bacterial cultures archived early in the twentieth century.[97] Bacterial plasmids also seem to have entered the human genome during evolution on scores of occasions.[92]

In certain disease-causing organisms, the set of genes responsible for the virulence of the disease, such as toxins or the facility to attach to the surfaces of human cells, can be present in transposons. Each of the strains of cholera responsible for the historic epidemics – we are currently on our seventh – has a distinct set of virulence factors that was introduced from another source to ignite each new epidemic.[98] The evolution of many human enteric disease organisms can be traced from their DNA sequences to episodes in prehistoric times when they acquired sequences from other organisms that conferred enhanced virulence to humans – usually called "pathogenicity islands." The dysentery organism, *Shigella*, has emerged in slightly different guises on at least three occasions by incorporation of novel virulence factors on pathogenicity islands.[99] *Salmonella* species (relatives of the typhoid organism) have evolved by acquiring many pathogenicity islands.[100]

How Life Experience and Heredity Affect Health

INFERENCES FROM IDENTICAL TWINS: Geneticists have words to describe the intrinsic genetic character of an organism (the genotype) and the appearance it gives through its interactions with the environment (the phenotype). The distinction is easy to imagine if we think of two lawns grown from genetically identical seed that has two phenotypes depending on whether it is mown or not.

Life experience profoundly affects certain human physical phenotypes, too, starting *in utero* or during childhood, and the consequences may not be reversible by a change of environment. Similarly, it is a reasonable expectation that psychological, behavioural, and mental attributes will be affected, perhaps irreversibly, by life experience. The Victorian polymath Francis Galton believed the importance of life experience in establishing the characteristics of individuals could be assessed by comparing the lives of identical twins. He was one of the earliest to appreciate that identical twins – monozygotic twins, we would say today – must have identical genotypes because they originate from a single fertilised egg. Consequently, the similarity of twins with respect to the onset of diseases and life span was an indication that these characteristics could be genetically controlled. Better information available today tells us that they may not be exactly identical, but they remain our best opportunity to evaluate the importance of life experience factors. Nonidentical twins – or dizygotic twins – originate from separate eggs or zygotes that are fertilised simultaneously and should be no more alike than any set of siblings, although they have shared the same uterine environment.

Registers of all the identical twins born in the Nordic countries have been maintained since Galton's time, and others have been established since. They are used to assess the importance of heredity in every facet of human life, from life span and risk of disease to intelligence and personality. Broad measures of human health, such as life span and the chance of contracting cancer or heart disease, suggest that life experience has a bigger effect than heredity. Only about 35 percent of the variation in life span between twins is attributable to heredity; the remainder originates in life experiences.[101,102] Similarly, 80 to 90 percent of human cancer is probably attributable to some aspect of life experience, although some types of cancer are clearly heritable.[103]

Clinical geneticists use data from identical twins to study the genetic contribution to individual diseases. Twins who have a health characteristic in common are described as *concordant*, and the degree of concordance in a group of twins is a useful measure of the importance of the genetic component. For type 1 diabetes (the condition in which no insulin is made and the patient is dependent on insulin; blood glucose levels are too high because the liver does not take up glucose), concordance is close to 100 percent. This suggests that susceptibility is almost entirely of genetic origin, but importantly, a crucial life experience is required to cause autoimmune disease. Concordance is also high but not so complete for a number of chronic health problems. These include type 2 diabetes (the condition in which the liver becomes unresponsive to insulin so that the blood glucose level becomes too high), asthma, and susceptibility to tuberculosis.[104] This kind of information explains the mystery of why diseases sometimes seem to have a substantial genetic element and yet vary dramatically across regions or environments.[105] We will return to this in Chapter 6, but we must remember that many chronic diseases originate in reactions to life experiences to which our bodies have inadequate defences owing to the lack of an appropriate gene.

REALISING OUR GENETIC POTENTIAL: Through much of human history, and particularly in certain strata of society, humans rarely grew as tall as modern people. Starting in the early twentieth century, what statisticians call a "secular trend" developed towards greater height in most industrial countries, a trend that has advanced too quickly to be of genetic origin. Indeed, adult height increased by one to three centimetres for each decade until 1980, although the trend was abruptly reversed in some countries subjected to food scarcities during the Second World War.[41,106] The secular trend was largely a consequence of increased growth of the long bones that overcame the tendency to "stunting" that was such a prominent characteristic of nineteenth-century slum dwellers and of children in the Third World. Endocrinologists suspect that improved foetal and postnatal nutrition makes long bones respond with greater sensitivity to growth hormone.[106] Scandinavia passed an historic milestone in the 1980s, when for the first time anywhere there was no significant difference in average growth rate between children from rich and poor homes.[107] The growth rate of children is now taken as a sensitive indicator of the quality of a child's diet. In Chapter 5, we will see that adult height and weight, as a genetic characteristic, is inherited in a rather peculiar way. Where stunted growth is endemic, several generations of improved nutrition must elapse before children attain their genetically determined height.

Accurate estimates of the height of historic populations are not easily obtained. Anthropologists believe that, compared to the height of Cro-Magnon man, the average height of their descendants declined sharply for many millennia during the first agricultural revolution. Only in the last two hundred years has the average intake of dietary protein reached the levels required for people to grow to the same average height as Cro-Magnon man.[6,108] Modern historians see in objectively recorded heights of individuals from previous eras evidence of the biological forces operating in the those times. Records of the height of recruits to the armed forces, with appropriate statistical treatment, give remarkable and important insights into the nutritional circumstances that existed in certain societies. For Britain, such records suggest that nutrition improved in the late eighteenth century, then plummeted during the "hungry 1830s and 40s" and recovered slowly as the century advanced. Similar sources indicate important regional differences in Europe, with clear indications that the relative affluence of the British was reflected in their stature.[108]

The age of menarche gradually decreased for about 140 years and then stabilised in the 1980s at about age thirteen in most of the industrial world. Although menarche is often associated with a body weight of about fifty kilograms, a deeper statistical analysis suggests that a critical event in early childhood related to nutritional status sets a developmental trajectory and the time of menarche.[106]

As the secular trend in height has stabilised in the developed world, we are now seeing another "secular trend" towards greater mass and, indeed, obesity. The abundance of high-energy foods is clearly important, but an increasingly sedentary lifestyle may well be the key. Certain ethnic groups may be genetically predisposed to respond to excess calories by becoming obese – a topic we will return to in Chapter 5.

Increased longevity is another secular trend that demonstrates our unrealised genetic potential. Improved longevity in the early twentieth century was due to reduced child mortality but since 1970 greater life expectancy in the industrial world has been, in part, the result of better management of cardiovascular disease and cancer in late middle age. Extrapolations from current demographic data suggest that the average and the maximum life span will continue to increase for the foreseeable future.[109]

Challenges, Risks, and Rewards – a Summary

Modern insights into heredity and the role of DNA tell us more about the human predicament than any single artefact of our history. The endless changes in sequence of DNA on the evolutionary time scale are wonderfully creative, but in the short term, the machinations of selfish DNA and mutation are almost wholly bad. Mutational events and chromosomal rearrangements transmitted in the germ line cause genetic disease and other distressing developmental defects; somatic mutations contribute to many other kinds of chronic disease and, indeed, to the ageing process (see Chapter 7). James Crow[86] believes that deleterious mutations may be accumulating in the human germ line at a faster rate than ever before, now that high child mortality no longer systematically removes the least robust individuals from the gene pool. This may increase the likelihood of genetic disease in the long term. If our environment ever becomes less benign, a cataclysmic selection process could eliminate individuals with genetic weaknesses. As our understanding of heredity becomes stronger, the challenge to find ways to resist these random and seemingly pointless forces becomes overpowering. A variety of new ideas, including gene therapy, gene product replacement therapy, and avoidance of genetically flawed births are being considered (see Chapter 14). The potent knowledge emerging from molecular biology carries with it great practical risks and the potential for serious misadventure; however, we have no moral obligation to accept the cruel fates that originate in DNA.

GENERAL SOURCES: 74, 85, 110, 111.

4 Cells in Sickness and Health

Hundreds of different types of cell, each based on a single design and all derived from one fertilised egg, make up our bodies. Their sheer variety, exquisite minuteness, versatility, and fortitude make cells truly extraordinary. Our entire existence depends on the actions of cells with precisely defined roles that are performed at very particular places in our bodies, governed by a vast network of chemical messages. Ultimately, a cellular clock determines how many times they can divide, and a program exists that can sacrifice cells for the benefit of the entire bodily organisation. When we are sick, it is really our cells that are sick, and it is in cells that scientists look for the immediate cause of most diseases. This chapter is concerned with the strengths and weaknesses of cells in health and disease.

The Architecture of Cells

Just as powerful telescopes revolutionised our perception of our place in the solar system, so advances in microscopy revolutionised perceptions of our material character. Early nineteenth-century microscopists realised that all living things are made up of cells and that only division of preexisting cells could generate more cells. The immensely influential Prussian biologist Rudolph Virchow was one of the first to absorb this lesson and to see that more was to be gained in studying pathology by looking at cells than by looking at the gross anatomy of cadavers. This outspoken political progressive, employed by a famously autocratic state, saw the body as a "republic of cells" in which the sickness of cells was the cause of the sickness of the body.[112]

Radio telescopes and the discovery of black holes and pulsars forced a reevaluation of the old cosmology; similarly, twentieth-century techniques

69

for examining cells have raised previously unimaginable questions about biology and provided some startling answers. The key discovery in this microscopic world was a highly developed organisation of tiny machines with specific tasks. In the next few pages, we will consider how, in the architecture of cells, we have discovered the source of our vitality and of our vulnerability to disease.

THE NUCLEUS: Every human cell contains about two metres of DNA, packed in a conspicuous nucleus whose diameter is about one-hundredth of a millimetre. The bald facts suggest a spaghetti nightmare, but the reality of cell division is that order usually prevails. Although no organisation can be detected between divisions, with either light or electron microscope, there are increasingly strong reasons for believing that the nucleus is highly organised.

Before every cell division the DNA replicates, the nuclear membrane disappears, and the chromosomes condense from a diffuse state – known as interphase – into exceedingly compact, tightly coiled structures that become visible to microscopists. The enormity of the enterprise is hard to convey. In humans, two sets of twenty-three largish chromosomes must be sorted accurately into two identical sets. The pulling-apart process happens in a medium of high viscosity without exterior intervention, so that the force required is considerable; it is generated within each cell by remarkable subcellular motors using chemical energy. Cell division is complete when a membrane finally makes a partition between the two sets of chromosomes; the daughter chromosomes are then packed up in new nuclear membranes. The chromosomes decondense and resume their busy inscrutable life as a template for transcription.

How are these extraordinary gymnastics achieved? Every cell division in the human body uses a similar apparatus. Two sets of daughter chromosomes are apparent once DNA replication has occurred, and these attach to a remarkable structure, known as the spindle, that separates chromosomes with extraordinary precision. It is not, however, perfect; it sometimes fails in spite of many built-in quality-control steps. One important checkpoint ensures that every chromosome attaches to the spindle, through a structure called the kinetochore, before "authorising" the spindle to pull apart the chromosomes. Should this fail, the daughter cells would receive either too many or too few chromosomes. A fierce and efficient system destroys miscreant cells of this kind by a procedure known as *programmed cell death*. Even so, there remains a small but significant chance that a cell with an abnormal chromosome complement may survive.

Scientists call cell division in somatic cells *mitosis* and distinguish it from the more complicated process of *meiosis* used to make sex cells or gametes for sexual reproduction. Both processes use a spindle apparatus. Meiosis involves two successive divisions – sometimes called a reduction division – to engineer the packing of only one set of chromosomes into egg and sperm cells (unlike the pairs found in normal somatic cells). After fertilisation, the two contributions are a double set again.

Many millions of potential eggs form in the developing ovaries of a female foetus. Many degenerate *in utero*, leaving perhaps two million at birth, and the attrition continues so that only two thousand remain at puberty, the number decreasing continuously until none remain at the menopause. Typically, only one is released per month and no more than five hundred in a lifetime. The chromosomes of these potential egg cells arrest at an early stage in the first meiotic cell cycle and remain in this state for many years. The final stages of meiosis occur as eggs ripen in the adult ovaries – a time when they are inherently susceptible, at a low probability, to errors in the chromosome separation process. This can lead to the grossly unsatisfactory and poignant consequences of a foetus carrying an extra chromosome, as discussed in the previous chapter.

The ends of chromosomes, the telomeres, have special features of great biological importance. They are composed of repeated units of DNA sequence arranged in an unusual loop. In normal human cells, the telomeres get shorter in each successive cell division through the erosion of a few units of telomere. As we will see in Chapter 7, this acts as a molecular clock that determines how many divisions a cell may make. A consensus seems to be growing that control of telomere extension and the activity of the enzyme that makes telomeres respond in critical ways to DNA damage and to chronic infection.[113,114]

Apart from being a venue for chromosomal gymnastics, cell nuclei conduct a lively import-export business with the surrounding region, the cytoplasm. A double membrane envelops the interphase nucleus, in which structures known as pore complexes are embedded that act as customs houses to regulate movement of proteins and messenger RNA into and out of the nucleus. This membrane is lined on the inside with a fibrous layer known as the nuclear lamina, which plays a part in organising the chromosomes in interphase and also in the regulation of transcription. We have much more to learn about this structure, not least because mutations in genes for the nuclear lamina cause mysterious genetic diseases, which indicates that the structure performs an important but unknown function.[115]

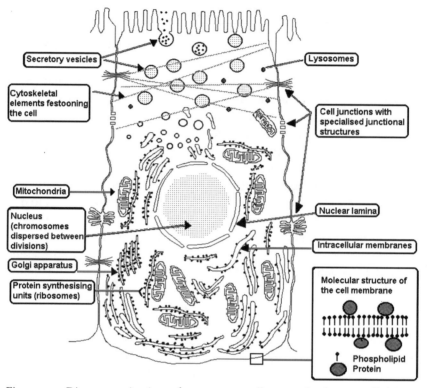

Figure 4.1. Diagrammatic view of a secretory cell at a point between divisions, illustrating the variety of organelles.

THE MITOCHONDRION – THE POWER OF LIFE AND DEATH: Generations of students have learnt to think of the mitochondria of an idealised cell as a discrete cigar-shaped structure outside of the nucleus, or perhaps lined up in impressive-looking phalanxes in muscle fibres. Their abundance is greatest in heart muscle and sperm, where great amounts of energy are consumed. Using better microscopic techniques, they appear, in most types of cell, as extended tubular networks that branch and fuse in the regions around the nucleus.[116] The outline of the mitochondrion is a membrane that encloses another extensively folded membrane system in which the important energy transactions occur (Fig. 4.1).

The mitochondrion is literally the power of life and death in a cell. It generates a chemical currency, called adenosine triphosphate or ATP, in reactions in which hydrogen combines with oxygen; if this process is ever interrupted, the cell commits suicide. Nothing better illustrates our precarious hold on life and the crucial importance of mitochondria than the consequences of suffocation. Just a few minutes of oxygen starvation irreversibly damages the brain, and a little longer will damage the heart. The underlying cause is a dramatic self-destruction of crucial cells brought about by the mitochondria. We will see that similar events underlie the pathology of many medical conditions, but also serve an important creative purpose during embryo development and in protecting the body against infection.

Damage sustained by mitochondria of a subtler kind causes cumulative damage to human health that contributes to ageing and to various chronic diseases. With advancing age, the key activity of mitochondria, the making of ATP, becomes perceptibly less efficient. As a consequence, a variety of compounds of oxygen are produced – such as superoxide – that are not the proper end point of the process. These extremely reactive substances – known as free radicals – can damage almost any biological material they encounter, causing a phenomenon known as oxidative stress. The DNA of mitochondria is easily damaged by free radicals because of their close proximity, causing mutations that impair the efficiency of the mitochondrion, leading to more oxidative stress and more mutations. Mutations accumulate in mitochondrial DNA at a rate about ten times greater than in chromosomal DNA.[117]

The existence of a genetic system associated with mitochondria, distinct from the chromosomes, was discovered first in the humble baker's yeast during the late 1940s, but it has a counterpart in all cells. In time, more than fifty rare and unpleasant human genetic diseases affecting mitochondria were found with an unusual pattern of inheritance – exclusively from the mother – that was eventually associated with mitochondrial DNA in 1988. Most diseases of this type start in middle age and get more serious as the damage from free radicals accumulates.[117]

Mitochondrial mutations that cause physical illness are rare, but about one in sixty males is affected by mutations that reduce the motility of sperm, and consequently fertility, with no apparent effect on general health.[81] A marginal reduction in ATP generation seems to affect sperm motility massively.

Each cell of the human body has about a thousand copies of the mitochondrial genome in which one might imagine a new mutation could hide and have no effect. However, that is not the case. For unknown reasons, the type of mitochondrial genome in a cell sometimes changes abruptly in a single generation. Particular molecules of mitochondrial DNA replicate more in one generation than another and seem to be passed on to the next generation with a much higher frequency than would be expected by chance.[118]

THE SHAPE OF CELLS – HOW AND WHY IT CHANGES: The boundary of the cell is marked by another miracle of biological architecture, the cell membrane – an entity visible only using the electron microscope, in which it appears as a thin double line. The key component is a double layer of small fatty molecules, called phospholipids, that have a natural tendency to form a film not unlike a soap bubble. The middle plane of this double layer is water repellent, with properties not unlike a fluid; scientists often describe it as a "fatty sea" in which other fat-loving compounds can move freely. This makes it an electrical insulator on which electrical signals can be transmitted, as in nerves. Crucially, it is impermeable to nonfatty materials unless they pass through special channels. Membranes of similar construction enclose every other unit of the cell, effectively compartmentalising each biochemical "factory." The surface of every membrane is studded with proteins, floating in the fatty sea, that act as receptors for information from their neighbours or the environment, or may protect the cell from the unwelcome attention of opportunistic microbes.

In tissues such as the lung, kidney, and gut, a continuous flow of fluid towards bodily orifices (e.g., in the lungs, towards the mouth) exists whose purpose is to wash away all infectious agents, leaving the tissue essentially sterile. The flow of liquid is sustained by co-ordinated beating of tiny molecular whips on the cell surface, called cilia, and the fluid contains anti-infective molecules that kill bacteria and viruses. A key element in this process is the flow of chloride ions and water through the membrane, facilitated by a cell surface protein called CFTR (cystic fibrosis conductance regulator). If this protein is defective, the failure of the flow of liquid leaves the lungs congested with thick mucus in which normally harmless bacteria prosper, eventually overwhelming the victim.[119] The entire syndrome is a serious disease (cystic fibrosis) that afflicts about one in two thousand Caucasians.

Microbes that cause intestinal (enteric) diseases usually exploit the CFTR channel to create symptoms of diarrhoea, a remarkably effective strategy for disseminating infection. The cholera organism, *Vibrio cholerae*, activates CFTR in the gut, which secretes water at such a rate that the ensuing diarrhoea is severe and life threatening. Historically, the high frequency of CFTR gene mutations in Caucasians may have conferred resistance to enteric infections such as dysentery and typhoid. The presence of a single copy of the mutation in humans seems to reduce marginally the effect of these organisms.

Membranes look fragile and vulnerable to an environmental knock, so the question of how cells on the surface of the body withstand wounds is an important question, though one that is not easily answered. It seems, however, that lesions in the surface membrane are rapidly filled with membrane derived from vesicles from inside the cell.[120] Membranes can also fuse in special circumstances, as when sperm meets egg. Viruses, which are packaged in membranes not unlike cell membranes, have also learnt to exploit this facility to unload their parasitic cargo into the cell. Scientists exploit the capacity of a virus to force membrane fusion to make cells of different character fuse, for creating hybridomas that secrete monoclonal antibodies (see Chapter 12). The remarkable feature of these cells is that the two components from which they are derived stay together forever.

Cells acquire their shape, not from their membranes, but from internal structures, in the manner of internal supports of a tent. This structure, the cytoskeleton, is a network of fibres that traverse the cell between attachments to the cell membrane. What a skeleton it is! Early electron microscopists failed to see it until a variety of ingenious tricks revealed its complicated structure in all its glory. At any one moment the camera captures a deceptively permanent-looking organisation in which ropes of fibrous protein festoon the cell, but the structure is highly dynamic and can be called into existence or packed away in a few moments.

Three kinds of protein rope make up the cytoskeleton. A thin fibre made from an extremely abundant protein called actin, akin to a component of muscle, is found radiating in all directions from the nucleus, with a firm anchorage in the cell membrane. Many kinds of cell make crawling movements by protrusion of the leading edge and a co-ordinated shift of the body of the cell in the same direction. Extension and contraction of the actin network, by addition or removal of subunits, drives these movements. Various kinds of cell extension known descriptively as ruffles, filopodia, lamellipodia, and micro-spikes are actin-rich membrane protrusions that locate new attachment points and contribute to intracellular communications by transmitting and receiving signals. Protrusive activity and mobility of the entire cell is an important feature of cell behaviour in healthy tissues, as in lymphocytes on the track of a microbial invader, as well as when malignant cells spread to new regions of the body.

Microbes, inevitably, learnt how to exploit the actin skeleton for their own nefarious purposes. *Shigella*, the cause of dysentery, penetrates cells of the lining of the intestine by means of an actin-based phagocytosis. Once inside the cell, the microbe hijacks the actin assembly system in order to propel itself across the cell and possibly into neighbouring cells. The gene that encodes the protein for attachment to actin is vital in establishing the pathogenicity of *Shigella*. Similar ploys are used by other bacteria and viruses of the smallpox family to enhance their disease-causing potential.

Intermediate filaments (intermediate in size between the actin filaments and microtubules) are the most permanent element of the cytoskeleton and play many roles in most cells. They make notable contributions to the character of skin cells and neurones because they provide mechanical strength. Skin is an armoured coat that is strong, flexible, and impermeable, originating in robust intermediate fibres known as keratin. Cells of the skin displace towards the surface as underlayers accumulate, and different keratins appear in the cells during this journey. As they reach the outside, the cells die, leaving tough, inert, and impermeable scales of keratin that are sloughed off in the hurley-burley of the environment. Certain rare genetic diseases (such as *epidermolysis bullosa simplex*) are characterised by severe blistering because of the fragility of the skin surface originating in mutations in these keratin molecules. The keratin filaments are anchored to the membrane through a complex structure called the desmosome, which makes a permanent attachment between adjacent cells not unlike a rivet.

Projections from neurones (the axons), which may extend for more than a metre, are very thin but have enormous mechanical strength conferred by the mass of intermediate filaments (usually called neurofilaments) that extend along their length. Subunits added to the filaments at the cell body push the filaments slowly down the axon together with membrane components that extend the tip of the neurone. The axons conduct messages away from the cell body, distributing the signal to many destinations simultaneously. Once again, we can see how the pathology of a genetic disease is rooted in cell biology. In some varieties of motor neurone disease and other neurological diseases, neurofilaments are produced excessively or in a disordered fashion and may be responsible for the death of the neurone and the paralysis of the victim.[121]

The microtubules, the third element of the cytoskeleton, participate in spindle formation, form the tips of growing neurones, determine the plane of division of cells, set up intermediate filament networks, and act as tramlines for intracellular transport.

PROTEIN SECRETION AND RECYCLING OF CELLULAR COMPONENTS:
Many specialised activities in cells are contained in membranes. One notable structure – called the Golgi body, in honour of a celebrated Italian pioneer microscopist – is a stack of membranes dedicated to preparing proteins for export. The same structure is found in many cell types, variously involved in secreting antibodies, hormones, neurotransmitters, digestive enzymes, and the extracellular proteins that hold tissues together. Proteins made on the outer surface of these membranes are squirted into long tunnel-like spaces where they are modified biochemically by the addition of carbohydrates and are then conducted to the cell surface. Finally, packaged in tiny membrane-bound bubbles or vesicles, they fuse with the cell membrane (see Fig. 4.1).

Another type of special vesicle, called a lysosome, plays an important role in breaking down and recycling the complex components of the cell surface. Over forty genetic diseases, affecting seven to eight thousand new-borns in Britain every year, can be traced to a defect in a lysosomal enzyme. These defects occur when materials that should be recycled accumulate in a toxic form in particular tissues. In Gaucher's disease, one of the commonest of these conditions, patients accumulate glycolipids – compounds of lipid and carbohydrate – in the spleen, liver, and bone marrow. This happens because the macrophages, the specialised blood cells that normally eat dead cells, fail to break down glycolipids. In the most severe form of Gaucher's disease, the brain is catastrophically damaged and the victim dies in childhood. As we will see, some progress is being made in treating these diseases by replacing the missing enzyme.[122] Genetic diseases that place insurmountable difficulties on lysosomes make the victim sick, but in old age, too, the lysosomal function is systematically damaged. This loss in function is marked by the accumulation of a brown-coloured pigment known as lipofucsin that is diagnostic of ageing cells.

Lysosomes recycle extracellular and membranous material, but another system of enormous subtlety degrades intracellular proteins as part of an organised recycling process. This process also removes proteins that are incorrectly folded or damaged by chemical modification in a protein degrading unit known as the proteasome – a delicate structure, easily inactivated by oxidative stress – after which the cell may undergo programmed cell death. In ageing mammals, this gradual inactivation of the proteasomes in neurones leads in different ways to several degenerative diseases of the brain, including Huntington's, Alzheimer's, and Parkinson's diseases and various prion diseases. In each case, brain cells fill with masses of fibrils that contribute to the death of these neurones. Although we are just beginning to understand the pathology of these diseases, they are the clearest illustration of how causes of disease are to be found in cells. Similarly, in tissues affected by oxygen deprival, proteasomes are damaged and contribute to the death of the cell.[123] The proteasome recognises and destroys proteins containing a "mistake" that causes a difficulty in folding into the normal structure. The proof-reading facility actually increases the severity of a genetic disease, because the mutated protein disappears completely. This explains why a single amino acid substitution in CFTR causes such a severe disease as cystic fibrosis. We will see in Chapter 14 that this insight may open a route to correcting many genetic diseases.

Long-chain fatty acids and other biochemicals are oxidised in another kind of vesicle that generates hydrogen peroxide and makes products that are potentially harmful to the cell. Mutations in these enzymes cause a substantial number of genetic diseases.

ANOTHER KIND OF CELL – THE BACTERIUM: Bacteria and multicellular organisms differ at every level of organisation. The small size of bacteria, their inability to form tissues, and their peculiar cell walls set them apart, but the most significant differences lie in the organisation of their genetic material and their arrangements to generate energy. Bacterial chromosomes are five orders of magnitude smaller than the human genome. They reside, not in discrete membrane-enclosed nuclei, but in an ill-defined blob in intimate contact with the matter of the cell. Energy generation in bacteria occurs in the peripheral membrane and not in a special organelle such as the mitochondrion. The bacterial lifestyle is characterised by their almost infallible ability to colonise any ecological niche as scavengers; as deadly, destructive opportunists; or as subtle parasites able to engage in the most promiscuous exchange of genetic material.

A Cell for Every Purpose – Specialists and Stem Cells

Hundreds of cell types have specialised roles in the human body. Muscle cells, with their massive blocks of contractile proteins arranged in rigid geometrical arrays that marshal substantial forces during contraction, are well known. Many other cells have a distinctive architecture reflecting their primary function. Secretory cells (Fig. 4.1) have conspicuously developed Golgi bodies. Dendritic cells of the immune system have tentacle-like protrusions for making contact with invading bacteria. Retinal cone cells, sperm, neurones, and red blood cells all have well-known distinctive features.

Some cells make just one product in enormous amounts. Reticulocytes, the cells from which red blood cells will form, make one major messenger RNA that encodes the protein globin to which haem will be added to make the red-coloured oxygen carrier. When these cells graduate from the bone marrow as red blood cells, the nucleus is ejected and their shape is transformed into the well-known convex lens shape in order to maximise the surface area for oxygen transfer. The gene that makes globin is active in just this one type of cell and is completely silent in every other cell in the human body. Behind this specialisation lies a powerful organising principle that provides cells with a "program" that gives them a character that usually cannot be reversed.

Most differentiated cells have little or no capacity to divide, but certain undifferentiated cells, known as *stem cells*, can propagate in an unlimited fashion. Some of the daughter cells of stem cells can differentiate into particular specialised cell types. Stem cells have great practical and conceptual importance because, unlike our bodies, they are essentially immortal. The value of bone marrow stem cells, for example, is plainly seen when transplants from a donor successfully cure blood disorders.

Most kinds of mammalian stem cells make a restricted range of differentiated cells. However, the daughter cells derived from a fertilised egg for the first few generations are a clear exception to this rule. Cells of this type, from the mouse or from human embryos, can be maintained indefinitely in the highly versatile stem-cell state using special culture conditions. However, with appropriate environmental cues, they can develop into any differentiated cell type – a facility that may have enormous therapeutic potential.

Since the end of the nineteenth century, biologists have wondered whether fully differentiated cells are genetically distinct from the fertilised eggs from which they are derived and whether they have lost genetic information. Isolated examples of such changes certainly occur in the animal kingdom, but how widespread they are was uncertain until recent times. Biologists now believe that mature differentiated cells are essentially genetically identical to newly fertilised cells of the same organism, with the important exception of antibody-forming cells. The evidence that differentiated cells and fertilised eggs are genetically equivalent comes from experiments in which differentiated cells are "reprogrammed" to behave like a fertilised egg that can develop into a complete animal. This is possible only because the essential genetic information necessary to direct the entire developmental program is still present in differentiated cells. Reprogramming of differentiated cells of a frog was first reported in the late 1940s. The result was confirmed and extended with many rigorous controls more than thirty years ago. The nuclei of eggs were replaced with nuclei derived from cells of various organs. In a small number of cases, the reconstituted cell could develop into a frog, and indeed a set of identical frogs was created – and "clones" of genetically alike individuals have remained a key part of science fiction ever since. This finding was published at about the same time as the first IVF birth and provoked the press to contemplate surreal fantasies, including the resurrection of dead dictators. The word "clone" is widely used in biology for any process that creates a colony of identical self-replicating entities. These can be organisms such as geraniums (raised from cuttings), bacterial colonies on a dish, or pieces of DNA amplified many times in a cellular factory.

Dolly the sheep is the best-known outcome of mammalian cloning.[124] Dolly's existence started when a nucleus from a differentiated cell of the mammary gland of an adult animal was "reprogrammed" by transfer to an enucleated fertilised egg. Proteins of the egg reorganise the incoming nucleus so that genes necessary for development of the animal are again "switched on." "Reprogramming" by nuclear transfer is technically difficult and rather inefficient. Nonetheless, the animal created by nuclear transfer seems more or less genetically identical to the animal that donated the nucleus, except in some details. The new animal would not resemble the donor with respect to its mitochondrial DNA, which would be characteristic of the recipient cell. Another difference arises because genes normally expressed in the original donor somatic cell are not completely silenced after nuclear transfer, and the converse – genes normally active in a developing embryo are less active in a cloned embryo. Scientists interested in these phenomena strongly suspect they are only at the start of understanding what is involved in "reprogramming" a nucleus.

Several mammals have now been cloned from a somatic cell, but the idea of starting the development of a child from a nucleus of a human somatic cell is regarded, almost universally, as ethically unacceptable. The procedure is illegal in most countries, not least because of the fear that the child might be harmed. We will return to this topic in Chapter 14, where we will consider the case for using nuclear transfer to create embryonic stem cell lines as a means to treat otherwise incurable degenerative diseases.

Cells Communicating – Growth Factors and Their Receptors

All cells of the mammalian body have the most profound sense of community. Their existence depends on chemical messages they receive from neighbouring cells and from far afield (e.g., insulin), which dramatically influence cell physiology. About one-tenth of all the proteins encoded by the human genome are probably secreted and therefore probably affect other cells. Every chemical message is matched by a class of cell-surface proteins called "receptors," which transmit the message into the cell. The interaction between these chemical messages and their receptors is extraordinarily discriminating. Biochemists compare it to a "lock and key" to convey the precise intimacy of the pairing that makes responses to the outside world so specific. Once the receptor is activated, a set of biochemical relays amplify the signal and pass it on to the nucleus for what could be described anthropomorphically as "executive decisions." These may include instructions that affect the timing of cell division, cell movements, metabolic changes, and secretion of other growth factors. Our defences against invading microbes also depend on the same kind of sophisticated sensing device that detects fragments of a foreign protein that the body has never encountered before.

The growth and division of cells is dependent on hormones and growth factors. The pioneer cell biologists used serum, the fluid in which blood cells exist, as the most promising medium to support the growth of cells, believing that it must contain potential growth factors. Today, the requirements of human cells cultured in the laboratory are better understood. In addition to vitamins, amino acids, and glucose, culture media include certain protein factors, such as insulin, to stimulate uptake of glucose and others with subtle and interesting effects on cell growth. Epidermal and fibroblast growth factors are two examples of factors that stimulate many cell types, but others have more specific roles. Survival and growth of certain classes of neurone depend on nerve growth factor (NGF). Proliferation of bone marrow cells that will become red blood cells depends on a factor called erythropoietin.

Nerve growth factor emerged from an inspiring long-term investigation that began in fascist Italy in 1944. Rita Levi-Montalcini was hiding from the Nazis in Florence after having been dismissed from the University of Turin.[125] In a makeshift laboratory, unpaid and in enforced seclusion, she continued her work on the nervous system of the chick, reflecting on a daunting question: "What makes nerves grow out of the spinal cord into developing limbs?" In these desperate circumstances, an answer emerged from her experiments. Nerve growth depends on a substance she called "Nerve Growth Factor," which is made in the developing limb and which encourages nerves to grow towards the developing limb from the spinal cord.

At the time, biochemical corroboration of this hypothesis was not feasible, but when the war ended an opportunity to continue the work materialised in St Louis. The news that certain tumours secreted material with similar properties alerted Levi-Montalcini to the possibility that better sources of the factor might exist from which a pure form could be purified. Stanley Cohen, a young graduate student who a decade later would emerge as a pioneer of gene technology (see Chapter 12), started on this path, working under her supervision. This led serendipitously to the venom of the moccasin snake and the salivary glands of male mice – the mammalian organ most comparable to the venom glands – as particularly rich sources of the factor. Once nerve growth factor was available in a pure form, its biological role became startlingly clear. In the absence of nerve growth factor, neurones failed to grow towards a target organ and actually died. Scepticism was the first response to reports of nerve growth factor, but that melted away under the steadily accumulating mass of data. The award of the Nobel Prize for Medicine to Levi-Montalcini and Cohen in 1986 was recognition of their part in opening an important new chapter in cell biology. Soon other aspects of nerve growth factor were discovered, such as its role in the maturation of the mast cells that release histamine in response to injury.

Epidermal growth factor emerged from the same investigation because of its idiosyncratic capacity to accelerate the formation of eyelids and teeth in mouse embryos. Cohen and colleagues found that the protein, once purified, had the capacity to accelerate healing of wounds and the renewal of cells that line the gut.

Growth factors discovered during this heroic period turned up in curious circumstances that illuminated their biological role. One, later known as platelet derived growth factor (PDGF), was identified because a certain strain of cultured cell would proliferate only in serum and not in blood plasma. Plasma is the liquid part of blood that remains when all cells are removed without forming a clot; serum is the liquid that remains when the blood clots first. The scientists reasoned that the growth factor appeared only during blood clotting and was therefore likely to originate from platelets – the unusual, short-lived cells, without a nucleus, that are activated when blood clots.

Growth factors contribute importantly to the organisation of the mammalian body, from the first to the last moment of life. Many thousands of these proteins exist, each with a special role and with enormous potency to affect cell behaviour. In Chapter 11, we will see how gene technologists believe they can be used in many kinds of therapeutic situation.

The Death of Cells

A human life ends when the heart no longer beats and electrical activity in the brain ceases, but the circumstance that provokes the collapse of these crucial organs is the sickness and death of individual cells. This section is about the death of cells and how cell death contributes not only to disease but also to normal healthy mammalian life. Fifty years ago, cell death concerned only cytopathologists and developmental biologists. The pathologists called it necrosis and attributed it to microbial infection, chemicals, or physical agents that made cells swell, burst, and initiate localised inflammation (Chapter 6). Developmental biologists were alert to the existence of cell death during embryo formation, although its significance was unknown.

In the early 1970s, scientists began to realise that cell death occurred continuously in most tissues, and dead cells quickly vanished. These cells were fleetingly identified by distinctive surface features that disappeared as the cells shrank and the nuclei condensed, in a fashion quite distinct from the swelling associated with necrosis. The phenomenon was called apoptosis (from the Greek for "falling away") to convey the quiet finality of the process compared to the melodrama of necrotic death and inflammation. (Aficionados of etymology struggle unsuccessfully to persuade scientists to make the second p silent.) Calculations suggested that at any moment, in any human tissue, perhaps one percent of cells die by apoptosis, a figure that translates into ten billion cell deaths each day.[126] Apoptosis involves a meticulously choreographed set of events that end with the complete disappearance of every trace of the affected cell. The cell begins to shrink, while inside the nucleus the chromosome is cut into tiny pieces very rapidly; finally, the cell material is bundled into small membrane-bound units. Phagocytes, the professional scavengers and undertakers of the body, then engulf the debris so rapidly and discretely that they rarely attract the attention of the immune system or stimulate an inflammatory response (unlike necrosis).

The importance of apoptosis in the development of all organisms was brought to a much wider audience by investigations of the nematode worm *Caenorhabditis elegans*. The mature worm contains only 1,090 cells, but in reaching this size, specific cells undergo apoptosis on 131 occasions at precise places and times.[127] The startling conclusion was that apoptosis is controlled by genetic instructions that cause certain cells to commit suicide. Consequently, the phenomenon became known as *programmed cell death*.

The existence of programmed cell death, operating on a massive scale, indicates that removal of unwanted cells is an important physiological function vital to the developmental and maintenance activities of all multicellular animals. An example known to virtually everyone is the disappearance of the tail of the frog tadpole at metamorphosis, in response to the thyroid hormone. However, as embryos develop, every aspect of their structure is remodelled by apoptosis continuously, of which a few examples follow. Limbs develop as a bud of tissue in which space is created between the prospective digits on the hands and feet by apoptosis. The embryonic kidney and the male and female gonads take their final form during embryonic development because of apoptosis. The shedding of the uterine lining during menstruation throughout adult life is caused by apoptosis. The cells that are destined to form the lens of the mammalian eye undergo apoptosis, leaving behind a transparent mass of a protein called crystallin. Similarly, apoptosis in the cells of the outer layer of skin leaves behind a skeleton of cytokeratin that becomes highly durable skin scales.

Massive programmed cell death occurs in the developing brain and immune system. Half of all neurones commit suicide during development. Mice that lack the principle gene for apoptosis dramatically illustrate the importance of this process because they die at birth severely burdened by a vast excess of neural tissue. Apoptosis is the means of ensuring that neurones exist only in appropriate places. To survive, neurones need the constant stimulation of proteins such as nerve growth factor, and the neurone undergoes apoptosis if this stimulation is lost.[127] Quite possibly, although we have no formal proof of this, any cell in the body might die by apoptosis if not sustained by an appropriate growth factor.

Apoptosis plays a crucial part in regulating the immune system. In the thymus gland, *all but two percent* of the lymphocytes undergo apoptosis because of a challenge that eliminates those lymphocytes that recognise, and therefore target, "self," thereby ensuring that only self-tolerant cells survive.[128] Lymphocytes with no useful recognition facility are also eliminated because they lack a survival signal. After an infection has come under control, apoptosis then eliminates antibody-making cells in order to conserve resources.

Apoptosis seems more like an execution than a suicide when a protein factor originating from another cell activates the death program. For example, a hormone originating in the adrenal gland prompts apoptosis of unwanted cells in the thymus. Many cells have a membrane receptor, known obscurely as *Fas*, which can be activated by a secreted protein called *FasL* that initiates apoptosis in certain places and at certain times.[129] T-lymphocytes recognise and attach themselves to virus-infected cells and then initiate apoptosis in the infected cell because of *FasL* binding to *Fas*. The same scheme guides the immune system to eliminate those lymphocytes that can recognise proteins that are normally part of the individual's body. A genetic disease that affects *Fas* makes child victims of the disease suffer from a nonmalignant proliferation of lymphocytes that leads to severe autoimmunity (an attack by the immune system on "self").[128,130]

Several tissues are exceptional in that they are not vulnerable to T-lymphocytes when they are transplanted; they are derived from tissues that are known as "immunologically privileged sites." This diplomatic immunity is most obvious in the cornea, the testes, and the placenta. An important consequence is that corneal grafts are not rejected – a benefit that made this procedure the earliest triumph of organ transplantation, because tissue matching and immunosuppression are not necessary.[131] Because of the presence of soluble *FasL*, activated lymphocytes that infiltrate the cornea undergo apoptosis without causing inflammation that might otherwise make the tissue opaque.[129] Immune privilege in the testes and placenta has great biological importance because it prevents immune rejection of the male contribution to conception. *FasL* in the seminal fluid kills T-lymphocytes of the female genital tract that might attack sperm. Then in the developing foetus, *FasL* in the placenta kills maternal T-lymphocytes that might attack the foetus.[132]

Apoptosis enigmatically controls the availability of eggs and sperm. Of the millions of prospective egg and sperm cells that are formed, most undergo apoptosis, for a purpose that is hard to ascertain.[133] Cells with chromosomal abnormalities may be selectively eliminated. The relentless attrition of eggs means that most women in their late forties have few eggs left and enter the menopause – a life stage that may have evolved to secure their role as grandmother in child rearing (see Chapter 2).

Apoptosis is vitally important for proper embryonic development and in defence against viral infection, but excessive or inadequate apoptosis contributes to many kinds of disease.[134] Episodes of apoptosis that destroy a particular group of cells mysteriously trigger – possibly by viral infection or a period of oxidative stress – scores of diseases, both rare and common. Autoimmune diseases in which "self" antigens are recognised by the immune system all involve apoptosis. They range from insulin-dependent diabetes (type 1), in which insulin-forming cells of the pancreas are destroyed, through multiple sclerosis, rheumatoid arthritis, and various blood disorders. Parkinson's disease is a crippling neurodegenerative disease that occurs when neurones of the *substantia nigra* are destroyed by apoptosis.[129] Another group of diseases – including genetic diseases, virus infections, autoimmune disorders, and cancer – involves decreased apoptosis.[134] One rare and fatal genetic disease is caused by a mutation in a vital part of the apoptosis machinery.

Apoptosis is a key weapon in the battle for supremacy between viruses and humans. T-lymphocytes attack and destroy virus-infected cells, but some viruses, such as the glandular fever virus and poxviruses, have antiapoptotic mechanisms that neutralise this offensive. The human immunodeficiency virus (HIV) has a subtle and exceedingly dangerous battle plan. The initial infection of T-lymphocytes creates no pathology of note and may persist in a latent form for many years, perhaps being spread far and wide. When full-blown AIDS starts, the virus provokes apoptosis in uninfected neighbouring cells so effectively that no immune response can be mounted, and the host then succumbs to an opportunistic infection that is not intrinsically dangerous. Uncontrolled apoptosis caused by bacteria results in the septic shock syndrome and the horrific haemorrhages caused by Ebola virus, which start with apoptosis of blood vessels.

Apoptosis is a formidable defence against any malfunction of the replicative mechanisms of cells. One gene – known enigmatically as p53 – is a key guardian of the genetic integrity of the genome and initiates programmed cell death as a way to eliminate damaged cells. This activity, sometimes called cellular proofreading, prompts spontaneous abortion of genetically damaged foetuses and apoptosis of genetically damaged sperm or egg cells. The gene also acts to suppress potentially malignant cells by eliminating genetically damaged somatic cells.[133]

In the Darwinian struggle between biological order and malignant anarchy, cancer cells that resist apoptosis emerge by selection. In more than half of all cancers, this is caused by mutations in the p53 gene, an event that marks a serious worsening in the prognosis because this crucial defence is no longer an option. Anticancer drugs become less effective because they cannot induce the apoptotic response.[135] Some leukaemias become resistant because genetic rearrangements create novel hybrid proteins that make cells resist apoptosis. Other tumours become completely impervious to an attack by the immune system because they are protected by *FasL* or have lost *Fas*, the receptor for FasL.[129]

Apoptosis or necrosis is a key response to oxygen deprival and causes the cell death that doctors call *ischaemia*. This may occur after a blood clot blocks the coronary artery of the heart (i.e., a heart attack) or blood vessels of the brain (i.e., a stroke). These tissues, which cannot be replaced, are easily damaged by ischaemia – the brain in just a few minutes, the heart in perhaps fifteen. For a short time, the damage done by oxygen deprival may be reversible, but over a longer period the damage progresses, becoming irreversible. Impaired oxygenation can initiate apoptotic or necrotic damage to any organ and may contribute to the pathology of many diseases. Paradoxically, the damage becomes more severe when oxygenated blood is restored to the affected tissue because of a sudden release of free radicals (called a reperfusion injury), a process that physicians routinely try to prevent.

Why is oxygen deprival so devastating? A period of oxygen deprival that significantly depletes the amount of ATP in the mitochondrion opens a pore in the inner membrane and abruptly halts the electrochemical process that makes ATP. From that moment, the cell is doomed; a process starts in which the mitochondrion swells and bursts, releasing a protein called cytochrome c that initiates the descent into programmed cell death. Beyond this, the detailed mechanism, though truly fascinating, need not concern us. The biological rationale of this self-destruction seems to be an attempt to localise damage, perhaps by reducing overall oxygen demand; but the stratagem is less than perfect, because the tissues may remain permanently damaged even if the victim survives.

Cells and Immortality

Some cells are immortal in the sense that they can propagate forever. Sex cells are derived from an unbroken lineage of cells that can be traced through every generation since the origin of cellular life (see Fig. 3.1). Stem cells have a similar capacity for unlimited propagation, in contrast to most cells that clearly have only a limited capacity to divide. Very few neurones, heart cells, or skeletal muscle cells can undergo cell division after birth, although they can probably manage more than the conventional wisdom of former times suggested. What about all the other hundreds of types of cell in the body? On average, to make all the cells in a full-grown human, about forty to forty-five generations of cell division are required.

In the early twentieth century, scientists found that cells from a variety of organs of human and animal bodies could be cultured to some extent in the presence of appropriate nutrients. As we will see, scientists were under the impression for many years that animal cells can grow indefinitely, as though they were immortal. This error was exposed in the 1950s by Leonard Hayflick, when he showed that normal diploid cells from human or animal bodies undergo only a finite number of divisions and then stop. When cells reach this point – widely known as the Hayflick limit – they are unable to divide further, remaining metabolically active in a state that is usually described as senescent.[136]

Such a singular phenomenon seemed to be saying something very important about the life of a cell. Early in our development, those cells that will not become germ cells are given an instruction that says, "You can divide n times and no more," exactly as in the biblical idea that "our days are numbered." Amazingly, we will see (Chapter 7) that this notion is underpinned by a molecular clock – embodied in the telomeres that we encountered earlier – that give the phenomenon physical reality.

The Hayflick limit probably acts to prevent cells from becoming malignant and thereby threatening the entire organism. Even so, under laboratory conditions some cells, at a very low frequency, succeed in bypassing the Hayflick limit, becoming capable of unlimited growth and potentially malignant. Cells cultured from tumours or transformed with certain cancer-causing genes are examples of potentially immortal cells that ignore the Hayflick limit.

Challenges, Risks, and Rewards – a Summary

The study of cells has revealed a frontier of biological existence that was invisible to earlier generations. This discovery, like our perceptions of the character of DNA, gives an important insight into the human predicament – human life is evidently contingent on the proper functioning of cells. The miracle of our existence depends on hundreds of types of cells with

hundreds of remarkable features that may develop weaknesses that ultimately betray us. Cells may lose control of the division process, may erupt
in necrosis, suffer from oxidative stress, slip away quietly in apoptosis, be
invaded by microbes, be poisoned by the accumulation of strange protein
aggregates, or just enter indefinite senescence. Each of these unregulated
fates has real and important implications for human life, often causing
disease and ultimately death. If an element of a cell is defective because
of mutation, we are vividly reminded of its importance to normal life by
the uncomfortable consequences for the victim of a genetic disease. Understanding cells better is therefore one of the supreme challenges we face
if we want to understand the real causes of ill health and ageing. We will
encounter therapies directed at specific kinds of cells (e.g., chemotherapy
at malignant cells; immunosuppressives at lymphocytes) that take a calculated risk that other cells of the body will not be seriously affected. If we
want to devise strategies to cure diseases where cells have been mysteriously
eliminated by disease (e.g., Parkinson's disease), we will need to replace the
deficit from another source, such as stem cells.

GENERAL SOURCES: 80, 110, 137–39.

5   Experiences *in Utero* Affect Later Life

A revolutionary nursery for their progeny in an internal organ is the immense evolutionary advance by which mammals escaped the desperate profligacy of lower vertebrate life. By abandoning yolk as a food for the embryo and becoming dependent on the mother for nutrition through the placenta, mammals made a huge step forward in reproductive efficiency. For human parents who invest such hope in every pregnancy, the possibility that a mother may carry to term a child burdened with an imperfection makes procreation a poignant lottery. The development of the foetus may be affected adversely by poor nutrition, premature birth, or genetic damage, and consequently the child's true physical potential dictated by the genes may not be realised. At birth, most of the genetic program is completed, but an infant's growth continues for many years after birth, following a trajectory set early in life.

The Perfect Nursery?

Imagine the reproduction of fish and frogs. Eggs are laid in thousands; the likelihood of fertilisation is low; the progeny are easily eaten, infected, or poisoned. However, if more than a few are fertilised and survive, the species prospers. Reproductive duties complete, the parents of these lucky few abandon their progeny to the lonely battle for survival. Mammals avoid this wasteful lottery. Eggs are fertilised with more certainty, and the resulting embryo is raised in an extraordinary high-grade "nursery," in relative safety. The nursery is the uterus, a uniquely mammalian organ in which the fertilised egg develops and where the embryo establishes its nutritional lifeline to the maternal bloodstream, the placenta. The succession of divisions that follow fertilisation creates a ball of cells. After eight days, the

conceptus breaks out of its coat, implants into the uterus wall, and bur-
rows vigorously into the lining. Before implantation, the ball of cells is
subdivided into an inner and outer layer that will become, respectively, the
embryo and the placenta with surrounding membranes. The placenta will
feed the foetus, and the membranes will envelop the embryo with a fluid-
filled chamber. The umbilical cord forms on about the twenty-third day to
become a conduit from the embryo to the placenta. Maternal and foetal
blood vessels lie very close together in the placenta, but the two systems
never mix. Nutrients and oxygen from the mother reach the foetus through
the placenta; carbon dioxide and the waste products of metabolism pass in
the opposite direction back to the mother for disposal through the mother's
lungs and kidneys. The placenta also makes hormones, such as insulinlike
growth factor and growth hormone, that sustain the embryo; and it contains the endocrine clock that determines the onset of childbirth.

The uterus is a safe haven for the conceptus, but there is a small chance
(about one in eighty) that the fertilised egg may implant elsewhere in the
reproductive system, in what is known as an ectopic pregnancy. This highly
undesirable outcome threatens the life of the mother by causing a haemorrhage. Such pregnancies sometimes abort spontaneously, but they are a
significant cause of maternal mortality and are usually terminated promptly
in order to avoid tragic consequences for the mother. A much more remote
possibility (perhaps one in 12,000 pregnancies) is that the conceptus bur-
rows into the abdominal cavity. If this is undetected, the risk of death for the
mother is extremely high, but very rarely the foetus may develop a placenta
attached to the liver and an otherwise normal amniotic sac in which it may
develop to term. In these cases,[140] obstetricians deliver the child through
the abdomen and allow the placenta to be reabsorbed by the mother, who
usually recovers with no long-term damage.

For almost four decades, the serene world of the human embryo was
known best through the remarkable images of Lennart Nilsson that first
graced the cover of *Life* magazine in 1965 and later appeared in his book.[141]
One image was a live fifteen-week-old embryo, photographed through the
cervix with an extraordinary wide-angle lens and a miniature fibre-optics
light source. This picture captures unforgettably the sleepy eyes and tiny
hands close to the mouth of an already almost perfectly formed foetus just
seven centimetres long, floating blissfully in amniotic fluid suffused in warm
pink and evocative light. In reality, most of his famous images were taken
outside of the body using products of miscarriage, but the one live picture
leaves us in no doubt about the character of the early embryo.

More than fifty years ago, Peter Medawar compared the situation of a mammalian embryo to an organ transplanted from a genetically dissimilar person. He raised the question of why the brutal efficiency that the body normally aims at foreign tissue is not directed at the foetus.[142] We know now that the danger of a direct immune response of the mother against the foetus is minimised because the maternal and foetal blood supplies are separated. The placenta is also protected against marauding lymphocytes by a locally produced protein called *FasL*. Furthermore, the system of innate immunity, which can destroy microbes on sight with a devastating barrage of biochemical weapons, does not attack the embryo.

Formation of a foetus raises the question, do the cells of the embryo and the maternal tissues remain completely unmixed and separate or could foetal cells invade the mother, or even vice versa? Highly sensitive probes based on the paternal Y chromosome have been used to track foetal cells, and show that some foetal cells migrate into the mother. Moreover, foetal cells from one pregnancy can survive in the mother and invade a second pregnancy. In the first such instance reported, cells of foetal origin were found in the mother twenty-seven years later, having survived because they propagated as stem cells. Although the subject is in its infancy, there is already evidence for both good and bad consequences; such cells sometimes repair damaged organs, but they may also cause autoimmune disease.[143] A small contribution from the mother to a child is often known as microchimerism, in distinction to the major chimerism found in rare instances after fusion of two fertilised eggs (see p. 56).

Threats from the immune system are not the only conflicts between mother and foetus. Another Darwinian struggle pits the three genetically interested parties against each other over matters of nutrition. The foetus is programmed to exploit the resources of the mother as much as possible; the mother must limit these demands to ensure her own survival. In times of undernutrition, the placenta enlarges and becomes more aggressive than in less challenging times.[144] We have seen how certain genes (known as imprinted genes), whose products are required for successful development of the foetus, contribute only if they are derived from one or the other parent but not both. Genes of this kind donated by the father (such as insulinlike growth factor) favour growth of the embryo; the mother's generally limit this tendency.

The most interesting and complicated phenomenon the human mind has ever contemplated is initiated when a fertilised egg starts to grow into a foetus. Its subtlety and complexity puts the topic far beyond the scope of this chapter, but the recent explosion of research in developmental biology is beginning to reveal important guiding principles. Lewis Wolpert captures the character of the process with a neat analogy, when he compares the development of embryos to origami, the Japanese art of paper folding.[145] In both, with no reference to any preexisting model, a figure emerges from a set of individual instructions that must be executed correctly, in the right place and at the right time. The biological programme originates from the genome and consists of tens of thousands of biochemical steps that affect the time and place of every cell division and the character of every new cell. The scale of the enterprise is evident when we consider that an adult probably contains about 10^{14} cells (1 followed by 14 zeros). This corresponds to about forty-six cycles of cell division and does not include all those cells that suffer programmed cell death. About forty of these divisions are carried out *in utero*, so that about 10^{12} cells are present at birth. Subsequent growth largely entails an increase in the volume and mass of cells.[41]

One month after fertilisation, the embryo is five millimetres long and already has small limb buds. After two months, the embryo is thirty millimetres long, with a flattish face, widely separated eyes, and clearly distinguishable digits on hands and feet. The first two months are a time of intense activity; cell division goes on at a cracking pace, while a subtle and extraordinary process is creating a body plan to be filled out by growth as time goes by. By fourteen weeks, the foetus is about 110 millimetres long, the heart is complete, and the body proportions are similar to those of a newborn.

Perils of Embryonic Life

At no moment in human existence is life more vulnerable than during the first three months of pregnancy. The embryo may have nothing to fear from predators, but a host of dangers can crucially affect it. Most human pregnancies end happily, but some produce children burdened with defects that arise from shortcomings in foetal nutrition, the character of the placenta, viral infection, encounters with alien chemicals from the maternal bloodstream, or immunological reactions. Development of the child may also be jeopardised by premature birth, inherited genetic defects, or mistakes in chromosome separation that happen before conception.

For humans who invest such hopes in their offspring, these are sobering eventualities, made especially poignant because the mother may carry to term an offspring burdened with a defect that may handicap the child for life. In nature's scheme, perfection for every pregnancy is not really the objective, although an alarm system provokes spontaneous abortion when certain malfunctions are detected in the foetus.[146] Possibly 80 percent of all human conceptions abort spontaneously and disappear unnoticed by the mothers during the first ten weeks of a pregnancy. Some embryos are lost because implantation fails; one-third of all implanted eggs abort because of mistakes in chromosome separation.[147] The key mechanism for terminating early pregnancy in mice seems to be the p53 gene, the master proofreader we encountered in the last chapter, which probably has a similar function in humans.[148,149] Embryos damaged by new mutations, genetic rearrangements, or mistakes in chromosome separation are often eliminated by spontaneous abortion. As a quality-control system, this is not completely effective, because as we saw in Chapter 3, children are born with extra chromosomes with a frequency of about one in seven hundred.

No system exists to eliminate inherited genetic damage, an intractable problem we will consider in Chapter 14. About three percent of babies are born with congenital disorders, many of which are inherited. Some disorders of the heart and circulatory system seem to be of genetic origin. The "hole in the heart" syndrome is a well-known congenital defect, caused by a hole in the septum separating the two ventricles, through which de-oxygenated and oxygenated blood mix together. Victims of this and related syndromes were at one time condemned to a short and sickly life, but today surgery at birth or even *in utero* can permit these children to grow up with little or no handicap.

Whatever perils lower vertebrate embryos face, nutrition is not a problem, because they start out with abundant supplies of egg yolk – a rich fuel, like the yolk of chicken eggs – that will sustain them until they can feed themselves. During mammalian evolution this was abandoned in favour of nutrition delivered to the embryo in a constant stream through the placenta. This scheme has many benefits but with serious consequences for the foetus if the mother's nutritional status is challenged by starvation or inadequately controlled levels of blood glucose.

HEALTH, LONGEVITY, AND EXPERIENCES *IN UTERO*: Rats reared on a calorie-restricted diet after weaning have a substantially increased life span. By contrast, rodents reared on a diet marginally deficient in protein produce normal, fertile offspring that breed through many generations, but they are smaller than control rats, and their lives are shorter.[150] This makes very good biological sense in a nutritionally challenging environment – the mother retains her reproductive potential, while the small body mass of the progeny means fewer resources are required for reproduction. The viscera of these rats – liver, kidneys, heart, and the insulin-producing cells of the pancreas – are scaled down compared to the brain. The brain depends crucially on the supply of glucose and evidently has a privileged status that protects it from any nutritional deficit. By avoiding brain damage, mating and reproductive processes are not compromised even when the body is just a shadow of its genetic potential. The organ systems of animals raised under protein-restricted conditions have little spare capacity to cope with the attrition of advancing years, and they develop disorders comparable to human diseases.[151,152]

Pregnant rats fed on a protein-deficient diet allocate fewer cells to the visceral organs. When fewer cells are allocated for control of insulin secretion (the beta cells), the animal has less capacity to respond to elevated blood sugar levels. The animal consequently develops a condition similar to the human disease type 2 diabetes – also known as insulin-independent diabetes, a form that cannot be corrected by injections of insulin.[153] Similarly, the number of kidney nephrons – the unit of blood filtration found in the kidney – may be reduced, and this may create circumstances that cause high blood pressure.[152] If the lungs are underdeveloped for the same reason, they are deficient in elastin and collagen and have reduced elasticity, and the animals develop a condition not unlike human emphysema. An underdeveloped liver affects cholesterol metabolism and increases the tendency to form blood clots and thromboses in the coronary artery. Malnutrition of the foetus in the later stages of pregnancy makes animals more susceptible to infection, because the size of the thymus gland is reduced.

Dietary restrictions affect the developing embryo very soon after conception. Four days of a low-protein diet from the moment of conception of a rat embryo, followed by a normal diet for the remainder of gestation, has serious consequences for the adult rat. Cell division in the embryo is affected even before the conceptus implants into the uterus wall, first within the inner cell mass that will become the embryo and later within the outer layer. This reduces the proportion of the body taken up by the visceral organs eventually reducing the birth weight and growth after birth, compared with a control. The four-day low-protein diet causes an immediate increase in the glucose level in the mother's bloodstream and a decrease in insulin and essential amino acid levels, and the animals eventually develop high blood pressure.[154]

ACUTE NUTRITIONAL DEFICIENCY AND THE UNBORN BABY: Human societies have always instinctively understood that the health of the mother must affect the health of an unborn child, but it was never easy to know exactly what was required. The vitamin and mineral requirements of laboratory mammals, and indirectly of humans, became known through experiment. The assumption has always been that the foetus obtains them from the mother through the placenta. We must assume that serious consequences for unborn children ensue if nutrients in the mother's bloodstream fall below certain levels.

An unknown eighteenth-century Dutch midwife was probably the first to realise that birth defects occurred in offspring of poor families more commonly in years with poor harvests, but the observation was corroborated many times in the nineteenth century.[155] The condition was almost certainly *spina bifida*, a birth defect that occurs when the embryonic neural tube fails to close properly, resulting, in the most extreme cases, in serious brain damage or even a stillbirth. The occurrence of spina bifida is geographically localised and usually occurs in a season of the year that suggests that victims were conceived at a time when fresh green vegetables were least available. At one time in northern Britain, neural tube defects were as frequent as seven per thousand births.[155] The long-held suspicion that spina bifida originated in poor maternal nutrition was finally attributed to deficiency of a specific substance called folic acid only in 1964. Folic acid was used to prevent pernicious anaemia in the 1940s, but the merits of taking folic acid pills from the time of conception to prevent neural tube defects were not proven unequivocally in a clinical trial until 1990. By fortifying flour with folic acid, the incidence of neural tube defects was substantially reduced in affluent countries.

Synthesis of DNA depends on folic acid. If our cells are deprived of it, the consequences are serious – and especially so for the foetus immediately after conception, when the mother is often unaware that she is pregnant and the embryonic rate of cell division is very rapid. Folic acid deprivation stalls DNA synthesis and causes mutations that may have far-reaching effects, including cell death and chromosomal rearrangements. Some women have a gene that makes them require a greater-than-average folic acid intake, which increases the risk of having babies with neural tube defects by three- or fourfold. Folic acid deficiency is now the commonest clinical vitamin deficiency syndrome affecting child development. Although contemporary diets contain enough vitamin B6 and niacin, deficiencies would probably cause birth defects, as these vitamins are required for DNA synthesis.

Maternal deficiency of micronutrients, unconnected with DNA synthesis, also affects the development of babies. Vitamin D is found in milk and oily fish or is made in the skin in response to sunlight. In excess, vitamin D's chief role is to mobilise calcium for bone formation; in deficit, children develop the characteristic bowed-leg condition known as rickets. In the industrial cities of nineteenth-century Europe, *most* children were disfigured by rickets, because dietary vitamin D was not readily available and circumstances conspired to prevent its synthesis in their skin. A fiendish combination of coal-blackened skies and long days incarcerated in factories – until the start of compulsory schooling – meant children saw very little daylight.[8] The consequences were far-reaching, because we know now that vitamin D is necessary for the efficient functioning of cell-mediated immunity, and widespread deficiency may have favoured greater virulence of tuberculosis during the nineteenth century.

The disorder has returned to Britain, amongst Asian immigrant families who do not consume milk and whose dark skins limit the synthesis of vitamin D in response to sunlight. Some authorities believe that breast milk does not contain sufficient vitamin D to confer maximum benefit to the immune system and that consequently children need some sunlight on their skins or nutritional supplements. Vitamin D may also play a part in preventing cancer.[156]

In Southeast Asia, millions of people suffer from a deficiency of vitamin A because the staple crop, rice, lacks vitamin A completely. The symptoms include night blindness and more serious damage to eyesight. Iron deficiency in the developing world contributes greatly to maternal mortality through anaemia, and iodine deficiency causes brain damage in babies. Taken together, these deficiencies affect the lives of billions of people.

The specific consequences to unborn children of a famine of known severity and duration were studied after an event usually called the Dutch hunger winter that occurred between October 1944 and May 1945.[157,158] In a brave attempt to assist the allied armies in the last campaign of the Second World War, Dutch railway workers staged a strike. The Nazis responded by prohibiting transport of food supplies by canal, immediately precipitating a serious famine throughout the western Netherlands that became a poignant experiment of history. At a precisely defined time and place, the dietary intake of a large population was reduced to less than one thousand calories per day. Accurately kept records indicated an immediate twofold increase in infant mortality and a reduction in the number of pregnancies. Surprisingly, although every mother was undernourished, the babies born had only slightly reduced birthweights. Much later, it became apparent that these babies conceived in the famine, although protected against acute malnutrition, also carried a significant burden of ill health in later life. Spina bifida, the congenital birth defect caused by folic acid deficiency, was common. Schizophrenia, caused by neurological damage, was more prevalent amongst male adolescent survivors than in the general population, and an elevated risk of heart disease emerged in middle age.

How Nutritional Experiences *in Utero* Affect Adult Health

Evidence connecting nutritional experiences *in utero* with particular kinds of ill health in later life emerged from another painstaking epidemiological investigation of the life histories of people born in the 1920s. The idea for this investigation came from an observed similarity between the geographical pattern of infant mortality in England and Wales between 1921 and 1925 and the pattern of coronary heart disease mortality in the decade after 1968. The cause of death of infants shown on the death certificate in the 1920s was often vaguely described as "low birthweight."[144] Infant mortality at that time was seventy-six per thousand – compared to less than six per thousand today – and varied by more than twofold across social classes. The nutritional circumstances of British people during that period were very poor, according to one famous survey.[159] About half of all families including pregnant women got by on a diet judged by contemporary estimates of requirements to be deficient in vitamins, calories, and essential minerals. Suboptimal nutrition in the lower socioeconomic groups was correlated with increased ill health and a higher death rate.

Starting in the 1920s, members of the newly created profession of health visitors, whose role was to attend women in childbirth and to encourage mothercraft, monitored the growth of babies, recording weight at birth and one year later. In some regions, measurements of body proportions were also recorded. Fifty years later, after more than 15,000 of these subjects had been traced, David Barker and his colleagues at Southampton University established an important link between weight at birth and the incidence of certain chronic diseases of middle age. The death rate from coronary heart disease was almost twice as high for people born at full term, weighing only 2.5 kilograms compared to individuals who weighed 4.3 kilograms at birth. The death rate was higher still amongst those who failed to grow after birth. Suboptimal birthweight was also correlated with an increased chance of developing high blood pressure, stroke, chronic bronchitis, and type 2 diabetes (a late onset form of the disease in which the patient becomes unresponsive to insulin). These tendencies were most obvious in babies who were below average weight and with small abdominal girth at birth. Boy and girl babies responded differently to undernourishment during gestation: boys were short in relation to head size; girls tended to be thin. A set of data from a hospital in Sheffield in which dimensions of newborn babies were recorded showed that males who were "thin" or "stunted" at birth were more liable to develop coronary heart disease fifty or more years later.

The important conclusions from this study have now been confirmed by independent surveys from a great range of social settings. Many kinds of optional life experiences – including smoking, consumption of animal fat, and not drinking red wine – also seem to be risk factors for developing coronary heart disease. However, a nutritional cause that originates *in utero* is at once more profound and more complicated to evaluate. Birthweight was evidently a crude proxy for an important attribute of foetal growth that was not readily identifiable with current knowledge. The question of whether people who were underweight at birth had other life experiences that might provoke heart disease was clearly an important issue but was discounted by statistical analysis.

One type of life experience unequivocally reinforces the damage done *in utero*. Underweight babies usually undergo an accelerated period of weight increase after birth, when nutrition improves. The risk of so-called catch-up growth implied by this phenomenon emerged from a study of Finnish men based on the detailed measurements made by school nurses in the 1920s and 1930s. Many of these subjects were traced more than sixty years later. The ones who were born thin and underweight but who became greater-than-average weight for height by the age of seven, emerged as the ones most likely to suffer from coronary heart disease.[160]

Coronary heart disease seemed to be an epidemic in the English-speaking world between 1920 and 1950. Men were more affected than women from the start, but after 1945 mortality of men continued to increase in the 35–55 age group, while it fell for women, until men were three times more likely to succumb to the disease. Paradoxically, the mid-twentieth-century coronary epidemic was associated with rising prosperity; today, the disease is loosely linked with the least affluent strata of society. However, the two are compatible if we suppose that the disease hits people whose lifestyle changes from undernourished and active to sedentary and overfed.

Doctors who practised medicine in the developing world forty years ago noticed how infrequent coronary heart disease and atherosclerosis were in hunter-gatherer peoples such as the Inuit and the !Kung of the Kalahari.[6,161] So-called Western diseases now occur commonly amongst people who migrate from a world of meagre food resources and intense physical activity to one where food is cheap and abundant and physical activity is greatly reduced. The abrupt change to a world of abundant high-energy foods makes "catch-up growth" possible, and individuals may become obese, with a tendency to develop type 2 diabetes, high blood pressure, and coronary heart disease.[144]

For the inhabitants of the Pacific island of Nauru, the Pima Indians of Arizona, and others, this had the most poignant consequences.[161,162] Within a generation, individuals with lean and fit physiques were replaced by individuals with marked obesity. The prevalence of type 2 diabetes changed from no recorded cases to approaching 50 percent of the adult population. In Nauru, the trigger was a delirious moment of enormous wealth that followed financial restitution for the island's phosphate mines, perhaps greatly exacerbated by a genetic predisposition towards diabetes.

In India, underweight babies are more common than in the developed world, are often born to very small mothers, and today frequently grow up to be overweight and to suffer from type 2 diabetes. The tendency to develop type 2 diabetes is greatly exacerbated if the children are sufficiently well fed that their weight increases to normal standard levels. Although markedly underweight, these babies have a relatively large amount of subcutaneous fat, which may contribute to their becoming resistant to insulin. Faced with the probability that a high proportion of Indian children will develop diabetes, public health officials would like to make a suitable intervention. Food supplements for children are likely to exacerbate the problem. As long as foetal growth is constrained by the small size of the mother, the risk of developing diabetes is likely to persist; several generations may elapse before babies are born at an ideal weight.[163]

We must ask, too, why the 1920s cohort in certain developed countries – particularly Japan and France – are much less affected by coronary heart disease than their contemporaries in the English-speaking world. In France, a large rural community may have had better food, but perhaps greater consciousness of the importance of maternal nutrition also played a part. Indeed, state-sponsored support of pregnant women in France started before the First World War.[164] No national group is more long-lived or has such a low incidence of coronary heart disease as the Japanese, even though the known risk factors (smoking, high blood pressure, and stress) are becoming more common in Japan. Japanese children of the interwar years probably had poor nutritional experiences *in utero* but remained unchallenged by energy-rich diets. Their peers who migrated to Hawaii were exposed to the American diet and started to develop coronary disease.[161]

American doctors were amazed to discover that atherosclerosis, the condition that precedes coronary heart disease, was present in three-quarters of the young soldiers on whom autopsies were carried out during the Korean War.[165] The precise significance of this is hard to know, but possibly this generation was born in underprivileged circumstances, suffered from suboptimal nutrition *in utero*, and was suddenly presented with the unprecedented abundance of America.

Barker's explanation of the foetal origins of coronary heart disease is that the foetus adapts to unsatisfactory nutrition by allocating fewer cells to the developing visceral organ systems in order to protect the brain.[144] The body tries to ensure that the supply of glucose to the brain is never compromised by excessive demands from other organs, in the fashion suggested by experiments with laboratory animals grown on protein-deficient diets. This happens because fewer cells are allocated to the developing pancreas, so that less insulin is produced. As insulin determines how much glucose cells use, this means that tissues such as skeletal muscle demand less glucose and become less responsive to insulin, and lengthwise growth of the foetus is also reduced.[154] In later life, such individuals become significantly predisposed to type 2 diabetes.

The liver and kidneys also receive fewer cells than normal. The reduced liver size seems to favour blood clot formation and therefore the tendency toward coronary thrombosis. For the kidneys, fewer cells mean the victim cannot resist the tendency to develop high blood pressure, and this is reinforced by a decreased elasticity of arteries originating in a relative deficiency of a protein called elastin.[166,167]

Poor nutrition *in utero* may also profoundly affect immunity to infection. In the rainy season of tropical West Africa – known locally as the "hungry season" – staple foods become seriously depleted, the workload reaches a peak, and women are probably in negative energy balance. Statistics collected over fifty years indicate that people born in the "hungry" season are ten times more likely to die prematurely – usually of infection – than people born in the dry season. The authors suggest that nutritional stress affects the development of the thymus gland and the capacity of the immune system to respond to infection.[168] The entire deal for the nutritionally stressed foetus *in utero* is a Faustian bargain.

Privileges bought for the brain *in utero* are exchanged for perpetually underdeveloped viscera, which in middle age become a source of weakness and poor health.

THE VULNERABILITY OF THE FOETUS: The unborn foetus is not protected from viruses and chemicals that may cross the placenta from the maternal bloodstream. Moreover, certain proteins in the foetus, of paternal origin, occasionally provoke the mother into an immune reaction that affects the foetus even though maternal white blood cells cannot readily attack the baby's tissues. Most viruses do not readily cross the placenta, but important exceptions include the human immunodeficiency virus (HIV), cytomegalovirus (CMV), and rubella (German measles). After infancy, rubella or CMV infection is either trivial or asymptomatic, but both viruses can seriously damage a foetus. The ability of HIV to pass from mother to child is one more poignant feature of the terrible epidemic.

Tobacco, alcohol, and narcotics crossing the placenta all affect unborn children adversely. About one in every thousand children born in the United States is thought to suffer from foetal alcohol syndrome. Alcohol provokes apoptosis of nerves of the forebrain whose connections are required in later life for learning and memory. These children suffer from facial abnormalities, stunted growth, and learning and memory deficits that in later life cause depression and psychosis. Alcohol has a chemical similarity to the neurotransmitters of the body and has the effect of overstimulating receptors in the developing brain. The same work raises the possibility that barbiturates and benzodiazepines (Valium® and related drugs) may also kill neurones in a foetus or newborn baby by a similar mechanism.

The sensitivity of embryos to strange chemicals became known universally forty years ago through the thalidomide affair. Physicians prescribed this notorious drug to huge numbers of expectant mothers as an antidote to morning sickness, in the mistaken belief that it had been tested adequately. The outcome was an epidemic of appalling birth defects. In the worst cases, limb bones were not formed and the victim's hands and feet were attached directly to the body. More than twelve thousand births were affected, and the number lost by miscarriage and stillbirth was unquantifiable. Later, it emerged that a single tablet taken between days sixteen and twenty-six of the pregnancy provoked damage to the developing limb buds. Legislation emerged from the tragedy that made exhaustive tests on animals obligatory for any medicines to be given to pregnant women.[169]

The thalidomide affair was not the only incident during that era that raised anxieties about the safety of unborn children. The publication of Rachel Carson's *Silent Spring* in 1962 drove the American public to reflect on a terrible omen: the insidious effect on innocent wildlife of chemicals in the environment. Enthusiasm for greater food productivity was constantly pushing the American agriculture industry to use more fertilisers, insecticides, and herbicides, with little thought for the ecological consequences. The publication of Carson's book dramatically alerted the public to an alarming possibility. If chemicals were damaging wildlife, they must surely threaten humans, too! Carson passionately denounced the insecticide DDT, arguing that it was interfering with reproductive hormones, decimating wild animal populations, and causing birth defects in humans and animals and an epidemic of human cancer. The book was essentially a polemic extrapolated from some shrewd observations – biological intuition rather than hard fact. However, Congress took the book's claims at face value and banned DDT in 1972. Accumulation of DDT in breast milk and body fat was also confirmed, and today suspicion is widespread that chemicals interfere with hormones in more subtle ways than Carson knew.

In 1991, a multidisciplinary group of scientists meeting at Wingspread, Wisconsin, published a kind of manifesto that drew attention to a set of man-made chemicals they called "endocrine disrupters." These compounds were defined as substances that could mimic hormones and disturb the normal endocrine functions of animals, producing effects that included thyroid dysfunction, growth deformities, behavioural abnormalities, feminisation of male animals, and compromised immune systems.[170] To establish unequivocally the same effects in humans is clearly almost impossible. However, the signatories of the Wingspread consensus were making the important claim that a proportion of human birth defects, declining fertility, and reproductive cancers could be caused by endocrine disrupters active on human embryos *in utero*. Experiences with two substances of this kind indicate the seriousness of this issue.

The synthetic oestrogen diethylstilbestrol (or DES), introduced deliberately to medical practice in the 1940s for prevention of miscarriage, now appears to have all the hallmarks of an endocrine disrupter. Although some doubts about its efficacy and safety were expressed in the 1930s, the FDA withdrew its approval of the treatment only in 1970. This move was provoked by an observant oncologist who noticed that an extremely rare vaginal cancer in seven young women was linked by a common factor: their mothers had taken DES in the third month of pregnancy. The situation looked even blacker when many of the daughters of DES-treated women were found to have developmental damage to the reproductive organs. The consequence was often ectopic pregnancy and miscarriage, the very outcome the medication was supposed to prevent.[171]

The second example concerns bisphenol, an endocrine modifier with well-established effects on experimental animals that could get into the human food chain because it is present in certain plastics. Environmental toxicologists fear endocrine disrupters of this kind could be a cause of cancer of the reproductive organs. Previously, experts doubted whether this compound ever reached levels sufficient to produce an effect in humans. However, recent reports suggest that it may have an important effect at low doses *in utero* by preventing meiotic chromosomes of mice (and, by inference, of humans) from packaging correctly.[172,173]

A number of layers of protection limit the ability of the mother's immune system to reject the foetus, but even so, some pregnancies are threatened. The best-known example occurs when the mother is "rhesus negative" and the foetus is "rhesus positive." The anachronistic name "rhesus," with its intimations of monkey business, dates from the 1930s, when a blood type was recognised that reacted with the blood of rhesus monkeys. If a rhesus-negative mother is expecting a rhesus-positive child, she may be immunised accidentally during childbirth by exposure to the rhesus antigen. In a second pregnancy, the mother's antibodies to this factor may then mount an attack against the baby, causing severe jaundice or even death *in utero*, a condition known as *hydrops fetalis*. Today, these dangers are just ancient history, because a vaccine against the rhesus factors is available. However, doctors often suspect that persistent miscarriage originates in immunological reactions. Maternal antibodies against phospholipid are a well-established cause of miscarriage.

The optimal development of the unborn baby is vitally dependent on the stream of nourishment through the placenta. One last example illustrates the vulnerability of the foetus to abnormal metabolic regulation in the mother. The unborn babies of women whose type 1 diabetes is inadequately regulated are at risk from excessive levels of glucose. This may cause excessive growth and unusually large babies or, for unknown reasons, may lead to a number of malformations, including defects of the heart and kidneys and spina bifida.

Premature Birth – the Danger of Arriving Early

For about one in ten babies, the tranquillity of the womb is shattered by a signal that activates birth prematurely and thrusts into the world an infant whose development is incomplete and whose hold on life is feeble. This tendency towards premature birth, not shared with other primates, can occur at little more than half the normal term. Unlike babies born at the normal term of thirty-eight to forty weeks, who attain a size and maturity that equips them to develop after birth without difficulty, premature babies require extensive support. Even so, one-quarter will die within the first month, and perhaps half of all premature babies will develop some kind of neurological handicap. Three-quarters of all infant mortality and most birth abnormalities occur amongst this group.[174,175]

What decides when a baby should be born, and why is the timekeeper so capricious? A protein called corticotropin-releasing factor, found only in humans, is the biochemical clock that determines the length of a pregnancy. Made by the placenta, it accumulates at a sharply increasing rate in the mother's blood during pregnancy, until it reaches the critical level at which the birth process is provoked. The alarm is complicated, subtle, and poorly understood, but it seems to be activated prematurely by circumstances that might harm the baby. Inflammation (see Chapter 6), infection, high blood pressure, maternal stress, and smoking all move the clock forward and provoke premature birth. Symptomless bacterial infections are a major cause of premature childbirth, but they can be eliminated by appropriate antibiotic therapy. Another cause of premature birth, established recently in sheep, is maternal undernutrition at the time of conception, which brings forward various hormone surges associated with the birth process. For obvious ethical reasons, the hypothesis of a link between undernutrition and premature birth cannot be tested in humans, but it seems likely that dieting around the time of conception may increase the chance of premature labour.[176]

When childbirth is initiated prematurely, the baby can be born naturally, but obstetricians tend to intervene using the caesarean section procedure in order to minimise any distress to mother or child. Preterm babies are not just early arrivals but may be incompletely developed, with only the weakest hold on life, and may still require nutrients that would normally come from the mother. Enormous care is required to help the smallest ones grow to maturity. Obstetricians have discovered that the survival of preterm babies can be greatly improved by antibiotics and certain nutritional supplements and by using steroid drugs to minimise inflammatory reactions. Even more remarkably, the baby's development can apparently be helped by touching the baby appropriately. Four out of five babies born with a birth weight under six hundred grams, usually at twenty-four weeks or earlier, are likely to die, and survivors often have neurological damage.[177] In Britain, however, both the earliest arrival (21.5 weeks and 482 grams) and the smallest live birth (350 g, born 1938, died 1983) survived.

Human lungs are not complete until thirty-two weeks, so that preterm babies must be held in a humid environment while the final development occurs, to prevent their becoming desiccated. Preterm babies find breathing almost too great an effort for their tiny bodies. Their lungs are deficient in a substance that reduces the surface tension of the lung (surfactant), minimising the effort required to inflate the lungs. Obstetricians have been looking for solutions to these problems for more than a century. Special incubators were invented in France and the United States in the late nineteenth century to provide a humid atmosphere and to stimulate better lung ventilation. The problem of the surfactant was solved by introducing into the lungs of the newborn, before the first breath, a preparation of surfactant extracted from calf lungs.[178]

Preterm babies may lack essential nutrients that would normally come from their mothers through the placenta, but there is no objective way to investigate such requirements with human babies. However, in the 1990s, evidence was obtained that supplements of long-chain polyunsaturated fatty acids, not usually present in artificial baby foods, improve eye function and cognitive development in premature babies. They are also essential for brain development.[179]

A foetus approaching full term is probably already developing the responses to touch that are so important in mother-baby bonding after birth. Indeed, the behavioural and emotional development of newborn monkeys is known to improve if they are stimulated by a gentle massage. Paediatricians have explored the same approach with preterm babies, and report that a massage with a gloved hand for a short period each day seems to stimulate weight gain, minimise stress response, and improve sleep patterns.[180] Experiments with preterm rat pups, which can be conducted with controls that are more robust, suggest that tactile stimulation by brush stroking was essential to stimulate growth so that they reached the weight typical of a normal pregnancy.[181]

Doctors remain greatly concerned about the right way to manage extremely low birthweight babies. Some years ago, a group of American doctors expressed their misgivings as follows: "*the prevailing moral and medico-legal climate dictates active intervention even when the prospects for survival are minimal and survival without severe impairment still smaller.*"[182] Emotional reports in the press are easily found, telling of a mother's joy that a baby described as "unviable" has proven a doctor wrong. However, doctors know objectively that babies whose birthweight is less than 700 grams may survive only with a lifelong handicap or not at all. To avoid futile and expensive intensive care, doctors are trying to define objectively the "limits of hope." Parental attitudes run the gamut from those who want to intervene whatever the cost to those who lack the fortitude to manage a baby with a handicap.[175] Success in caring for extremely premature babies means that late-stage termination of a pregnancy is no longer ethically appropriate, and in 1991 the legal limit for abortion was lowered to twenty-four weeks in Britain (see Chapter 2).

The solution to the biological difficulties caused by preterm birth has always been to find better ways of prolonging gestation, because every extra week in the womb makes the baby bigger and stronger. Treatments are being developed to help suppress premature contractions and to remove some of the causes of premature birth, such as infection.

From Birth to Maturity

Soon after birth, the growth pattern is set on a "trajectory" that can be predicted fairly accurately. The adult height is usually close to twice the height at the age of two, and the anticipated full adult height of well-nourished children is usually midway between the heights of the parents. Babyhood is a time of rapid lengthwise growth followed by the steadier pace of childhood and then the final adolescent spurt. People differ substantially, but physiologists can define the average growth pattern and identify abnormal variations. The crucial events that determine our adult height seem to occur when we are just tiny embryos.

All mammals, including humans, respond to a period of malnutrition by stopping growth in order to conserve resources. When better conditions are restored, growth resumes at a somewhat faster rate than before – called catch-up growth – so that within a reasonable time growth is back on track.[41] Even babies exposed to malnutrition *in utero* during the Dutch famine (see p. 97) eventually attained a normal height.[158]

Chronic malnutrition is another matter. The normal growth period may be extended even with moderate undernutrition if it continues throughout childhood, and this can be exacerbated by chronic enteric infections. The deficit in height can be made up, but a large percentage of children in certain developing countries – sometimes greater than 50 percent – are said to be "stunted" owing to lack of food, chronic enteric infections, or both.[183] The World Health Organisation uses the comparative growth rate of children as a sensitive and objective indicator of the quality of nutrition. Scandinavia passed an historic milestone in the 1980s when, for the first time in any country, the average growth rates of children from affluent and poor homes were not significantly different.

Immediately after birth, babies from every social setting grow at the same rate, on average, because the nutritional value of breast milk is a more or less universal constant. Children reach a critical moment around the time of birth, when receptors for growth hormone appear on cell surfaces; the number of receptors determine how fast the child's long bones will grow. Recent research suggests that a child's length at birth is the best predictor of final height, and that the overall growth trajectory is set at or before birth. Subsequently, the quality of nutrition affects how responsive the long bones will be to growth hormone.[184]

Many historical peoples were shorter than modern people. Archaeologists can deduce from the skeletons of ancient peoples their age, sex, height, and probable social status. Cro-Magnon man, whose hunting exploits 30,000 years ago vividly decorate the caves of Lascaux, was quite tall and almost indistinguishable from modern *Homo sapiens*. During the Ice Age that followed this period, human welfare declined, but after the ice receded, agriculture eventually provided a critical lifeline. In the good years, population growth must have been boosted, but sustained periods of nutrient deficiency, particularly in early childhood, and persistence of infection would have retarded growth in height. Measurements of Old World skeletons of this period suggest that average height fell by almost fifteen centimetres and did not return to the old level again until two hundred years ago.[6] New World skeletons from the fifth millennium B.C. were often the same height as modern people, but in more recent developing agrarian societies stature has often declined, a trend that was greatly exacerbated during the Spanish conquest.[185]

The stunted growth of children in the workhouses and slums of Victorian Britain is burnt unforgettably into the British national imagination through the pen portraits of Charles Dickens. However, it is the reports of Edwin

Chadwick, the apostle of sanitary reform, that tell us objectively that British factory children in 1833 were on average twenty-three centimetres shorter than the mean height of modern children of the same age.[185]

Statisticians speak of a "secular trend," meaning a movement that occurs once in an era – for example, a trend towards greater height. Such a trend started in the early nineteenth century in most industrial countries, largely as a consequence of increased extension of the long bones of the legs in each succeeding generation brought about by improved nutrition. This continued until the 1980s, adding one to three centimetres to adult height for each decade in almost every industrial country.[41] This trend has sometimes been attributed to genetic changes generated by increased out-breeding consequent to the greater mobility of modern society, but most experts believe that nutrition is the key, because the trend reversed during major wars.

The key nutritional factors that affect growth are not easily determined, and in any case may vary from place to place, but cow's milk has clearly affected the growth of children in many societies.[185] The introduction of cow's milk to the diet of Japanese children in the 1960s probably explains the abrupt acceleration in their rate of growth. Similarly, differences in height between traditional pastoralist communities of Central Asia and Africa and the rice-eating communities of Asia are usually attributed to the nutritional effects of cow's milk. Whether calcium, vitamin D, or protein is the critical component of milk is uncertain, but the betting seems to be on the effect of calcium on bone growth.

Nutritional experiences in early life that affect adult height also seem to have a far-reaching effect on life expectancy. In a massive survey of physical characteristics and life expectancy of Norwegians in the 1960s and 1970s, researchers found short people were more than twice as frequent in cohorts of people who died prematurely. No indication of such a bias was found amongst those who lived at least seventy years. Short people are evidently more susceptible, on average, to cardiovascular and respiratory disease but not to other kinds of ill health, an observation reminiscent of the consequences of low birth weight.[108]

Infection exerts a powerful negative effect on growth in many parts of the Third World. Children with enteric infections for many days a year have a negative energy balance during those periods that retards their growth and is not easily overcome if they fail to shake off the infection or are not better fed.[41] Better nutrition takes several generation to improve stature to reflect the true genetic capability. The mother's weight determines the growth trajectory of each generation and remains a strong predictor of adult height even under an improved nutritional regime. The first generation of children under the new regime will still be undersized, and the girls will have children who will also be slightly undersized, but stature will improve to some extent in each succeeding generation. The statistics show quite strongly that each mother's own birthweight is a better predictor of the eventual height of a child than her actual weight. Similarly, the size of babies born to surrogate mothers is correlated more with the size of the mother than with that of the egg donor.

An ominous new secular trend is the advance of obesity. Ethnic groups that carry genes better adapted to the feast-or-famine existence of the hunter-gatherer period seem to be predisposed to obesity.[36] However, in affluent societies, being overweight is becoming more common, and some individuals have a tendency to extreme obesity, fuelled by the abundance of food and the decline in necessary physical activity. A secular trend seems to be developing in which overweight mothers have overweight children, perhaps because the predisposition towards being overweight is established *in utero*. Some nutritionists believe that the "fast food" industry contributes to the problem by promoting food with a high energy content that does not satisfy appetites adequately, so that the customers consume excessive numbers of calories. The consequences could be serious, since being overweight is often associated with a greater risk of developing type 2 diabetes, high blood pressure, heart disease, and serious leg ulcers.

Challenges, Risks, and Rewards – a Summary

Human reproduction may be remarkably efficient for the species, but for individual couples the process is often fraught with difficulties. The greatest practical lesson of twentieth-century embryology may be the recognition that experiences in the womb have far-reaching effects on adult health and life expectancy. The challenge, now, is to understand these influences better and to devise ways of avoiding experiences that damage the prospects of the unborn foetus. Inadequate nutrition arising from specific vitamin deficits or a more general deficiency of calories and protein clearly damage the prospects of an unborn baby, although the consequences may not be seen for perhaps five decades. A reciprocal danger probably originates in overnutrition, which predisposes an unborn child to obesity in later life. Unborn children are also adversely affected by poisons that cross the placenta – such as tobacco smoke, alcohol, and narcotics – and by viruses such as HIV and rubella that take the same route. The babies of mothers whose diabetes is not adequately controlled also have a 10 percent chance of being born with development defects. Factors affecting length of gestation, such as bacterial infections, also crucially affect the health and prospects of offspring. Physicians have responded to this challenge by devising strategies that attempt to simulate the nurture normally conferred by the womb. Although this may facilitate the survival and proper development of premature babies, extremely low birthweight babies still have a low probability of thriving in the long term. Doctors worry about the correct ethical response to this problem, particularly as not all parents have the fortitude to manage a baby with a handicap. The prospect for avoiding poignant misfortunes is steadily improving, but the understandable wish for perfect babies cannot be granted by anyone – threats of litigation notwithstanding. Today, premature and underweight birth can be managed better than at any time in history, and, with appropriate vigilance, congenital defects and chromosomal aberrations may be identified *in utero* early and plans made to manage the consequences. Surgery after birth or even *in utero* can correct some of these problems with startling success.

GENERAL SOURCES: 41, 141, 144, 145, 185.

6   Infection, Nutrition, and Poisons: Avoiding an Unhealthy Life

The Israelites believed all ill health was caused by the Almighty's displeasure, and the idea persisted in the Judeo-Christian tradition for more than two millennia. Today, we understand that the good health, and indeed the survival, of any mammal depends on three things: resistance to infectious disease, an appropriate diet, and avoidance of poisonous substances. Our defences against microbes are complicated and powerful, and although they may protect us, they can also create symptoms of disease. We can define our dietary requirements as never before and can recognise that deficiencies of these nutrients have blighted human societies throughout history, although we still have some way to go to ensure the optimum micronutrient intake. The human body is vulnerable to poisons in plant foods and to environmental pollution, in spite of the extraordinary powers we have to detoxify natural and man-made poisons. This chapter tells the story of how, in the twentieth century, we came to understand some of the requirements for a healthy life.

Microbes – Historic Enemies

Microbes are nature's greatest opportunists; their almost infallible capacity to exploit any chemical as a nutrient allows them to take advantage of virtually any ecological niche, using their astonishing speed of multiplication and genetic adaptability. Every external and internal surface of the mammalian body is constantly under siege from microbes. Exterior surfaces are effective physical barriers to almost all microbes, but dynamic multilayered defences of a more subtle kind also protect against this perpetual threat. A skin lesion of any kind becomes highly vulnerable to such an enemy, and all multicellular organisms have evolved remarkably effective systems for rapidly sealing wounds and excluding microbes. Mammals lack the re-markable capacity of the newt to restore amputated limbs, but in adapting to terrestrial life they have acquired a more effective system of skin repair. Even so, microbes and parasites still have the potential to enter the body and create serious mischief.

Mammals have two main lines of defence against infection. The first is perpetually on guard, a kind of hair-trigger reflex that biologists call innate immunity; the second is intelligence-based and involves the body's learning to recognise new enemies. Innate immunity is encoded in our chromosomes and is inherited from our parents. The second kind is "acquired" – in the language of immunologists – by a complex reshuffling of the chromoso-mal cards in specific cells of the immune system and cannot be passed on to our offspring. Our understanding of the mechanism of self-defense, which is probably still far from complete, is a remarkable achievement of twentieth-century science. It also provides profound insights into the origin of symptoms of disease – which, paradoxically, are often the reactions of our bodies to the invaders.[186]

The mammalian immune system is the culmination of an evolutionary process by which animals have learnt to recognise potential aggressors by the foreignness of their chemistry. The innate system identifies accurately certain classes of microbe and never responds to "self"; the adaptive system "learns" to recognise almost any previously unseen chemical but may attack "self" motifs in error.

INNATE RESISTANCE TO INFECTION: A microbe appearing in a wound will probably be recognised and destroyed by the innate immunity sys-tem. Once the skin is broken, the key element in recognising microbes is an unusual-looking and extraordinary cell that has many tentacle-like projections that penetrate between cells without forming intimate connec-tions that restrict movement. Although originally described in 1868 by the Prussian cytologist Paul Langerhans – who also discovered the famous pan-creatic "islets" that bear his name – their function remained unknown for more than a century. Today, they are usually called dendritic cells. Their branches sample their immediate environment using sensors or receptors that can recognise the chemical attributes of at least nine different groups of microbes that betray their foreignness. These receptors have a common design that probably evolved from a protein that existed at the dawn of multicellular life and that was engaged, even then, in repelling dangerous advances from microbes. In the absence of a better word, they are known in the idiosyncratic language of modern genetics as "toll-like receptors." This is an allusion to their similarity to an insect protein known as *toll*, the first gene of this class to be described. These receptors enable dendritic cells to act as scouts that can pick up the scent of a microbe. The information is passed to other lymphocytes to start the immune response that, together with its inflammatory dimension, always follows a wound.[186]

A complicated substance called lipopolysaccharide, found on a class of gut-inhabiting bacteria that include *Escherichia coli* and disease-causing visitors such as typhoid, dysentery, and cholera is one of the most typical bacterial chemicals recognised by the human innate immunity system. Most organisms respond to lipopolysaccharide, in some way, but humans have the most exquisitely sensitive reaction. The tiniest dose injected into the bloodstream provokes an immediate increase in body temperature, and bigger doses produce symptoms comparable to the onset of a severe enteric infection. Dendritic cells and macrophages – the white blood cell types that are professional eaters of dead or infected cells – recognise lipopolysaccharide on the bacterial surface and then quickly engulf the bacteria. Mutant mice that cannot recognise lipopolysaccharide die of septicaemia when they are infected by otherwise harmless strains of gut bacteria.

In a similar way, the innate immunity system recognises at least eight other classes of chemical – too many to mention here. Invading microbes are greeted with a robust broadside from a class of powerful antimicrobial peptides called defensins. These are delivered by a battery of special granules found in many kinds of cells of the blood, on the surface of the intestine, the kidney and uro-genital tract, and in the skin.[187] Their importance is particularly obvious when they are absent. In severely burnt patients, the skin becomes highly vulnerable to infection because the front line of innate immunity – the cells that make the defensins and the macrophages – is destroyed. Patients with the genetic disease cystic fibrosis lack a chloride ion pump, which means that the surfaces of the lung and other tissues have an abnormal salinity that disables the defensins. This leads to the peculiar pathology of the disease, in which the victim becomes chronically infected with otherwise harmless bacteria.

INFLAMMATION – THE TWO-EDGED SWORD: "Inflammation" is an evocative and venerable name for the set of symptoms that characterise reactions of the human body to many insults. Redness caused by dilation of blood vessels, localised heating from increased blood flow, swelling from accumulated fluids exuding from blood vessels, and pain are all universally recognised symptoms of inflammation. These symptoms are provoked by trivial scratches and major wounds, burns and exposure to radiation, heat and corrosive chemicals, and also by systemic catastrophes such as deprival of oxygen to the heart muscle or brain.

Inflammation has evolved as part of our innate immune system; it kills microbes with devastating efficiency but with weapons that can be extremely damaging to us. Consequently, while the response to a cut or a bruise is usually limited, the inflammatory response to certain kinds of damage may far exceed the insult. A bee sting, to a person already sensitised to bee venom, can evoke an extreme and sudden swelling of the airways or of the brain that may be lethal – anaphylactic shock. When the heart muscle is deprived of oxygen in a coronary heart attack, an inflammatory reaction kills the affected tissue. More surprisingly, we find the inflammatory response underlies the symptoms of most chronic diseases of middle age, such as atherosclerosis and rheumatoid arthritis.

What happens after an injury? Cells at the site respond to the trauma by summonsing an "emergency team" of assorted white blood cells within seconds of the injury. Some of them start making chemicals, such as histamine, with a special role in transmitting the inflammatory response to surrounding tissues. Platelets – strange little fragments of a cell that lack a nucleus – accumulate in wounds where blood clots form. Their special role is to release proteins called cytokines that affect neighbouring cells powerfully and to lure towards the wound phagocytes – the cells that eat other cells. Once engulfed by a phagocyte, microbes die under a fierce onslaught of chemicals not unlike household bleach (hydrogen peroxide and the free radical called superoxide) plus powerful destructive enzymes. The importance of this process is tellingly illustrated by chronic granulomatous disease, a rare human genetic disease in which the blood cells that normally kill microbes are unable to make the chemicals (superoxide, etc.) necessary to kill bacteria. Patients therefore suffer from a succession of localised infections and frequently die from septicaemia at an early age.

Acute pain often accompanies inflammation, an evolved response that keeps us conscious of the dangers of injury. Bradykinin is a pain-inducing chemical released by affected cells that binds to receptors on nerves and signals "distress" to the brain. Bradykinin has the curious distinction of being the most potent pain-inducing chemical known, measured in a test based on the excruciating pain inflicted when applied to open blisters. Its other roles include making blood vessels constrict, inducing secretions from gut epithelia, and acting as a chemical attractant for the releasing cell.

Eventually, inflammation subsides and the wound heals, infection free. However, following tuberculosis, leprosy, and certain viral infections, the microbes that provoked the inflammatory response sometimes persist in the phagocytes.

TUMOUR NECROSIS FACTOR (TNF) AND THE INFLAMMATORY RE-SPONSE: Lipopolysaccharide, the calling card of enteric bacteria we met earlier, activates the innate immune response and a cascade of agents – too numerous to mention here – that contribute to inflammation. The first of these to become well known was blessed with an evocative name – tumour necrosis factor (TNF) – that seemed to promise some kind of victory over malignancy. The discovery was foreshadowed in the work of a late nineteenth-century surgeon from New York, William Coley, who noticed how some malignancies regressed during recurrent erysipelas infections. He deduced that the microbe that caused this nasty creeping inflammatory disease of the skin released a substance that provokes regression of the cancer. To test his hypothesis, he injected patients with extracts from cultures of *Streptococcus* and found tumour regression; he claimed, remarkably, that ten percent of his patients were cured. At a time when cancer was becoming more common and the treatments available were the despair of patients and doctors alike, his approach seemed to herald an intellectually elegant cure that would make surgery and X-ray therapy seem rather crude.

Sadly, Coley's toxin, as it became known, never developed much further, but eight decades later the dramatic effect of the inflammatory response on tumours was confirmed. A protocol similar to Coley's was reconstructed in mice in 1975, using the endotoxin of the tuberculosis bacillus.[188] The serum of endotoxin-treated mice contained a factor that made tumours regress by what pathologists call necrosis – death of cells by bursting. This factor, known ever since as tumour necrosis factor, began to look like a potential weapon against cancer of some importance. Ten years later, the genes for TNF were cloned and the protein was made on an industrial scale. Unhappily, it seems to be too toxic to be a useful cancer therapy, but once its importance in inflammation was established, it found other medical uses in controlling the worst effects of inflammation (see Chapter 12).

The crucial part played by TNF in serious inflammation first became apparent in 1980 in the aftermath of an epidemic caused by *Staphylococcus* infection that took the lives of thirty-eight women and made thousands very sick. These infections, which became known as toxic or septic shock syndrome, started during menstruation and were associated with a new design of hyperabsorbable tampon in which bacteria multiplied during prolonged use. Substances released by microbes into the bloodstream of the victim (so-called superantigens) provoked inflammation and production of TNF by the victim's body. TNF has devastating effects; it makes blood vessels permeable and causes loss of blood pressure, cardiovascular collapse, and the failure of other critical organs, with death ensuing within seven days. The key part played by TNF in creating this damage was confirmed when the syndrome was brought under control by antibodies directed against TNF.[189]

Toxic shock hit the headlines as a new disease, but in reality it has probably always existed, associated with death by blood poisoning, and it will probably recur in other forms in the future, whenever hygienic safeguards fail. A *Streptococcus* species, gruesomely celebrated as the notorious "flesh-eating bug" that causes necrotising fasciitis as well as toxic shock, is a major cause of death in intensive care units.[189]

TNF causes a serious and dramatic wasting in certain other diseases, such as the sleeping sickness of African cattle infected by *Trypanosomes*, and in the pathology of AIDS. In the terrible last days of the disease, the HIV virus provokes the body into making TNF and the characteristic wasting of the victims ensues.

ATHEROSCLEROSIS IS AN INFLAMMATORY DISEASE: Atherosclerosis, the cause of most cardiovascular disease, is a kind of inflammation. In the previous chapter, we saw how coronary heart disease became the leading cause of death in most industrial countries, affecting at least one in three people but remaining rare in the less-developed world. The root cause remains uncertain, although events *in utero* relating to foetal nutrition, diet in later life, smoking, and other factors all play their part. The most important clue to the immediate cause of coronary heart disease emerged from a famous American survey known as the Framingham Heart Study carried out in 1954. This demonstrated a correlation between amounts of a chemical known as low-density lipoprotein (or LDL) circulating in the bloodstream and the chance of coronary heart disease, unrelated to other risk factors. LDL is a biochemical complex containing certain proteins, the steroid cholesterol, and other fatty substances whose role is to transport cholesterol from the liver to other tissues to make and repair cell membranes. A similar complex known as high-density lipoprotein (HDL) transports cholesterol to the liver for excretion and recycling. LDL can enter cells of artery walls and may accumulate there, excessively, in certain individuals. The trouble starts when LDL is oxidised by free radicals – the notorious nuisances that initiate the classic inflammatory reaction, just described, that are products of imperfectly functioning mitochondria (see p. 73).

The next phase is the arrival of the emergency team of specialised blood cells called monocytes, whose normal role is to fight infection. These cells attach to the blood vessel walls, where they transform into macrophages and start gobbling up materials, becoming replete with fatty droplets. Cells lining blood vessels then grow over this so-called fatty streak or atherosclerotic plaque, and eventually constrict the blood vessel. Later, the plaque may rupture, releasing factors that can initiate blood clot formation, and the clots may be big enough to block the flow of blood through the heart muscle. If this happens, cells of the heart muscle may die because they are deprived of oxygen, provoking a coronary heart attack in the victim. Coronary heart disease has a fiendishly complex origin, rooted in many kinds of life experience from foetal life onwards. In this and later chapters, the variety of ways in which medical science has sought to overcome the problem will be addressed.[190]

LEARNING TO RECOGNISE THE ENEMY – ACQUIRED IMMUNITY: The innate immunity system cannot master every kind of microbe. A second line of defence literally learns to recognise microbes by experience, so that our bodies can produce an overpowering response that completely halts infection. The key players are two kinds of white blood cells, the B- and T-lymphocytes, which start life in the bone marrow and the thymus gland, respectively, and which together can recognise most kinds of microbe that enter the body. The role of B-lymphocytes is to gain global control of the bloodstream using molecular policemen – the antibodies – that circulate everywhere in that vital thoroughfare. These have the singular ability to recognise villains accurately, by their foreignness, and to "tag" them. Microbes marked so distinctively then quickly attract the attention of a dedicated crime squad – specialised cells called phagocytes – that promptly eat the tagged individuals.

T-cells recognise and kill cells infected with viruses or bacteria because they display on their surface fragments of the invader. They have a robust and ruthless fighting style that would appeal to human military strategists, as they engage the enemy extremely closely and take no prisoners. Using the same ruthless principles, T-lymphocytes attack and reject grafts or organ transplants from other individuals.

Immunologists make the extraordinary claim that every peptide – a small chain of amino acids that can be permuted into billions of alternatives – that might enter the bloodstream can be recognised by an antibody made by a specific lymphocyte. This supremely ingenious evolutionary invention enables most vertebrates to trump nearly every infectious agent that might ever arise through evolution.

How is it done? The genes that make the antibodies are not present, as such, in the DNA of human sperm or eggs but are created after conception through rearrangements of the chromosomes, generating a vast library of different sequences. The process is too great a challenge to describe here in detail, but briefly, all of these sequences are available in the bone marrow as if in a library. Each sequence is in a different lymphocyte, and on first exposure to a novel peptide, a "book" containing the recipe for a particular antibody is withdrawn from the bone marrow library. Initially, the amount of antibody made by the specific lymphocyte is not sufficient to be of practical value, but these few cells start to divide and become quite numerous in several weeks. However, continuous exposure or a second exposure to the same peptide provokes a massive surge in the level of antibody circulating in the blood. The reason why children catch diseases of childhood only once is that a single exposure is sufficient to train the immune system to fight every subsequent infection successfully. This is the essential principle of the immune system, recognised for the first time by Edward Jenner when he devised vaccination against smallpox.

OVERREACTION OF THE IMMUNE SYSTEM: Sophisticated and powerful as the immune system may be, it occasionally overreacts, with no discernible biological benefit and with consequences that run the gamut from annoying to lethal. The overreaction or uncontrolled inflammatory response may be a physiological reflex to pollen, fungal spores, or dust mites. It takes the form of allergies and asthma, with symptoms such as sneezing and breathlessness, although more serious reactions are possible. A special class of antibody known as IgE responds to these agents in the skin, nose, lungs, or gut of sensitive individuals. These activate special blood cells – known as mast cells – that initiate inflammation by releasing histamine and other molecules. Fortunately, antihistamine medications control such reactions fairly effectively. Rare but alarming episodes of acute allergic reaction (known as anaphylactic shock) may follow a second exposure to allergens such as bee venom. Muscles of the lung and gut contract, the blood pressure falls, and skin and airways swell dangerously. The victim faints, and may even die if a physician does not intervene with a shot of adrenalin.

When John Bostock first described hay fever in 1819, the condition was very rare, but it now affects one in two people in the developed world. Similarly, alarming acute reactions such as nut allergies have been increasing very sharply in the last three decades.[191] Why this is happening is a mystery that makes the public suspect a wide range of possible causes. Medical opinion mostly favours the idea that stimulation of the immune system by a minor infection in early life is necessary to confer lifelong immunity to allergens. A striking comparison of children who grew up in the former East and West Germany illustrates how this might occur. The two groups are likely to be genetically equivalent, but their lifestyles and life experiences differed sharply at many points. The children from the East lived in a town with substantial air pollution and attended day-care centres where, most probably, they were constantly exposed to germs. The children from the West, at home with mothers who received paid maternity leave, were less likely to be exposed to infection and air pollution. The surprising outcome was that children from the West, brought up to avoid infection and grime, suffered most from allergies and asthma.[192]

AUTOIMMUNITY: The acquired immunity system would be seriously flawed if it created antibodies that recognised our own proteins. This is clearly not generally the case, so there must be a mechanism that eliminates from the repertoire antibodies that might react against ourselves. Occasionally, disorders flare up in which the immune system attacks a component of the victim's body. Once again, inflammation is the culprit and results in autoimmunity – a condition that is widespread and intractable. In some individuals, a few anti-self lymphocytes that should have been eliminated in childhood seem to survive and suddenly start a sustained attack on certain tissues in later life. These diseases are not infectious but start in unaccountable ways, as though triggered by an environmental cue. Neither are they genetic, although genetic predisposition plays a part; and, very mysteriously, they are more common in women than men.

The idea that diseases might be caused by autoimmunity was new in the 1950s, but now at least forty autoimmune diseases are known. This dismal list includes insulin-dependent diabetes (type 1), rheumatoid arthritis, multiple sclerosis, inflammatory bowel disease, and lupus. The lives of perhaps one in thirty people will be adversely affected by an attack by lymphocytes or antibodies against an organ.[193] Such insults activate the inflammatory response, unleashing powerful forces against important structural proteins. In rheumatoid arthritis, hyaluronic acid – a complex carbohydrate crucial to the smooth operation of joints – is damaged. In type 1 diabetes, T-lymphocytes destroy the capacity of the islets of Langerhans – the insulin-making cells of the pancreas – to make insulin. In Crohn's disease and some related disorders, inflammatory attacks damage the gut.

The entire inflammatory system contributes to this damage, including agents with far-reaching effects, such as the tumour necrosis factor we encountered earlier. Victims of rheumatoid arthritis and inflammatory bowel disease seem to be responding in the most encouraging way to monoclonal antibodies that can "neutralise" TNF (see p. 254).[194]

Autoimmune diseases get started under mysterious circumstances. Infection is always a suspect, because it could expose otherwise hidden aspects of "self" to the immune system or introduce a component that mimics a "self" protein. Indeed, microbes have a well-established role in inflammatory diseases such as rheumatic fever and stomach ulcers (*Helicobacter* – see Chapter 9), but some people believe – albeit rather controversially – that other chronic diseases originate with microbial infections. A genetic component probably confers a crucial sensitivity to something in the environment in all autoimmune diseases. The genetic predisposition to Crohn's disease, for example, seems to reside in the inability of a defective protein to protect the victim from inflammatory reactions provoked by certain gut bacteria.[195]

Immunologists believe allergic disease and autoimmunity are less likely in individuals who had a robust well-regulated anti-inflammatory response soon after birth. Two sets of T-lymphocytes organise responses to bacteria and viruses (T_H1) and helminth infections (wormlike parasites) (T_H2). Uncontrolled strong responses of the T_H1 system cause autoimmunity; uncontrolled responses of T_H2 cause allergy. However, if infants are exposed to infections early in life, their responses to later challenges rarely produce pathological inflammation. Helminth infections are so rare in the industrial world that the T_H2 arm of immunity is rather undeveloped there, in contrast to the developing world, where such infections are commonplace and allergy is almost unknown.[191]

Table 6.1. *Twenty micronutrients essential for human health*

Vitamin A and/or β-carotene	Pantothenic acid
Vitamin B1 (thiamine)	Biotin
Vitamin B2 (riboflavin)	Folic acid
Vitamin B6 (pyridoxine)	Calcium
Vitamin B12	Iron
Vitamin C	Zinc
Vitamin D	Selenium
Vitamin E	Magnesium
Vitamin K	Iodine
Niacin	Chromium

How Can We Know What We Should Eat?

Our remote ancestors were tree-dwelling apelike creatures whose diet was fruit and foliage but who learnt to hunt herbivorous animals and acquired a taste for meat. Almost a million years elapsed between that crucial step and the emergence of the Cro-Magnon people, who lived in Europe about thirty thousand years ago. Their enthusiasm for big game hunting is well known from cave paintings, but their artefacts also indicate a partiality for some vegetable material. As the climate cooled and big animals became less abundant, the successors of Cro-Magnon man explored a wider range of foodstuffs, including fish, molluscs, and small game. The tools that survive tell us that plant-based foods eventually became a large part of their diet.

The experiment of self-sustaining agriculture, while hugely important to the development of civilisation, was a substantial nutritional difficulty for a species with omnivorous habits. As we saw in the last chapter, we know from European and New World skeletons dating from the agrarian revolution that people often died showing signs of protein and vitamin deficiency and indeed of starvation. Only in the last two hundred years has the intake of dietary protein reached the levels required for humans to grow to the same average height as Cro-Magnon man.[6]

Our penchant for exploiting available resources has allowed us to explore many kinds of diet, but such experiences rarely gave any indication of what is optimal for our health. The subtlety of our dietary requirements did not become apparent until the early twentieth century, when the effects on rodents of diets of defined composition were investigated by Frederick

Gowland Hopkins and others. These studies showed that mammals require – in addition to protein, carbohydrate, and fat – miniscule quantities of certain accessory foodstuffs that are present in small amounts in milk. Using Hopkins's rigorous methodology and the ascendant discipline of natural-product chemistry, an explosion of research in the United States and Britain revolutionised our understanding of nutrition, discovering the chemical identity of more than twenty micronutrients.

The concept of dietary deficiency was not entirely new; the value of citrus fruits for preventing scurvy in humans was known for more than a century. In the late nineteenth century, a disease widespread in South and East Asia called beriberi was also established as a nutritional deficiency disease. This could be avoided if the rice that was the staple crop in the region was cooked with the husks. Once the major vitamins were known and available as food additives, the extent of nutritional deficiency in many communities became apparent. The historic dependence of poor subsistence farmers throughout the world on single staple crops was recognised as an inadvertent blunder of monumental proportions. Most staple crops (except maize) seemed to be inherently deficient in vitamins C, D, and A, and, more poignantly, traditional food preparation methods generally removed the traces of B vitamins that were present. In Britain, green vegetables were commonly cooked long enough to destroy vitamin C. Rice, the staple crop of Southeast Asia, was prepared by a mechanised process that discarded the only source of vitamin B6, which is present in the outer husks of the grain. Rice is also deficient in vitamin A, an important nutrient for development of the retina, so that the estimated four hundred million people who still live almost solely on rice are at risk of vitamin A deficiency. Consequently, night blindness and *xerophthalmia* – a disease that affects the cornea that can lead to incurable total blindness – are widespread.[196] Wheat, milled to produce white flour by traditional methods, is devoid of vitamin B6 and contains little protein or fat. In the cotton monoculture districts of the American South, during the early twentieth century, poor sharecroppers lived almost entirely on corn mash – a staple food made from maize – and suffered from a disease called pellagra. After four decades of research during which three million Americans contracted the disease, this potentially lethal complaint, characterised by severe dermatitis and extreme lethargy, was eventually attributed to a deficiency of nicotinic acid (niacin). Throughout southern Africa and South America, maize was and remains a staple food, but pellagra never arose because the traditional alkaline-processing step used in those regions released niacin in a form that was useful nutritionally.[197]

The vitamin content of most foodstuffs was known by the late 1930s. For the first time in history, experts in nutrition emerged who could make rational dietary recommendations and who could objectively assess whether the diets people actually consumed were adequate to support a healthy life. Although nutritional circumstances in Britain were better than at any time in history, a survey by John Boyd Orr revealed widespread malnutrition. Possibly one-fifth of the children and one-tenth of the population were chronically undernourished, and perhaps one-half of the population suffered from a specific nutritional deficiency.[159] Defenders of the political status quo thought the survey had used excessively severe standards. However, the conclusions were soon confirmed and extended by more surveys that linked higher death rates and ill health to poor nutrition amongst lower socioeconomic groups.[53] Nutritionists were now able to promulgate the importance of a balanced diet containing milk products, meat, eggs, fruit, and vegetables, with considerable conviction, as this would provide the vitamins required for good health. The survey also gave a poignant meaning to the demands of organised labour for a "living wage." Later on, strategies would be devised, at very little cost, to fortify staple foods with the stable vitamins B1, B2, B6, A, and D, niacin, and later folic acid in the hope that vitamin deficiencies would never recur. During the 1940s in Britain, nutritional supplements for pregnant mothers and children provided at no cost by government agencies ensured that wartime children were in many ways healthier than previous generations.

Other advances in the biological sciences came to overshadow the activities of nutritionists later in the century, but their important work was certainly not finished. The importance of folic acid and vitamin B12 in preventing pernicious anaemia was established in the late 1940s, but the part folic acid played in preventing birth defects was not appreciated until the 1960s. Essential fatty acids or polyunsaturated fatty acids, another class of micronutrient discovered in the 1920s, assumed greater significance when their physiological role became known more than fifty years later. In the late twentieth century, the role of vitamin D as a nutrient for a class of lymphocytes important in conferring cell-mediated immunity to tuberculosis, as well as its possible role in preventing certain kinds of cancer, became apparent. Awareness is growing that we need a small daily exposure to sunlight to generate sufficient vitamin D to confer this anticancer benefit.[156]

By careful monitoring in every country, the WHO has established that serious deficiencies of the micronutrients iron, iodine, and vitamin A still exist in many regions. The WHO believes that the most common preventable cause of brain damage in the world today is iodine deficiency, which, in its most serious form, causes an irreversible mental retardation known as cretinism. Iodine is needed to make thyroid hormone, which in turn is required for brain development during foetal and early postnatal life. Although the deficiency can be brought under control with iodised salt, the scale of the problem is huge – two billion people are probably at risk. Anaemia (iron deficiency), greatly aggravated by parasite infections, also affects two billion people and causes about twenty percent of all maternal deaths; it is being tackled by the distribution of iron supplements and antiparasite treatments. Vitamin A deficiency remains common in Southeast Asia in spite of attempts to distribute vitamin A capsules and to encourage the use of more green vegetables containing beta-carotene. Nutritionists also occasionally discover instances of people who suffer from deficiencies of zinc and chromium, a situation that could only arise through long-term dependence on poor-quality food.

WHAT DO MICRONUTRIENTS DO? Every vitamin has a particular biochemical role that we need not consider in any detail except to indicate the significant consequences of deficiency. Most important is the effect of deficiency of the group that includes folic acid, vitamins B12 and B6, niacin, vitamins C and E, and the elements iron and zinc. Deficiencies of these micronutrients have specific consequences but also indirectly damage synthesis and repair of DNA, which can start cells on the road to malignancy and birth defects.[157]

Another group of four micronutrients (vitamins C and E, the pro-vitamin of vitamin A called beta-carotene, and the element selenium) has a far-reaching role in removing free radicals, highly undesirable products that appear during an episode of inflammation. Absolute vitamin deficiencies – for example, vitamin C deficiency, which causes scurvy – are now quite rare, but marginal deficiencies of antioxidants, causing a steady accumulation of oxidative damage to DNA, have serious long-term consequences. The damage is a major risk factor for cancer, and chronic inflammatory diseases contribute importantly to the ageing process. Vitamin C has special importance as an antioxidant because it can regenerate vitamins E, A, and possibly the correct form of selenium. Lipid-soluble vitamin E can penetrate fat deposits where vitamin C may be deficient, and therefore may shield low-density lipoprotein (LDL) from oxidation and protect against atherosclerosis and coronary heart disease.[198]

Humans and a few other primates are the only mammals that require dietary vitamin C. The exact amount needed for healthy life is controversial, but conservative authorities suggest that sixty milligrams a day is enough to prevent scurvy, the most obvious symptom of deficiency. A widely quoted, if vague, recommendation is that people should eat at least five servings of fruit or vegetables each day to obtain sufficient antioxidants.

Although the so-called essential fatty acids were known in 1928, another half-century elapsed before a clear idea of their role emerged from investigations into why Greenland Inuits only rarely suffered from cardiovascular disease. Their traditional diet was extremely rich in fat and deficient in vitamin C, but, consistent with their low risk of thrombosis, the time required for their blood to clot was much longer than for a Danish control group. Their platelets, the curious cells that play a special role in blood clotting (see p. 83), contain unusually high amounts of one of the two groups of essential fatty acids (linolenic acid, a member of the n-3 group).[199]

Another "experiment of history," enacted during the Nazi occupation of Norway, suggested how a fish-rich diet might protect against atherosclerosis and coronary heart disease. Because meat was scarce, fish consumption increased threefold, and within a short time the death rate for coronary heart disease fell sharply.[199] The explanation of both phenomena is that cold-water fish and marine mammals are notably rich in essential fatty acids of the n-3 variety, which they obtain from plankton. While fatty acids of the n-6 and n-3 varieties are both essential components of the human diet, they have opposite effects on blood clotting and on inflammatory processes – that is, they are pro- and anti-inflammatory, respectively. Nutritionists now believe that our health would be substantially improved, and a variety of inflammatory conditions reduced, if we consumed more n-3 fatty acids, preferably as fish. In particular, there would be a significant reduction in the risk of heart disease and stroke, lung disorders, high blood pressure, type 2 diabetes, eczema, and autoimmune diseases such as Crohn's disease, rheumatoid arthritis, and lupus.[200] Moreover, these fatty acids are required for the development of the retina and brain in the developing foetus. Newborn babies obtain them in breast milk, but premature babies require baby milk supplemented with linolenic acid (n-3) to ensure proper development of cognitive functions.[179]

NON-NUTRIENT DIETARY FACTORS: Probably all the micronutrients absolutely required for human life are now known, but the idea that non-nutrient dietary factors exist that favour better health is being investigated.

This argument is used to explain the differences in the incidence of disease between Japan and the rest of the industrial world. Japan has a notably low incidence of cardiovascular disease and of cancer of the breast and prostate gland. Japanese women have about one-sixth the risk of breast cancer of American women, but the effect disappears in the second generation after emigration. Clinical prostate cancer is ten to fifteen times more frequent amongst American white men than amongst Japanese men. The difference reflects the progression of the disease rather than the cause, because latent prostate cancer is perhaps only 50 percent greater in the United States. Many cultural and genetic differences between the two populations could explain the difference. However, attention is focussing on the soya bean products used in Japan and specifically on iso-flavones, a class of chemical structurally similar to steroids, which can antagonise steroid hormonal responses in rats. One representative, called genistein, suppresses tumours of the rodent mammary and prostate glands and also hastens development of the rodent mammary gland – an observation of unknown significance that has not been corroborated in humans.[201]

LEARNING FROM DIFFERENT LIFESTYLES: The spectrum of chronic disease encountered in different countries associated with different lifestyles has important lessons for us all. Although rampant infectious disease drastically affects average life expectancy in developing countries, the bald statistics of disease tells another story. Western society, for all its advantages, suffers from maladies that were formerly uncommon in the developing world. In a statistically informal compilation of the experiences of doctors who practised in the developing world published in 1981, remarkable differences in the incidence of chronic ill health between different regions emerged.[161] Conditions that were uncommon in the developing world included high blood pressure, coronary heart disease, peripheral vascular disease, type 2 diabetes, gallstones, varicose veins, colonic disorders, obesity, dental caries, and cancer of the breast, ovary, prostate, and lung. On the other hand, stroke, kidney and heart failure, and different kinds of cancer connected with chronic infection were more common in nonindustrial regions, in addition to their much greater burden of infectious disease. These differences are not easily explained, but a relative lack of exercise, excessive calories consumed, a relative deficiency of dietary fibre, and an excess of salt and fat in the Western diet are all thought to be relevant. Sadly, as the less developed world adopts Western culture, its people are being affected by Western diseases.

Other kinds of comparison suggest different lessons. In the last fifty years, for example, coronary heart disease has affected the English-speaking world much more than it has affected Japan or France, but it is virtually impossible to identify the key factor with certainty. Inadequate foetal nutrition, excessive consumption of animal fat, excessive dependence on dairy products, and insufficient consumption of vegetables, olives, and red wine have all been suspects, but nothing conclusive emerges.[202]

Toxicological Dangers – Living in a Chemical Zoo

Vegetarians often suppose that their lifestyle is intrinsically virtuous because they have no sentient victims, but the evolutionary history of flowering plants indicates that they have resisted predators of every kind with a sustained ferocity since the dawn of time. Plants must have provided an unforgettable chemistry lesson for the first primates as they learned to negotiate the unpalatable and the exquisitely poisonous. Millennia of selection have given us cultivated plants that are relatively free of natural toxins, although traditional crops of the Third World need elaborate processing to remove high levels of toxins before consumption. An Indian lentil-like crop (*Lathyrus sativum*) used to make dahl contains a nerve toxin, and the cassava root, an important food throughout Africa and South America, contains large amounts of cyanide. Most cereal and legume crops contain large amounts of phytic acid (inositol hexaphosphate), a compound that can sequester calcium, iron, and zinc, preventing them from entering the body. In largely vegetarian societies, dependence on certain cereals probably causes a partial deficiency of these minerals. On the other hand, phytic acid is also an antioxidant; it probably neutralises free radicals derived from iron in the gut and protects the bowel against cancer.

During evolution, powerful generic systems have evolved to combat acute poisoning by natural chemicals, but they are equally effective against manmade poisons never encountered by animals. These systems are based on liver proteins that attack foreign chemicals, using the oxidising enzyme, cytochrome P-450, or an enzyme called glutathione transferase that adds a water-loving molecule to toxins, rendering them soluble, excretable, and detoxified. The livers of alcoholics are able to metabolise ethanol so efficiently that they tolerate quantities of alcohol that would be lethal to the uninitiated, but they can convert other industrial solvents into dangerous poisons.

Detoxification genes have multiplied steadily during evolution, and in humans they vary remarkably between individuals. We have at least fifty representatives of the cytochrome P-450 family and at least seven families of glutathione transferase that have an important bearing on the efficiency of medicines. Certain plants, such as broccoli, contain substances that stimulate animal detoxification systems to dispose of potential toxins more efficiently and may explain why the mysterious non-nutrient dietary factors are beneficial.[203]

It is providential, indeed, that our detoxification systems are so effective against strange chemicals when our species has invented and carelessly dumped so many novel chemical poisons. Chemicals have been accumulating in accidental emissions from factories since the beginning of industrialisation. More recently, weed killers and insecticides have been spread copiously and deliberately to increase crop yields, under the misguided impression that they are harmless to all living things except the target species.

A family of polychlorinated biphenyls (PCBs) is the most notable worry, because they have important biological effects as endocrine disrupters and also, when partially incinerated, generate highly toxic cancer-causing chemicals (or carcinogens) known as dioxins. These spread through the atmosphere, enter the food chain, and are detectable in the tissues of virtually every person tested.[204] One member of this class is the most toxic substance ever created by man (2, 3, 7, 8-tetrachlorodibenzo-p-dioxin, or TCCD) and a proven carcinogen. Investigations of sediments indicate that dioxins have been entering the atmosphere in Europe for two centuries, initially from burning wood and seaside peat and later from coal. The trend apparently peaked between 1930 and 1970 and is now declining.[205]

The extent to which dioxin can cause human disease has been investigated in long-term studies of people contaminated during accidents. When a herbicide factory exploded in Seveso, Italy, in 1976, dioxins were liberated on a massive scale; the accident's long-term consequences were investigated in detail. Many victims had acute symptoms, and children were born with permanent neurological damage sustained *in utero,* but in addition there was a long-term risk to people exposed to lower levels of dioxin. Exposures several orders of magnitude greater than the levels of dioxin found in the general population increased the risk of developing cancer and heart disease.[205] The sex ratio of babies born in the seven years following the incident increased to 1.64 females for every male born because of the selective mortality of boy babies. The people of Viet Nam still count the cost

of the American military's efforts, during the 1960s and 1970s, to defo-
liate forests using the herbicide Agent Orange, which we now know was
contaminated with dioxin.

Dioxin is evidently a serious cancer risk for those who suffer occupational exposure, but what effect will dioxins accumulated from the environment have in human tissues? They are present in the breast milk and
body fat of humans and in animals living far from possible sources. Early
estimates of the amount in human tissues seemed to be well below levels
likely to cause cancer.[206] However, even if the risk to an adult is slight,
these levels may be a significant risk to the developing reproductive organs
and endocrine glands of an unborn child (see p. 103).[172,204]

Challenges, Risks, and Rewards – a Summary

The secret of a healthy life is to avoid infection, follow a sensible diet, and
avoid poisons. While we have an extraordinarily versatile armoury to fight
microbes, we remain vulnerable to them and to the unpredictable overreac-
tions of the immune system that cause inflammation. Even as we celebrate
progress in controlling infection and inflammation, new challenges are arising from our *lack of exposure to microbes* as infants, from allergy, and from
autoimmunity. The academic study of nutrition has identified many compounds and elements essential for human health of which our ancestors
were unaware, but nutritional deficiency still exists, particularly amongst
children in the developing world. Knowledge of the dietary requirements to
avoid malnutrition is widespread in the developed world but is not always
matched by appropriate behaviour; chronic deficiencies of micronutrients,
particularly of antioxidants, contribute significantly to the health problems
of late middle age. Through countless millennia, the predominant human
experience was Adam's curse: the desperate struggle to feed a family. Suddenly blessed with an abundance of food and a physically less demanding
lifestyle, we are challenged by the insidious problem of obesity and type 2
diabetes, which will shorten the lives of most affected people. Ever since
the industrial revolution, we have been contaminating the environment
with endocrine-disrupting chemicals that have spread through the atmosphere to the remotest parts of the planet. We have remarkable powers of
detoxification of poisonous chemicals, but these pollutants seem to diminish fertility, cause cancer and birth defects, and interfere with endocrine

function and the immune response. An unanswered question is: how much do environmental pollutants affect developing embryos and contribute to reproductive cancers in adult life? We are increasingly conscious of the ways in which we contaminate the environment, but whether we can stop is quite uncertain.

GENERAL SOURCES: 6, 190, 191, 202, 207.

7 Signs of Ageing: When Renovation Slows

In an era when premature death is relatively rare, people eventually show the well-known signs of ageing and are afflicted by one or more diseases of old age. How we fare depends on our genes and our life experience, but even if our genetic constitution promises long life we may be robbed of it by the strenuousness of our lifestyle or by misfortune. The signs of ageing reflect inexorable chemical change to important proteins acquired in childhood that never renew, making some kind of decay in old age inevitable. A major culprit responsible for this chemical damage, and probably the major risk factor for age-associated disease, is oxidative stress caused by "free-radicals" that leak from our mitochondria and other subcellular particles. Reduced calorie intake, exercise, and adequate intake of antioxidant vitamins and essential fatty acids all militate against free radicals and arguably might slow physical ageing. Ageing at a cellular level seems to be governed by a molecular clock that limits the number of divisions a cell makes in a lifetime, which may have a role in reducing the danger from potentially malignant cells. This process, too, may respond to free-radical damage.

Life Span and Ageing

"We spend our lives as a tale that is told." This notion of human existence, transient and predestined, is attributed to Moses and apparently dates from about 1500 B.C. (Psalm 90). He goes on to tell us that three-score years and ten are mankind's allotted life span. Although much greater than the estimated life expectancy of those days, it was not an entirely improbable age for a few notable individuals in each generation. After more than three millennia, average life expectancy at birth has now exceeded this; indeed,

it is forecast to be ninety-one in Japan and eighty-three in the United States in the year 2050.[208]

Death from the diseases of late middle age is not related in any precise way to the advance of the well-known signs of ageing: frailty, white hair, cataracts, loss of skin texture and muscle tone. Some individuals survive into their tenth decade, robust and relatively free of chronic complaints but always marked by the signs of ageing. We must suppose that as we age, our susceptibility to disease increases, but whether we contract them depends on a bad-luck lottery in which lifestyle risk factors and genetic predisposition skew the odds. Apart from a greater susceptibility to infectious diseases, people who live to a great age succumb to much the same diseases as younger people.

Animals reared in a protected environment, where accidental death is unlikely, have a characteristic life span. Whether or not this is true of humans and whether this life span is changing are less easy to determine. The most extensive data from which this can be judged comes from a unique archive in which the dates of birth and death of every Swedish subject have been recorded with undisputed accuracy since 1751. The age of the oldest person alive of either sex at any one moment during this period is also known with absolute certainty. Within a fluctuating pattern, the maximum age at death has increased steadily, by about 0.44 years per decade, from 101 years in 1861. Around 1969, the rate accelerated to about 1.11 years per decade, and it seems set to continue.[209] No Swede, according to these records, has ever exceeded 113 years old, but the world record human life span of undisputed authenticity is the 122 years of Jeanne Calment, who died in 1997 in France. The Swedish data also charts the improved survival of the cohort of people over seventy years old who have benefited from better preventative medicine for age-related conditions and possibly from better nutrition earlier in the century. With better-managed lifestyles, improved medications, and no drastic change in the economic climate, the ranks of the very old are likely to continue to swell.

Scores of theories endeavour to explain why humans and animals age, but one proposed by Tom Kirkwood in 1977 accommodates the key phenomena in a particularly convincing way.[209] The physical state of all mammals, he suggests, deteriorates with age because they devote insufficient resources to maintaining their bodies in the pristine state of their youth. Kirkwood argues that the universally recognised signs of ageing are therefore a relatively trivial reflection of the same process.

But why should maintenance be flawed? Why should evolution, which produces such marvels, be less than perfect when it comes to maintaining aged life? The first and obvious answer is that if organisms had not died, the world would be overrun with every creature that has ever lived; there could be no natural selection and no evolution. However, Kirkwood makes the much more subtle point that the biological cost of efficient maintenance *beyond the years of reproduction* is an uneconomic luxury. Such maintenance would seriously encumber any animal running in the survival stakes because novel and costly biochemical arrangements would be needed.

Different mammals live their lives according to quite different strategies. As a rule, highly fecund species are short-lived, while those that have few progeny are long-lived. Long-lived species such as *Homo sapiens* have better systems of maintenance than short-lived species and have a much greater capacity to repair damaged DNA than short-lived species.[210]

If our cells deteriorate progressively with each succeeding division, what happens to the sex cells? We need not worry: children of aged parents are not born prematurely aged. We must assume that sex cells are protected from the ageing process and have remained juvenile throughout the history of our species, although male sex cells tend to be more susceptible to acquiring mutations than female sex cells.[89]

Do all organisms show signs of ageing? In a group of animals that includes the lobster, some cold-water fish, and a few amphibians, an ageing process is hard to discern because, unlike mammals, they continue to grow throughout their lives; but they certainly die.[211] Geraniums and many plants have a sort of immortality that enables them to be propagated year after year from cuttings that regenerate roots and leaves. They contain cells similar to animal stem cells that can make new shoots and roots whenever the opportunity arises. Lowly invertebrates such as *Hydra* also propagate from stem cells in established tissues that can create a new organism that buds off. Seeds and spores, though apparently immortal, are an entirely different phenomenon because their life processes are suspended or dormant to facilitate survival. A finite life span is a fundamental characteristic of all living things, with the possible exception of bacteria, which may be differently organised. Even the lowly yeast, which makes wine and bread, has a finite life. Its single cells propagate by budding, produce a fixed number of daughter cells, and thereafter divide no more.

Demographers emphasise that each human generation has a characteristic mortality pattern with respect to age and that, as life-threatening hazards come under control, each generation survives a little longer.[108] We are all curious to know how much human life span can increase and what changes to our lifestyles will be necessary to attain this.

What Happens as We Age?

Sooner or later, everyone wonders why joints get stiff, why skin wrinkles, why sight deteriorates, and why hair goes white. The question is not rhetorical; each aged feature represents inexorable chemical modification of components of our bodies that have been with us since early life. The effect of advancing years is written as a kind of chemical signature on three important molecules: collagen, elastin, and crystallin. The long chains of collagen make tendons and connective tissue. In middle age, they start to change chemically into something less supple, so that tendons become stiff and skin less resilient. Elastin gives arterial walls elasticity but in the same way becomes stiff with age because of chemical cross-linking, making arteries harden and contributing to high blood pressure. Crystallin, the translucent material that makes the lenses of the eye, gradually becomes opaque through oxidation of the protein, causing the condition known as cataracts. The most tangible support for Kirkwood's idea is an important truth; the body cannot repair this kind of damage. Probably many other proteins change irreversibly with the advancing years, but their significance is mostly unknown.

In parallel with chemical transformations of our bodies, complicated changes in our physiological parameters are occurring. The levels of athletic performance attained at the age of twenty-five soon decline, an uncomfortable notion sadly attributable to ageing. The same process makes our hearts pump blood more slowly, makes kidney filtration work more slowly, and causes our lung capacity to decrease, our muscular strength to wane, and the speed of nerve conduction to decline. At the same time, neurones, muscle fibres, and nephrons that can never be replaced just disappear, reducing the capability of each organ by a small amount in each succeeding year.[212] For most purposes, the capacities of our golden youth were more than we needed, but after more than forty years what remains may seem insufficient to manage the activities of former times. What is cause and what is effect in this slow decline is not a simple problem. We need to consider far-reaching changes in cells, in hormonal activity, in immunity, and in the transmission of genetic information.

OXIDATIVE STRESS – POSSIBLY THE ROOT OF ALL EVIL: A lot of evidence suggests that the root of the ageing problem may be the oxidative stress that originates in the activities of individual cells. Slightly inefficient respiration generates so-called free radicals (superoxide and hydrogen peroxide are just two of them) that cause oxidative stress. These chemicals are highly reactive and can damage almost any biochemical they encounter. The significance of free-radicals for the ageing process was first recognised in 1956 by Denham Harman. As a veteran of an atomic radiation project concerned with the importance of free radicals in biological damage, Harman's mind was receptive to the dangerous potential of free-radical by-products of respiration. These, he argued, would be a potent and cumulative threat to the chemical integrity of the cell that could underlie the pathology of ageing. In support of this notion, he looked back to some famous investigations of longevity performed in the 1930s by Clive McCay at Cornell University, using rats reared on a diet that was calorie-deficient but containing adequate vitamins. These rats were healthier and longer-lived than control animals fed with unlimited supplies of food. Harman boldly suggested that the calorie-deficient animals produced fewer free radicals than the control animals, consequently suffered less cellular damage, and therefore lived longer.[213] Although a dietary supplement of chemical antioxidants known to consume free radicals could significantly extend the life span of rodents, the idea languished for another decade for want of more detail and better experimental evidence. However, Harman's argument suddenly acquired new force when an enzyme, superoxide dismutase, was discovered that could consume the superoxide free radical.

Independent evidence for the role of free radicals in reducing longevity emerged from a completely different source: the genetics of simple animals. The life spans of the fruit fly, *Drosophila*, and the nematode, *Caenorhabditis elegans* (a tiny worm with only one thousand cells), are crucially affected by the free radicals they produce. Long-lived mutant varieties of either animal have a more powerful system for consuming free radicals than normal individuals. Conversely, mutants of the nematode unable to consume free radicals have shorter lives than control animals, but their life span can be restored artificially by chemicals that mimic these enzymes. Is this true of mammals? Scientists find the variety of mechanisms of antioxidant defence in mice too complicated to make interpretation of experiments easy. However, the longevity of mice can be increased by treatments that reduce free radicals.[214,215]

The principle cellular victim of free-radical damage is mitochondrial DNA. The mitochondrial DNA of brain, heart, and skeletal muscle cells accumulates genetic deletions as we get older. These inactivate critical proteins necessary for mitochondrial respiration, so that the appropriate products are not present in every cell. In skeletal muscle, different fibres accumulate different deletions of mitochondrial DNA, creating regional inefficiency in energy generation.[216] This becomes a vicious circle in which mitochondria become progressively less efficient as oxidative stress becomes more severe, more free radicals accumulate, and the individual cells die through apoptosis or necrosis. Damage to cells causes wasting of skeletal tissue, exacerbates heart disease, and contributes importantly to the frailty of the elderly.[79] Free radicals also affect nuclear DNA and can start a cell on the road to tumour formation. Why, one might wonder, have mitochondria remained throughout evolution so delicately poised between the constructive and the destructive? Kirkwood's answer would probably be that free radicals serve a vital role in preventing infection in the early life of mammals; in later life, the cost of controlling free radicals becomes an evolutionary liability.

AGEING OF OUR BODIES REFLECTS AGEING OF OUR CELLS: Just as the sickness of the individual reflects the sickness of the cells, the ageing process reflects malfunctions at a cellular level. Increased oxidative stress originating in the mitochondria is an obvious manifestation. In cells that are no longer dividing, such as muscle cells or neurones, subcellular components such as mitochondria and lysosomes (the vesicles that recycle old proteins; see Chapter 4) undergo conspicuous age-related changes. Mitochondria characteristically swell, suggesting some indefinable damage that is not correctable.

The lysosomes of actively dividing cells break down damaged intracellular proteins as well as extracellular debris brought into the cell by an engulfment process (known as endocytosis) that participates in recycling dead cells. In old age, this recycling system gradually fails, and an undegradable oxidative product of unsuccessfully processed junk, called lipofuscin, accumulates. This brown-coloured pigment is diagnostic of ageing cells and a permanent marker of lysosomes.[217] In laboratory studies of growing cells, lipofuscin reduces their growth and metabolic activity and probably permanently damages the lysosomal system of recycling. The damage done by lipofuscin is particularly evident in damage to the retina, where apoptosis, caused by lipofuscin, impairs the vision in a condition known as age-related macular degeneration that affects one-fifth of individuals over sixty-five.[218]

As cells age, the litter of history accumulates within them, but several alternative fates are open to them; their lives can end with either a bang or a whimper. As we will see shortly, this means that scientifically speaking, they either undergo a dramatic programmed cell death (apoptosis) or their capacity to divide just dwindles away according to a schedule determined by a molecular clock. Both the characteristic wasting of skeletal muscle in old age[219] and the loss of neurones in disorders such as Parkinson's disease[129] originate in the suicide of cells. Less obviously pathological, but indubitably contributing to the ageing process, is the progressive loss of cells from tissues as they age, presumably also by apoptosis. This includes the loss of neurones from the brain, the loss of cells from the liver (possibly a third by middle age), and the loss of nephrons from the kidney (possibly a third by middle age).[212] Survival of most cells is probably dependent on a cell growth factor; if or when this influence diminishes, for whatever reason, apoptosis may ensue. One other fate is possible – to be reborn as a cancer cell (see Chapter 8). Such cells escape from all the normal constraints that exist in a tissue and characteristically undergo uncontrolled cell division.

HORMONES AND AGEING: Chemical messengers, or hormones, carried in the bloodstream are of major importance in early life in regulating growth, sexual development, and the energy transactions of cells. We can hardly be surprised to discover that their decline in middle age heralds the descent into frail old age. Why this decrease should occur is the mystery at the heart of ageing that raises the question of whether the endocrine system is a master timekeeper. Some people consider old age to be a state of hormone deficiency that can be corrected by "replacement therapy." It is an appealing idea, but one so full of practical complications that we must be wary.

Some reductions in hormone function are really consequences of ageing-associated diseases. About half of all people over eighty suffer to some extent from type 2 diabetes and respond poorly to insulin. Five to ten percent of women suffer from an underactive thyroid gland and thyroxine deficiency, possibly a result of autoimmune disease.[220]

Growth hormone and insulin-related growth factor play a mysterious but important role in the functioning of tissues. In old age, the level of these hormones tends to decline, so that muscle mass, skin thickness, and immune function are lost, contributing importantly to the frailty of the elderly. Growth hormone now attracts attention because of its apparent ability to rejuvenate the elderly. In one study, a group of men in their seventh and eighth decades who had received injections of growth hormone seemingly became more muscular and lost body fat.[221] The subjects were chosen because they had depressed levels of a hormone called insulin-related growth factor that is often taken as a proxy for growth hormone. In late middle age, this is known to decline by about one-third compared to the level found in young men. A mysterious alchemy suggests to certain people that growth hormone could be an elixir to reverse the normal ageing process – to turn the clock back twenty years, reverse osteoporosis, and prevent strokes and heart attacks. Today, it is easy to find advertisements on the internet, for expensive courses of treatment with the hormone or with mysterious dietary supplements that are supposed to trigger growth hormone release. Whether or not such treatment is effective, the research is evidently at an early stage, and it suggests that growth hormone supplements might have unpleasant consequences.[222]

Production of sex hormones begins to decline in both sexes in middle age. Oestradiol is dramatically reduced in women at menopause and may cause loss of bone mass in some women (osteoporosis). Hormone replacement therapy (HRT) can correct these effects and reduce this symptom of ageing, but it clearly requires clinical supervision because of the associated dangers, including a small but significant risk of breast cancer. On the other hand, the onset of Alzheimer's disease and atherosclerosis may be delayed by HRT.

In older men, a more subtle decrease in testosterone secretion contributes to the decline of bone mass, muscular mass, and strength and is sometimes called the andropause, by analogy to the menopause. Replacement therapy is sometimes proposed but is rarely prescribed because of the unpredictable consequences for the prostate gland.

Levels of a steroid of uncertain function made by the adrenal gland, called dehydroepiandrosterone (DHEA), decline sharply in later life.[223] DHEA has attracted particular interest because rodents, which normally have none, respond to it as though to an "elixir of youth," boosting the immune system and preventing obesity, diabetes, cancer, and heart disease. In monkeys reared on a reduced-calorie diet that increases the life span, the decline in DHEA levels is postponed.[224] DHEA may be a significant alternative source of progesterone and testosterone. Its supposed ability to improve skin texture, bone formation, and lean body mass may result indirectly through boosting levels of insulin-like growth factor.[220] However, no data conclusively supports the conclusion that old age in humans is a DHEA-deficiency syndrome – but this has not impeded a brisk trade in replacement therapy. Moderate doses are probably harmless but of unknown value. Nonetheless, DHEA is readily available in the United States, without prescription, as a widely touted "cure" for ageing.

DECLINING IMMUNITY: The well-known vulnerability of the elderly to infection reflects a decline in the performance of the immune system and physical changes in one of the relevant organs. The thymus gland – a small organ found in front of the larynx that plays an important part in the making of T-lymphocytes – reaches its peak development in adolescence and thereafter slowly atrophies during adult life. Once this decline starts, an adult probably cannot extend his or her range of responses to microbial challenges. Consequently, increased susceptibility to influenza, tuberculosis, and the *Staphylococci* that cause sepsis after surgical operations may be inevitable in the elderly.

We will see later that an intrinsic part of the ageing process is the limited capacity of somatic cells to divide. This means that immune memory cells, whose resources have been called on repeatedly, may become less responsive in old age. Certain phagocytic white blood cells become less fierce in old age, apparently losing their appetite for eating bacteria and infected cells and also becoming less potent sources of superoxide, the free-radical bactericide.[225] Immunity to diseases caught in childhood can persist until very old age. However, paradoxically, the system that prevents reaction to self-antigens seems to lose its stringency in later life, leading to the unwelcome appearance of troublesome autoimmune diseases.

Free radicals are enormously important in the highly aggressive inflammatory response. This famously two-edged weapon can destroy our infectious enemies but can also create the symptoms of a number of diseases. In rheumatoid arthritis, for example, the immune system starts to recognise "self" components of joints, and the tissue is then damaged by an inflammatory attack.[226] Formally, these are not strictly symptoms of ageing, but the loss of control of inflammation is surely an example of inadequate maintenance that increases with age. The damage caused by most autoimmune diseases is usually confined to one tissue, but very probably all tissues suffer some oxidative damage with increasing age.[214]

IS GENETIC INFORMATION TRANSMITTED FAITHFULLY THROUGH-OUT OUR LIVES? Can we really expect the information that shaped our bodies after conception to remain uncorrupted after eight decades? Could the processes that make proteins become error-prone, so that in time ineffectual proteins compromise the smooth running of a cell? In the early days of molecular biology, scientists predicted that the accuracy of RNA and protein synthesis would be jeopardised by errors. The notion was that errors in the machinery of protein synthesis would propagate into a cascade of errors that would damage a cell, causing the phenomena we associate with ageing.[210] Errors occur with a finite frequency, but powerful systems recognise and destroy incorrectly folded proteins, preventing more errors from accumulating. The most interesting case is oxidative damage to mitochondrial DNA, which propagates into cellular damage in a positive feedback loop. Mutations occur sixteen times more frequently in mitochondrial DNA than in chromosomal DNA, owing to its proximity to the energy-generating machinery.[214] These mutations can make respiration less efficient, increase the production of free radicals, and consequently create worse damage.[216] Defective mitochondria also seem to exacerbate the problem because they have a selective advantage over sound ones.[217]

Longevity, Life Experience, and the Genome

GENETICS AND HUMAN LONGEVITY: Some families clearly have a predisposition for a long life, but investigations of pairs of identical twins suggest that in practice, genetic influences on the human life span are disguised by the stronger influence of life experience.[227] Registers of identical twins are kept in many countries, but the Scandinavian registers, maintained since the mid nineteenth century, have had special importance to geneticists investigating differences in life span between identical (monozygotic) twins. This data suggests that the variation in life span attributable to heredity (between 23 and 35 percent) is less than the part played by life experience. Put another way, our lifestyle or misfortune may deprive us of the theoretical maximum life span our genetic constitution suggests. The early demise of one of a pair of twins from an infectious disease or a road accident would disguise genetic factors affecting health and mortality. The health records of identical twins also tell us that the chance of contracting cancer, an age-related disease, is not strongly affected by genetic factors.

However, many specific diseases affect pairs of twins equally, indicating an important genetic component.

Many variants of human genes are likely to affect the rate of ageing without contributing to a particular disease syndrome. The ApoE gene makes a cell-surface protein that affects the amount of cholesterol circulating in the blood by a small but very important amount. The variant E2 is most effective and is significantly more frequent amongst centenarians. E4, the least common variant, causes the smallest change in cholesterol and is associated with premature atherosclerosis, Alzheimer's disease, and cardiovascular disease. E4 evidently shortens the life span, but it may confer a selective advantage in youth, as fifteen percent of the population carry this variant. E3 is intermediate between the two.[228]

Werner's syndrome, an extremely rare genetic disease of humans, causes premature ageing. During adolescence, the victim grows slowly, and the signs of ageing – grey hair and wrinkled skin – appear in the fourth decade. Atherosclerosis, osteoporosis, and cataracts appear prematurely, and the victims die at half the normal life span. Tumours occur prematurely compared to the general population, but other age-related conditions, such as Alzheimer's disease and high blood pressure, are infrequent. Victims of another, even rarer, premature-ageing disease, the Hutchinson–Guilford syndrome, develop an intense progressive atherosclerosis in childhood and die of circulatory disease in their early teens.[229]

WHAT WE CAN LEARN ABOUT AGEING FROM SIMPLE ORGANISMS: Organisms such as the nematode, the fruit fly, and the yeast provide opportunities to study life span objectively in the laboratory, as it is possible to identify mutations that increase or decrease their life spans. Two important rules about the life strategies of organisms have emerged from this approach. First, mutations that modify life span favour efficient reproduction or great longevity but not both, just as different species of mammal favour one or the other strategy. Second, genes that help organisms survive nutritional stress favour longevity at the expense of reproduction; those that are important for reproduction in the good times carry a biological cost that must be paid for in reduced life span.

Geneticists were surprised and immensely excited to discover a nematode gene that affected longevity that was also similar to the human insulin receptor. In humans, when insulin combines with its receptor, it stimulates use of glucose by cells. In the worm, the equivalent receptor is a target for a biochemical messenger from the gonads that urges every cell to use more energy for reproduction. If the gonads are destroyed using a laser microbeam, the signal is no longer transmitted, reproduction is reduced, and the life span of the animal increases by 60 percent.[230] A reduction in the food supply diminishes the same signal naturally and increases the life span of the worm. A *mutation* in the receptor has the same effect, simultaneously favouring longevity and restraining reproduction. Why should a flow of energy reduce the life span of the worm? Again, the connection is free radicals. They are released when energy is abundant and shorten the lives of nematodes, just as they do flies. As we have seen, a parallel phenomenon is seen in rodents when restricted calorie intake extends life span and reduces free-radical production.

HOW LIFE EXPERIENCE AFFECTS LONGEVITY: Apart from chance events such as infection and accidents, nutrition probably affects longevity more than any other aspect of life experience. Nutritional experiences *in utero* have immense consequences for health and longevity as an adult. The undernourished foetus tends to develop into an adult who will be especially susceptible to cardiovascular disease, high blood pressure, and non-insulin dependent diabetes in later life. In a foetus threatened by nutritional stress, the brain is protected from glucose starvation by a stratagem that allocates fewer resources to the viscera, with far-reaching consequences for most aspects of physiology. Such a foetus may have a range of apparently small imperfections, such as a slightly underdeveloped liver, a deficiency of insulin-secreting cells, a raised level of insulin resistance, and reduced numbers of nephrons in the kidney. These individuals are born with less reserve capacity for a variety of functions, which in middle age predispose them to chronic complaints such as high blood pressure, coronary heart disease, atherosclerosis, and type 2 diabetes ahead of more fortunate individuals.

After birth, the calorie and nutrient content of the diet continues to affect health and longevity. Nothing in our social history suggests, objectively, what an appropriate calorie intake might be. The outcome of Clive McCay's experiments with calorie-deficient rodents sparked widespread astonishment, because the animals were healthier and longer-lived than control animals reared with unlimited supplies of food. The latter lived for about twenty-four months; the calorie-restricted group lived between eight and twelve months longer, and they lived twice as long as their well-fed cousins when fed on the absolute minimal diet. The long-lived rats developed tumours less frequently but were also much less fertile. The result has been replicated in many investigations and with many species. Inevitably, people wonder if humans are affected in the same way: an idea that appeals to dreary puritans who relish unrelenting austerity.

What does this tell us about the biology of ageing? In calorie-restricted rats, the loss of skeletal muscle fibres (sarcopenia) that characterises ageing is greatly delayed. Early in life, damaged proteins are removed by the proteasome – the cell's machinery for recycling proteins – and quickly replaced by new ones. Later in life, this process becomes less effective, but the decline in proteasome activity is delayed in calorie-starved rats, and the proteins that characterise old age appear more slowly. Chemical signs of oxidative stress in protein and DNA are less common in calorie-restricted animals than in normal animals, and their mitochondria make less oxidant in laboratory tests.[214,231]

Similar investigations on humans are not possible, but experiments with monkeys are now in progress. They will take another twenty years to complete, but the midterm reports suggest the same outcome.[232] So far, calorie restriction seems to have prevented age-associated increases in blood pressure, blood glucose, and cholesterol.

The clear implication seems to be that if we consumed fewer calories, we would be healthier and probably live longer; but life is not that simple. For adults, a dietary restriction may pose no special risks, but children and possibly the elderly may become more susceptible to infection. Similarly, young people indulging in high-energy competitive activities for prolonged periods almost certainly need a high-calorie diet. McCay's half-starved geriatric rats would have little survival value compared with their well-fed cousins in the real world, one suspects.

LIFE EXPECTANCY ACROSS THE GLOBE: One grotesquely unbiological statistic reveals a key factor determining life expectancy in every country on the planet. This is national wealth measured as gross national income (GNI) per capita.[48] For countries with a very low GNI – just a few percent of the wealth of the richest countries – life expectancy can be under forty years. The reasons for this are very obvious: in these countries, infants in particular but also every stratum of society are prey to infectious disease and nutritional deficiency. Most of the poorest nations are in Africa, where malaria, AIDS, and tuberculosis are fearsome threats that restrain economic growth, and in the case of AIDS actively reduce life expectancy.

Quite small increments in GNI generate substantial increases in life expectancy (Fig. 7.1). If a country is to approach the life expectancies of the richest nations, a critical threshold exists, evidently between $5,000 and $10,000 of GNI. As James Riley pointed out a few years ago, a handful

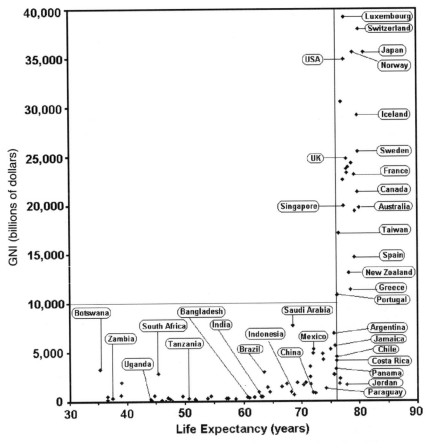

Figure 7.1. Life expectancy and gross national income – data for eighty-seven countries for 2002 (compiled from the CIA *World Fact Book*).[48]

of countries have managed remarkable increases in life expectancy in the recent past with quite modest GNI and health expenditures.[9] This data shows that in fourteen of the eighty-seven countries with GNIs of less than $7,000, people reach on average seventy to seventy-six years old. On the other hand, in a few cases countries with a promising GNI are not matched by the life expectancy that could reasonably be expected (e.g., Saudi Arabia, South Africa). The key to preventing premature death, particularly of children, and therefore to improving life expectancy – easily said, but not easily delivered – is obviously public health infrastructure. It must provide clean water, ensure safe disposal of sewage, ensure an equitable distribution of food, and provide primary health care, particularly for infants. An excessive birth rate obviously means the national wealth is divided among more individuals, negating any increase in economic growth.

For countries at the top end of the range of life expectancy (i.e., greater than 76 years), GNI varies over a twentyfold range, with little correlation between GNI and life expectancy. Figures for expenditure[48] suggest that the proportion of GNI spent on health is roughly constant, with a few countries getting notably good value for their money (Japan, Singapore) and a few paying very dearly for much the same life expectancy.

How Cells Grow Old

Early in the twentieth century, biologists learnt how to take cells from certain organs of human and animal bodies and culture them in appropriate nutrients. During this period, the disastrously wrong notion that animal cells could grow in an unlimited way, like bacteria, became widespread. This idea emerged from the laboratory of Alexis Carrel, a Nobel laureate who made important discoveries in vascular surgery and who performed the first person-to-person blood transfusion. The misapprehension arose because a culture of cells taken from a chick heart was apparently grown continuously for more than twenty years. The absurd origin of the mistake was not discovered until after Carrel's death, but in the meantime, his ideas had diverted attention from the real and much more interesting behaviour of animal cells in culture.[136]

This perspective changed utterly in 1961, when Leonard Hayflick and Paul Moorhead reported that most types of cells taken from human donors had a finite life span, or underwent a fixed and reproducible number of divisions in culture. This was quite different from cancer cells, such as the famous HeLa cells – cultured from a tumour for the first time in the 1940s – which could grow forever.[136] Widespread agreement emerged that the phenomenon itself was the real and typical behaviour of cells from many sources, and the idea was soon corroborated by other information. Cells from patients with premature-ageing diseases had less than half the normal capacity to divide, and the number of divisions made by cells from different animals seemed to reflect their average life span.

However, while the idea that the ageing of an organism might reflect the behaviour of individual cells was bold and original, it was very difficult to prove. The revolutionary implication was that cells behaved as though controlled by a timekeeper that allocates a set number of divisions before senescence strikes. The theory did not explain how ageing occurred in nondividing tissues such as brain and muscle, where cells undergo very few divisions after birth and function perfectly, sometimes for a hundred years. On the other hand, as we shall see, extraordinary support for this far-reaching conjecture came from a completely different source that suggested a molecular basis to the clock.

Does the number of divisions permitted by the Hayflick limit correspond to the number of divisions needed by the body to reach "old age"? The answer is probably no, and although the idea is appealing, there is no reason to believe that the life span of cultured cells reflects the life span of people. We saw in Chapter 5 how it takes about forty-six cycles of cell division to make an adult human being, of which about forty occur *in utero*. This does not include the cells that will die, but even making allowances for them, the eighty cycles suggested by Hayflick seem too many.

For some years after the discovery of the Hayflick limit, the available data seemed to suggest that the replicative capacity of cells taken from a donor reflected the donor's age. More recent investigations, of cells from undisputedly healthy donors of different ages, show no correlation between the number of divisions made and the donor's age, except in the case of foetal cells that underwent many more cycles. There is evidently no simple relationship between the replicative life span of cells and the life span of the donor. In one case, cells taken from the same donor at different times before he reached the age of seventy-eight all seemed to undergo the same number of divisions.[233]

Cells that reach the Hayflick limit are active metabolically but differ from normal growing cells because they do not respond to agents that stimulate cell division in younger cells. Hayflick described such cells as senescent – with no implication that they reflect the senescence of the organism – and noted that their appearance and biochemical character was changed. Forty years later, a tentative consensus still exists that senescence of cells in culture is related to the ageing of cells in living mammals. A number of examples illustrate the idea that cultured senescent cells seem to be similar to cells found in tissues of aged animals and humans. Skin cells of elderly people produce a secreted enzyme, called collagenase, that breaks down the collagen found between cells, and makes their skin thinner and wrinklier. Cultured skin cells as they approach the Hayflick limit begin to secrete this enzyme, too, suggesting that the processes seen in culture reflects the ageing process seen in people. Similarly, cells that line blood vessels, when cultured to the Hayflick limit, start to make the characteristic biochemical markers that scientists know are found in atherosclerotic blood vessels found in aged subjects.[234]

Another line of enquiry is to ask whether cells of tissues that are manifestly ageing undergo a Hayflick type of restriction. A number of examples indicate that the answer is yes.[234] Wound healing in the skin and cornea of elderly people is slower than in younger people because of reduced proliferation of the cells that pioneer the wound-healing process. Similarly, the declining immune response in older persons probably reflects the reduced capacity of immune memory cells to proliferate. In individuals suffering from osteoporosis, a condition in which bone making becomes less active in late middle age, the bone-making cells are much closer to the Hayflick limit than in nonosteoporotic patients.[235]

A MOLECULAR CLOCK – THE TELOMERE: Three decades ago, the Russian theorist Alexei Olovnikov envisaged, with extraordinary insight, a molecular clock that could explain the origin of the Hayflick limit.[236] Replication of chromosomes right to the tip – usually known to cell biologists as the telomere – seemed to Olovnikov to be a conundrum that could not be explained using the known system for DNA replication. In order to explain this and to accommodate the idea of the Hayflick limit, he suggested that DNA sequences would be lost in every replicative phase until they reached a critical level, at which cell division would stop. We now know that the ends of chromosomes (the telomeres) are made of repeated units of a simple DNA sequence, ending in an unusual terminal loop. At each cell division, exactly as Olovnikov so astutely anticipated, a few subunits are eroded.

Telomeres are assembled by a special DNA-replicating enzyme that scientists call telomerase. This enzyme is most active in those cells that have the longest telomeres, such as in fertilised eggs. As the embryo develops after fertilisation, the enzyme gradually disappears, and it seems to be completely absent from most cells at birth. Important exceptions to this rule are stem cells – the immortal cells that renew blood, the intestinal lining, and the skin, which characteristically contain the full amount of telomerase. Until very recently, the consensus was that all cells capable of division after birth lacked telomerase and that because of this, telomere units were lost at a constant rate at each cell division until the Hayflick limit was reached. New information suggests that an active telomerase is present in all dividing cells in a rate-limiting amount. The onset of replicative senescence is unlikely to be a passive process; it seems likely that telomere length responds dynamically to all sorts of exterior influences.[113] Free radicals damage the telomere more readily than the rest of the chromosomal DNA. This suggests that the telomere could be a sensing device for oxidative stress and that the ageing process could be driven by free-radical damage that reflects life experience rather than a simple clock.[237] Chronic infection of T-cells by HIV or hepatitis, which also involves excess production of free radicals, causes a reduction in telomere length.[114] Genetic diseases that damage the DNA-repair system also accelerate the erosion of telomere subunits.

Cancer cells, which are also immortal, contain excess telomerase.[238] If telomerase is introduced into a typical somatic cell, what happens? The technicalities of how this is done need not concern us here, but the outcome is clear. Cells continue to divide for more than twenty divisions beyond the normal Hayflick limit – providing support for the notion that telomerase might govern the life span of cells.

The phenomena of ageing, the shortening of telomeres, the Hayflick limit, and senescence raise far more questions than we can answer. Scientists suppose that the telomere clock and the Hayflick limit are devices that have evolved in long-lived creatures to reduce the risk of wild, uncontrolled cell division (or cancer) before their reproductive days are over.[234] About twenty division cycles are required for a mutated cell to become a life-threatening tumour. If such a mutation occurred, in the early part of human life, there would be no brake to resist the descent into malignancy, but fortunately, the probability of such a mutation is initially quite remote. The probability of developing a malignancy increases continuously during human life, and *senescence* – in the special sense used by Hayflick – is a barrier to malignancy. In the laboratory, senescent human cells can be made to divide again if they are infected by certain DNA tumour viruses. At the same time, telomerase is made once again and the tips of the chromosomes begin to extend through addition of telomere subunits. These cells are now equipped to become malignant. Whether DNA viruses play a part in the cancers of elderly people whose cells have reached a senescent phase is unknown. We will return to this in the next chapter.

We have already alluded to the rather surreal idea that budding yeast undergoes cellular ageing comparable to the Hayflick phenomenon. Remarkably, yeast also has a gene that is similar to the normal version of the human disease gene that causes Werner's syndrome. When this gene is removed entirely from the yeast, its genome undergoes a catastrophic reorganisation reflecting the importance of this gene product to the welfare of the cell. The yeast can be genetically engineered to contain the human premature-ageing gene instead of its yeast counterpart, and the life span of the yeast is reduced by sixty percent.[239]

Stopping the Clock

Paradoxically, although elixirs that confer everlasting youth are the stuff of fairy tales, a market for schemes that promise to stop the clock in the real world always seems to exist. The most hilarious of these was used to relieve the affluent of their wealth, if nothing else, for half a century. It started in 1889, when an old and distinguished French physiologist, Charles Brown-Sequard, claimed he was "rejuvenated" by a testicular extract from a monkey. Word spread, and by the 1920s gullible people in Europe and America were submitting themselves to fantastic surgical interventions that promised to restore youthful looks and vigour. Preposterous "quacks" such as Serge Voronoff in Paris and John Romulus Brinkley in the United States were inserting testes from various animals into their male customers, promising to restore their waning sexual powers.[240] A vasectomy operation, invented in Vienna by Eugen Steinach and offered as a treatment to make "old age bearable," attracted scores of patients, including Sigmund Freud and W. B. Yeats.[241] In mysterious Alpine clinics, Paul Niehans was "reversing the ageing process" by injections of foetal cells or extracts of "monkey glands" and claiming that Pope Pius XII and Winston Churchill had benefited from his ministrations. Incredibly, successors to these clinics still exist. Our own era is not deficient in hilarious novelties. An American corporation is prepared to freeze our bodies when we die so that when a cure comes along we can be thawed, mended, and put on the road again.[242]

Today a rich assortment of "elixirs" promoted to counter the ravages of time are still available for the desperate and neurotic, if they are undeterred by lack of clinical evidence or endorsement by reputable scientific authorities. They range from "royal jelly" – the chemical that extends the life of a queen bee tenfold compared with her genetically identical sisters, the worker bees – to DHEA. Clinicians are still evaluating some of these for efficacy and adverse effects (growth hormone, melatonin, carnitine, and creatine). Some could be harmful if taken in excess, but the effects of many are essentially unknown. Serious professional scientists are becoming increasingly angry at the presence of preposterous quackery in contemporary nonspecialist literature, masquerading as scientific journals.[243]

One recent development is the biotech company dedicated to finding compounds that might delay ageing. This can be done rationally by examining the effect of compounds on lowly organisms, such as yeast, that demonstrate the ageing phenomenon. One compound has already come to light, called resveratrol – a polyphenol coincidentally found in red wine and grapes. Whether this is the explanation of the low prevalence of heart disease in Mediterranean countries is unknown.[244] The compound is certainly interesting, as it dampens the activity of $p53$ – the cellular damage-control gene – and could conceivably be a brake on the rate of cell suicide. Companies will almost certainly look for more compounds that extend the life of lowly creatures. A compound that is safe and that confers even a small benefit probably has a promising commercial future. Compounds with antioxidant properties that extend the lives of mice are already known.[215]

Challenges, Risks, and Rewards – a Summary

The human life span increasingly seems like another frontier to be explored and tamed. We see the prospect of further increases in life span as a realistic possibility, but our ancestors were never in any doubt that ageing was inevitable and unavoidable. Moreover, we know that in evolution, the biological cost of bodily maintenance much beyond the years of reproduction is a price rarely paid by other animals. On present trends, we might see an increase in life expectancy of five or ten years in the next fifty if avoidable causes of mortality are eliminated. We can see, from studies of both animals and humans, indications of how we can improve our prospects. We have already passed an important evolutionary landmark when middle-aged people are counselled to take aspirin, statins, and blood pressure pills as precautionary measures that will extend their lives. Free radicals seem to be a key biochemical factor, but proposals to minimise their effects are inevitably vague, although a low-calorie diet containing adequate antioxidants is clearly beneficial. Our new understanding of the importance of the uterine environment suggests that further improvements in life span may be possible if foetal nutrition is made optimal. In the not-too-distant future, some kind of intervention affecting endocrine hormone levels may find routine use in reducing frailty among the elderly.

The prospect of a steadily increasing healthy life span usually receives universal approval as long as officious efforts are not made to keep alive people who are permanently comatose or in extreme pain. In particularly difficult cases – such as the persistent vegetative state – the medical profession seems to find it easier to continue a hopeless treatment than to decide to stop. The problem is already widely recognised, and people in full possession of their faculties facing serious medical interventions at an advanced age increasingly reject, explicitly, the possibility of resuscitation. Another milestone in perception of this aspect of ageing is growing public sympathy for those with incurable and unbearably painful diseases who wish to end their lives by assisted suicide.

The character of society will inevitably change if the cohort of people living healthy and active lives in their ninth and tenth decades continues to increase. The much-discussed economic consequences of an ageing population would be less problematic if steps were taken to gradually increase the existing retirement age and to make arrangements for retirement pensions workable. Grandchildren may be important beneficiaries of the extended life spans, if grandparents come to play a more important role in their lives than they did in former times.

GENERAL SOURCES: 136, 210, 212, 214, 234, 238, 245.

8 Cancer and the Body Plan: A Darwinian Struggle

Cancer, one of the defining personal anxieties of the twentieth century, is also one of the greatest biological puzzles that confront humanity. Perhaps a single cell, out of the trillions of cells in the body, starts a kind of rebellion against the great plan that gives each cell its identity. More genetic changes follow in the same cell or its descendants, to establish a dangerous malignancy. Biologists see the disease as a Darwinian struggle between the anarchic and outrageous stratagems of malignant cells and the organisational forces of the body. Probably only 20 percent of the risk of cancer is inherited; the remainder originates in life experiences that cause the genetic changes that start these insidious chain reactions. Like the ageing process considered in the last chapter, cancer originates in the absence of effective maintenance. Cancer is the most complicated and scientifically difficult disease we face, but our understanding is advancing sufficiently to allow hope for really effective control of ongoing cancers and the identification of environmental cues that initiate the disease.

Twentieth-Century Nightmare

In the generation that reached middle age in the early twentieth century, the family was frequently faced with cancer for the first time in its collective memory. No remedies or palliative care could ease a terrible illness that often struck when its victims were still in their prime. Until the 1920s, when family magazines began to publish articles about the disease for the first time, a kind of taboo inhibited public discussion of the experience and of the victims' medical prospects. Doctors knew that cancer existed in many forms that often took several decades to develop. They knew that the disease became incurable once it had spread to other organs; the best hope for treatment was early discovery and swift surgical excision of the tumour. Driven by desperation and unwillingness to accept this grim outlook, pressure groups began to emerge, particularly in America, whose aim was to find better solutions to the problem.

One of the earliest responses to this pressure was a bill drafted by U.S. Senator Matthew Neely in 1927 that proposed a five-million-dollar reward to the first person who discovered a practical and successful cure. Although hugely popular with the public, the proposal was rejected by the Senate, which feared an expensive charade.[246] Ten years later, President Roosevelt opened the National Cancer Institute. At about the same time, in Britain, the ill-fated Chamberlain government introduced a cancer bill that addressed similar anxieties. The bill included a proposal to organise better treatment and access to surgery and radiation therapy, but it deliberately excluded provisions for research. The Royal Colleges would not recommend support of "laboratory workers" because they believed such a development would undermine the status of clinicians. In response to a question, a government spokesman revealed that all research would be left to voluntary organisations.[15]

Meanwhile, in the United States, a lay group of activists was gradually wresting control of the American Cancer Society from the conservative physician-scientists who had managed the society since its inception. The key figure in these events and in similar later developments was Mary Lasker (1900–1994), who, with her husband, Albert, founded the prestigious Lasker Awards for Medical Research in 1942. Albert died of cancer ten years later, and Mary devoted the rest of her life and their immense fortune to securing more funds for research and improved treatment of the disease.[246]

The cause of cancer was not obvious in the mid twentieth century, and the danger seemed to be growing as the death rate from cancer seemed to be increasing steadily. The public, witnessing the tumultuous material changes of the mid twentieth century, found it easy to put the blame on chemical pollution, radiation leaks, or food additives. In the Delaney Amendment to the American Food, Drug and Cosmetic Act of 1958, Congress said the FDA could not "approve for use in food any chemical additive found to induce cancer in animals *at any concentration.*" To the public, the legislation was common sense, but in reality it created an impossible burden for the FDA arising from mandatory tests of chemicals on animals of an extensive and unrealistic kind. More than three decades later the difficulties remained, and the proposal was finally abandoned in 1996.[206,247]

By the late 1960s, Mary Lasker's movement was profoundly dissatisfied with the progress of cancer research. Using her legendary diplomatic skills and unrivalled access to officials, she managed to persuade Senator Yarborough to set up "a citizen's inquiry" into how effective cancer research was. Yarborough's enthusiastic response was to suggest that if Uncle Sam could put a man on the moon in seven years, the problem of cancer could be solved by 1976, the two hundredth anniversary of the Declaration of Independence. After several months of deliberation, the committee concluded that investigator-led research had failed. Progress was slow, the committee said, because efforts were not firmly directed at the real objective. Research should be done outside of NIH in an independent agency with a greatly increased budget and strong management.[246]

The scientific community was horrified. Most believed the reason for the perceived lack of progress was that scientists were concentrating on understanding fundamental life processes, a necessary preliminary before any insightful contribution could be made. The "moon shot" approach was doomed to be an expensive failure, they said, because it could not be focussed on a clear work plan. Others prophesied a catastrophic implosion of NIH if funds were diverted to cancer. Meanwhile, just as a Senate committee was starting to hear testimony, the eponymous agony-aunt Ann Landers encouraged her huge readership to bombard Congress with letters supporting a "War on Cancer." After a few more obscure twists and turns, President Richard Nixon, in December 1971, signed a National Cancer Act that proposed a massive increase in funding for cancer research, accelerated trials of available drugs, epidemiology studies, and cancer care.

Ironically, the bicentennial – the year in which cancer was to be finally nailed – was the year of the most significant breakthrough in cancer research for decades, but one that could not be credited to the "crusade." This was the discovery of the first *oncogene* – a gene with a proven capacity to set an ordinary cell on the road to malignancy.[248] This vital breakthrough was the most salutary triumph for the molecular biology that critics had derided as irrelevant and for the creative, investigator-initiated research they had so deplored. It came as a culmination of more than a decade of introspection about whether certain viruses might cause cancer. It also unearthed the startling enzyme that could make a DNA copy of an RNA virus – reverse transcriptase – which played a crucial role in founding the new gene technology industry.[249] Researchers were ecstatic, but in truth the discovery revealed the magnitude of their ignorance and powerfully endorsed the need for a more profound understanding of cell growth.

The chief positive outcome of the "war against cancer" was an immense boost to cancer treatment facilities, with little benefit to cancer research. President Nixon put the affair in perspective when he remarked that the money spent between 1962 and 1969 was only one-twenty-fifth of the amount spent in sending a man to the moon.[246] Later still, it was clear that no lasting damage had been done, and indeed many more scientists became interested in cancer research.

WHAT CAUSES CANCER? French scientists were the first to suspect that some viruses could transmit cancer, but it was Peyton Rous, in 1911, who showed that a filtered extract of a chicken tumour could cause a tumour in another chicken. The initial enthusiasm for this idea then died down, when similar experiments with rodent tumours failed to produce the same result, and it remained dormant until the 1950s, when new viruses were discovered that caused leukaemia in various mammals. The first human cancer to be attributed to a virus was Burkitt's lymphoma, a disease found principally in East Africa. This disease is caused by the Epstein-Barr virus (normally associated with glandular fever in adolescents) in persons suffering from malaria. Much later, a retrovirus (HTLV-1) was discovered that caused a rare human leukaemia known as hairy cell leukaemia. Twenty years ago, it seemed unlikely that viruses might cause more than a small number of cancers. Today, many specific examples are known, and scientists suspect that other cancers are caused in part or wholly by viruses.[250] In most cases of viral cancer, the disease is attributed to the effect of particular oncogenes carried by the virus. Certain chemicals and radiation may also cause cancer with a high probability, usually several decades after the primary exposure; other subtle elements of life experience may have the same effect. From this evidence emerged the important idea that most cells have the potential to become cancerous when exposed to certain triggers. The next section describes a widely held view of how this might occur.

Six Steps to Chaos

Six crucial changes in the behaviour and physiology of cells are necessary for a cell or its descendants to form a life-threatening malignancy.[251] The first event is almost certainly DNA damage that may occur at any time in our lives, but five more genetic changes must occur for descendants of the first maverick cell to become a dangerous malignancy. This means that, with the exception of some hereditary cancers and childhood leukaemias, the risk of developing cancer increases as we age.

In general, it is hard to attribute any cancer to a particular mutation, with the major exception of certain heritable malignancies. Individuals who carry a recessive gene for a disease such as retinoblastoma – a childhood cancer of the eye – are at risk of sustaining a mutation in the good version of the gene. In other words, individuals born with these mutations are already some way down the path towards developing cancer and consequently contract diseases of this type in early life. In the case of retinoblastoma, the normal unmutated version of the key protein has an important role in DNA synthesis in each cell cycle and in ensuring that chromosome separation is carried out correctly.

In the next few pages, we will examine the steps by which a normal cell or its descendants is transformed into a dangerous malignancy. The process starts by blind chance and develops into a chain reaction that ends in unlimited and anarchic cell division.

MAVERICK CELLS EMERGE: Genetic changes drive the development of cancers from the first fateful mutation that starts the process to the massive loss of whole chromosomes or major rearrangements of DNA sequence that occur later. Every normal cell sustains a certain number of spontaneous mutations each day. This number may be elevated by several factors, to be explored later, including radiation, environmental chemicals, vitamin deficiencies, and oxidative stress. The cell responds either by attempting to repair them or by undergoing suicide (apoptosis) in order to rid the body of a dangerous cargo. Not every mutation has serious consequences, but probably every cancer starts with a cell in which a mutation has conferred a growth advantage over its neighbours or a genetic instability that favours the appearance of a fast-growing descendant. The mutation rate is very low; indeed, for every potentially serious, cancer-causing mutation there are about ten trillion cells in the human body.

Once on the road to malignancy, the cell's DNA becomes a melting pot in which the program that established the character of the cell is lost or rearranged. In normal cells, chromosomes are stably maintained by processes that correct damage in DNA or that permit cell division only if cell cycle events are executed correctly. In the development of a malignancy, the mutation rate may accelerate if a DNA-repair gene mutates. The descendants of the founder cell will then compete in a kind of Darwinian struggle in this genetic turmoil, from which the fastest-growing and most undisciplined cells emerge as victors.

Chromosomal rearrangements that are well understood occur in lymphomas and leukaemias when parts of one chromosome join to another. In Burkitt's lymphoma, a cancer of antibody-forming lymphocytes (the B-lymphocytes), chromosome 14 becomes fused with chromosome 8, putting the sequences that drive antibody formation (the promoter) next to a gene called c-*myc* that drives cell division. This rearrangement results in catastrophic uncontrolled production of c-*myc* that stimulates excessive cell division by white blood cells. In at least two-thirds of leukaemias and in many kinds of lymphoma, rearrangements of the chromosomes initiate the disease by placing important genes under a new and irresponsible controller. Cells of the developing immune system are particularly vulnerable, because genetic rearrangement is part of the process involved in forming antibodies. In addition, the cells that make the new chimeric proteins are arrested at a stage at which they can reproduce forever, like stem cells.[252] Some malignancies originate in a stem cell population and retain the character of stem cells at later stages of the disease, compromising the effectiveness of chemotherapy, probably because they are less affected by the agents.[253]

CELL DIVISION GETS OUT OF CONTROL: Every cell of the human body is probably engaged with its neighbours in acts of mutual stimulation or inhibition through receptors on their surfaces, using secreted proteins known as cytokines or "growth factors." Because of changes in crucial genes that are known as oncogenes, cells may make abnormal amounts of a growth factor or start to behave as if they were under permanent stimulation by a growth factor.

The word "oncogene" entered the language in the late 1960s, to conjure up the hypothetical cancer-causing genes that scientists were beginning to think might be activated by viruses or carcinogens when cells became malignant.[249] This idea reached its definitive form with the great bicentennial discovery of 1976 – that the cancer-causing gene in the Rous sarcoma virus was similar to, but not identical with, a gene found in uninfected chickens.[248] The normal version of the chicken protein (*src*, in the vocabulary of molecular geneticists) transmits a signal from outside of the cell into the cell nucleus that affects the rate of cell division in a subtle, low-key fashion. The authors proposed that during evolution this chicken gene, which could make cells divide faster, had become part of the viruses' genetic material. Scientists soon began to wonder just how many genes could become true oncogenes if they were genetically modified by the forces of evolution. By the time Michael Bishop and Harold Varmus were awarded the Nobel Prize for Medicine in 1989 for their seminal discovery, more than forty oncogenes were known, each associated with a particular type of cancer.

Activation of oncogenes is just one of many kinds of offensive activity associated with malignant cells. Genetic changes also inactivate so-called tumour-suppressor genes. The retinoblastoma protein mentioned earlier plays exactly that role in a normal cell; once this protein is inactivated by a mutation, unregulated cell division is permitted. Arguably, the most important tumour-suppressor protein is the enigmatically named p53, because in more than half of all human tumours, this protein is mutated. We can think of p53 as a damage-control protein that appears suddenly and in great abundance whenever a cell is damaged by radiation, oxygen deficiency, or a restriction on DNA synthesis. What does the damage-control protein do? When activated, p53 arrests cell division, participates in quality-control steps necessary for chromosome separation, and can initiate a programme of cell suicide. If p53 is disrupted by a mutation, none of these useful security functions can be performed and the cell is free to undergo uncontrolled division, because the quality-control function and the self-destruction safety net is gone.[135] We have seen how, in the early mammalian embryo, p53 detects genetic damage and is usually instrumental in terminating pregnancy.

PROGRAMMED CELL DEATH IS EVADED: All mammalian cells can commit suicide when they sense stress by a process known as programmed cell death or apoptosis. A sensing device reports on many cellular activities that affect the welfare of the individual cell, including damage to the genome, levels of growth factor stimulation, and oxygen availability. If the situation is serious, the signal to self-destruct is given and, in quick time, the genome is rapidly cut into tiny fragments and the cell membrane ruptured. The cell contents leak out, and neighbouring cells engulf and destroy the cell debris – a dramatic "scorched earth" tactic that elegantly solves a tricky problem. The offending cell disappears, leaving no trace to attract the attention of the immune system; adjoining cells divide to replace their deceased neighbour. The crucial component in this scheme is, again, the damage-control protein, p53.

Apoptosis is a vital safeguard, and the loss of this facility is an essential step on the road to malignancy. Our lymphocytes have survived a test of competence in which all those that did not make proper antibodies were ruthlessly eliminated by apoptosis (95 percent of all that are made). Occasionally, one of these survives this giant pogrom because of a mutation in a critical apoptosis gene that allows it to continue propagating. Much later, the same cell may sustain a second mutation that activates an oncogene, but now cannot protect itself by self-destruction and is more likely to become a malignancy known as a lymphoma.

MALIGNANT CELLS CAN DIVIDE FOREVER: Normal cells of the human body make only a fixed number of divisions before entering a phase called "senescence." One way of thinking about this phenomenon is to regard it as a device for limiting the potential for uncontrolled cell division in long-lived creatures before their reproductive days are over.[254] A cell damaged by mutations needs about twenty division cycles to make a life-threatening tumour. This means that in the latter part of human life, when the cumulative risk of cancer-inducing mutations is greatest, the cells that enter senescence or reach the "Hayflick limit" could be protected from the risk of becoming malignant. Even so, cells that had started earlier on the route to malignancy would be unconstrained by the Hayflick limit. Once more, the damage-control protein, p53, is the key player. In normal cell division, p53 is absolutely necessary for a cell to enter senescence, but many malignancies ignore the Hayflick limit because p53 has already been lost by mutation.[255]

TUMOURS DEVELOP A BLOOD SUPPLY: In order to receive sufficient oxygen to ensure their survival, every cell of the human body needs to be within one-tenth of a millimetre of a blood vessel. The moment a tumour acquires a blood supply, the oxygen and the nutrients needed for growth are available, and consequently the tumour becomes more dangerous. The blood vessels are also a conduit through which malignant cells migrate to new sites. The key event, which can happen early or late, occurs when yet another mutation enhances the production of a protein called vascular endothelial growth factor (VEGF), which stimulates nearby blood vessels to make new capillaries that grow towards the tumour.

END GAME – THE CANCER SPREADS: At the culmination of its loathsome existence – a stage known as metastasis – a tumour acquires the sinister capacity to infiltrate neighbouring tissues and to migrate to distant sites to set up another focus. A tumour cell on the point of embarking on its voyage of metastasis will have lost its cell-to-cell contacts that might restrain cell division and will be free to migrate, systematically destroying organ function en route. If cancers had no capacity to spread, surgery would usually be successful, but unhappily that is not the case, and 90 percent of deaths from cancer occur in patients in whom metastasis has occurred.

Life Experience and Cancer – Six Major Risks

Eighty to 90 percent of the risk of developing cancer originates in life experience. Surveys of more than ninety thousand pairs of Scandinavian identical twins born between 1870 and 1930 indicate that the presence of cancer in one twin was not a strong predictor of cancer in the other, whether they were brought up together or lived separately.[103] Similarly, people who migrate from countries of low incidence to high-incidence countries gradually develop the risk pattern of their adopted country, as though life experience overrides any genetic predisposition. Prostate and colorectal cancers are exceptions to this rule and seem to involve a substantial genetic predisposition.

Gross differences in life experience affect the prevalence of different types of cancer in different geographical regions and across microcosms of any one society. Economic progress seems to burden people with cancers of the colon, pancreas, lung, and reproductive organs; the mouth, oesophagus, stomach, liver, and cervix are affected by cancer more commonly in the preindustrial world.[161] The specific cause is often hard to find, but deficiency of certain dietary factors (e.g., insufficient vitamins or dietary fibre) or the presence of toxins or poisons may increase the risk of cancer. We will see how, in the industrial world, earlier exposure to reproductive hormones may be a key influence in increasing the incidence of reproductive cancers. Cancer in preindustrial societies originates in diverse causes such as chronic inflammation caused by infection, food spoilt by fungal toxins, arsenical contamination of drinking water, chewing tobacco, and alcoholic concoctions.

Statistical analysis of the incidence of cancer is an exceptionally complex undertaking, not least because decades may separate the outcome from the original potentially hazardous experiences. Moreover, the disease is intrinsically unpredictable. The chance of contracting a second cancer in paired organs that are obviously genetically identical and share the same life experiences is very small (0.8%).

CARCINOGENS IN THE ENVIRONMENT: The idea that substances encountered in our environment may trigger cancer is not new. In 1775, the British surgeon Percival Potts ascribed the high incidence of scrotal cancer in chimney sweeps to soot remaining on their bodies, at a time when bathing was a rare luxury. Others noted the prevalence of cancer in pipe smokers and snuff takers. One of the earliest recognised chemical causes of lung cancer, reported in Saxony in 1879, was arsenic inhaled in mines. Recently, cancers have been discovered in Bangladesh that were caused by arsenic in drinking water obtained from new wells, sunk to relieve the chronic unavailability of clean water.[256] Early in the twentieth century, occupational cancers were found in chemical industry workers that had taken two decades to develop after the primary exposure. In parts of Africa and China, a high incidence of certain cancers is attributed to stored grain in which fungi make highly potent carcinogens (i.e., cancer-causing substances).

Not surprisingly, the sharp rise in the number and quantity of synthetic chemicals accumulating in the environment has worried ecologists and thoughtful members of the public for many years. In the early twentieth century, many chemicals were introduced into food and other consumer products with little thought for long-term dangers. The environmental health community is now struggling to assess the magnitude of these conjectural hazards. The problem of establishing the risks associated with environmental chemicals that might cause cancer was laid out by Bruce Ames of the University of California in a highly influential publication in 1979. Noting the huge worldwide increase in production of old and new chemicals, Ames reviewed disturbing instances where suspect carcinogens could already be found in significant amounts in human body fat and human milk.[257] He also mentioned how routine testing of new substances for carcinogenicity by the response of rodents was manifestly impractical even three decades ago, because so many were appearing each year. Ames's proposal was to assess the risk of potential carcinogens using their capacity to cause genetic damage (i.e., mutations) in a very simple, inexpensive, quantitative test of high sensitivity that he had developed in 1964 using bacteria. Based on fifteen years' experience of the test in his own and many other laboratories, he concluded that most known carcinogens cause genetic damage. Moreover, most known mutagens (i.e., substances able to cause genetic damage), with some important exceptions, are carcinogenic. In time, it became clear that asbestos, for example, is not intrinsically mutagenic but provokes cells to become cancerous indirectly by inducing serious chronic inflammation, a process that is intrinsically mutagenic.[258] Other substances became mutagenic only after being modified in the liver by the so-called detoxification process. Disturbingly, the test identified a few suspect carcinogens in commercial products introduced during the 1960s and 1970s. Examples included a fire retardant used on children's clothes, a hair dye, a petrol additive, and a Japanese food additive called furyl furamide. The mutagenicity of the food additive was particularly disturbing because two carcinogen tests carried out on rats had been negative and the substance had been in use in Japan for nine years. New synthetic chemicals were not the only suspects. Disturbing evidence emerged that carcinogens appeared in charred fat during the roasting or grilling of fish and meat. Similarly, many food plants contained mutagens.[257]

These were salutary observations. The Delaney Amendment, in force since 1958, stipulated that any chemical known to cause cancer, at any dose, must be eliminated from the food supply. There was clearly a practical problem of how long-established customs could be regulated. About half of all chemicals tested at that time caused cancer in rodents, but the tests were conducted at concentrations close to a lethal dose and were probably meaningless, as the animals developed pathologies that provoked cancer indirectly. The Ames test provided the first rational system for assessing and prioritising the risk associated with the level of chemicals actually occurring in foodstuffs or the workplace. The recent discovery that acrylamide, a suspect carcinogen and nerve toxin, is created when potatoes are deep-fried indicates how apparently innocent activities can have unanticipated chemical consequences; but the scale of the problem still cannot be gauged.

We saw in Chapter 5 how the publication of Rachel Carson's *Silent Spring* in 1962 suddenly provoked Americans into awareness of the potentially insidious effects of agricultural chemicals. Carson's particular anxiety – she was suffering from the breast cancer that killed her in 1964 – was that chemicals in the environment might unleash an epidemic of cancer in humans and animals of unprecedented proportions. Anecdotal evidence that the insecticide DDT did cause cancer already existed, and there was abundant evidence of the ecological harm done. Later analyses showed a minor metabolite of DDT was responsible for its carcinogenicity. On the other hand, few people were ever contaminated to an extent that presented a significant risk; indeed, using the Ames test, the risk appeared lower than for natural pesticides. The intake of polychlorinated biphenyl pesticides (PCBs) by humans (about one-sixth of a milligram per day in the United States) is also said to be far below the level at which it is likely to be carcinogenic.[206] Nonetheless, the possibility of an above-average dose from unwashed fruit and vegetables or accidental exposures is a realistic anxiety. More recently, environmental toxicologists have raised the issue of whether unborn babies are more sensitive to PCBs than adults (see Chapter 6). We also considered, in the previous chapter, the anxiety caused by dioxins – the highly toxic and probably carcinogenic products of incomplete incineration of PCBs. The idea that dioxins played some part in sporadic cancers during the last century is plausible, as they have clearly been accumulating for two centuries.[205] Mutagen tests and a variety of newer tests for potential genetic damage are now part of routine public health surveillance aimed at identifying environmental hazards and new products containing unacceptable amounts of carcinogens.

SMOKING – THE EIGHTH DEADLY SIN: One more sin should be added to the deadly seven. The link between cigarettes and lung cancer was the first instance of a malignancy being attributed to a mundane, avoidable life experience. The key to why smoking persists is that nicotine, delivered so effectively to the bloodstream as a vapour, is also highly addictive. Bizarrely, doctors in Nazi Germany had essentially proven that smoking caused lung cancer in the 1930s and were hoping, with the support of the Führer, to make the Third Reich a healthy paradise by banning smoking. Understandably, this made little impact elsewhere. In Britain, a fifteenfold increase in deaths from lung cancer between 1927 and 1947 spurred the Medical Research Council to investigate the cause of lung cancer. At that time, few British men were nonsmokers. Virtually nobody suspected smoking, and indeed one American brand, Lucky Strike, carried the claim that it improved breathing.[259] As early as 1950, Richard Doll and colleagues had demonstrated an unequivocal increase in lung cancer among heavy smokers. However, it took twenty years, using methods now widely seen as the gold standard of epidemiological analysis, to define the risk precisely. These studies show that persons who smoke twenty-five cigarettes a day have a twenty-five-fold greater risk of developing lung cancer after twenty years. Moreover, if the subject stops smoking, the risk falls sharply. Faced with overpowering evidence of the folly of smoking – a one in six chance of dying of lung cancer for a lifelong heavy smoker – one might have predicted the rapid demise of smoking and the tobacco trade.

Lung cancer was not the only evil consequence of smoking. Smoking was evidently a major risk factor for cardiovascular disease and twenty-three other conditions,[259] but the evil was not just confined to health. Knowing their customers were addicted to the product, the tobacco industry could have done nothing. However, to find ways of reducing adverse publicity, they recruited senior scientists to discredit the connection between tobacco and lung cancer. The director of the Sloane-Kettering Cancer Institute of New York in the early 1960s, Frank Horsfall, accepted financial support from Philip Morris in return for attempting to silence Ernst Wynder. The latter's offence was his persistent efforts to publicise the dangers of cigarette smoking while employed by the Institute.[260]

Smoking is a truly astonishing example of the human capacity for self-harm. About three million of the world's population die every year from smoking, and this will increase steadily for the foreseeable future as the habit grips the developing world.[261] The circumstances it creates are bizarrely corrosive. Some governments, like that of Britain, obtain vast tax revenues from tobacco that would have to be found elsewhere if smoking ceased. Philip Morris has told the Czech government that the economic benefit of smoking outweighs its harmful effects because the premature death of smokers spares the country substantial expenses incurred in later life.[262]

SURGES OF REPRODUCTIVE HORMONES AFFECT CANCER RISK: A remarkable epidemiological investigation in Verona by Domenico Rigoni-Stern, in the early nineteenth century, showed that chastity protected nuns from uterine cancer, while child bearing protected married women from breast cancer.[263] The risk of breast cancer is now more precisely defined in developed countries as being proportional to the time interval between menarche and first birth. The risk of ovarian cancer is lowest in women who reach the menopause early.[250] These cancers probably initiate during the bursts of cell division that occur in every menstrual cycle, provoked by surges in the concentration of steroid hormones. By contrast, when these surges are controlled, as in oral contraceptive users, the incidence of uterine cancer is lower. Cancers of the reproductive organs are notably more frequent in certain industrial countries than in agricultural ones. The explanation is quite uncertain; there are many differences in lifestyle between the two. Late puberty, the physically demanding lifestyle, bigger families, and dietary differences may all be significant. We saw in Chapter 6 how the rate of breast and prostate cancer is lower in Japan than in other industrial countries and may be associated with differences in nondietary nutrients.

RADIATION – BITTER EXPERIENCE INSPIRES GREAT CAUTION: One of the earliest warnings of the danger to health of atomic energy was cancers amongst people who painted watches and clocks with radium-containing luminous paints. However, in the terrible aftermath of the atomic bomb blasts at Hiroshima and Nagasaki, the dangers of radiation hazard were revealed in the most excruciating detail when survivors of the blast succumbed to cancer. As only one bomb fell on each city, the probable exposure of every survivor could be calculated fairly accurately, and in later years the incidence of cancer amongst the survivors was monitored and correlated with the calculated exposure. The types of cancer found were no different from those seen by physicians in normal times. Radiation-induced leukaemia began to appear two or three years after the blast, reaching a peak six to eight years later and thereafter slowly declining; solid tumours continued to appear for many years.[264]

Meanwhile, as nuclear power was harnessed for peaceful purposes, workable standards of acceptable exposure to radiation were urgently needed for the industry. The decision was taken to design nuclear installations to prevent emissions exceeding the level of natural radiation (four to five milliSieverts per year, the unit of exposure used by scientists). Natural radiation comes from two sources – cosmic rays and radon, the radioactive gas found in certain types of rock.

A number of important questions arose. What was an acceptable standard of exposure? Was damage cumulative with exposure? Would natural repair processes correct damage below a certain threshold? By 1960, the data for the first twenty-five years after the Japanese blasts was available. The evidence seemed to suggest that the incidence of cancer amongst the survivors was roughly proportional to the dose received. About one to three cases of leukaemia and three to six cases of breast cancer occurred each year for every ten thousand people exposed to one Sievert. The industry's conclusion, however, was that a dose of one-tenth of a Sievert (one hundred milliSieverts) was an exposure below which genetic damage would be corrected. By contrast, cell biologists of the 1950s, studying the response of cells to radiation in laboratory experiments, concluded that DNA would be damaged in proportion to the dose of radiation received and that a safe threshold was unlikely.

The practical appeal to administrators of a threshold was obvious, as it suggested that people working on their premises would not accumulate radiation-induced damage below that threshold. However, information from an entirely different source suggested the level chosen was set too high. This data was collected by Alice Stewart (1906–2002),[265] a physician-epidemiologist working in the 1950s in Britain. At that time, pregnant women were commonly X-rayed to determine the position of their babies. Stewart carefully examined the incidence of childhood leukaemia amongst children born to mothers who had received a single diagnostic X-ray. She concluded that a foetus exposed to ten to twenty milliSieverts had a 40 percent increased risk of developing leukaemia, when the supposedly safe threshold adopted by the atomic power industry was five times that value. Professional reaction was muted, sceptical, and even hostile, but hospitals quickly abandoned X-rays for pregnant women once the complete study was published in 1970. Other investigations of low X-ray exposures administered to the thyroid gland during childhood corroborated Stewart's findings, again showing increased leukaemia. The nuclear power industry, however, struggling to compete in the conventional energy market, was not impressed. Meanwhile, in America, Thomas Mancuso enlisted the help of Stewart – now officially retired – to reexamine some epidemiological data originating from an accidental leak of radiation from the Hanford Atomic Weapons plant in Washington state from 1942 to 1947. Together, they concluded that subjects who were exposed to very low amounts of radiation because of the leak had a higher incidence of cancer than would be expected based on a "threshold" theory. The authorities were less than happy with this interpretation and responded by dismissing Mancuso and withdrawing their co-operation from Stewart. Eventually, a congressional enquiry forced the U.S. Atomic Energy Commission (AEC) to make the information public and to admit that the level of permitted exposure was set too high by a factor of ten.[265]

By 1991, the International Commission on Radiation Protection (ICRP) went one step further by accepting that damage from radiation is proportional to dose. The total damage incurred in a lifetime – and consequently the total risk of contracting cancer – would be the sum of small exposures to radiation. Today, the maximum permitted exposure to radiation, set by the ICRP, for nuclear industry personnel is twenty milliSieverts per year; for the public, it is one milliSievert. The wisdom of this decision was apparent in 1996, when the latest mortality data from the Japanese A-bomb survivors showed that the risk of cancer was still rising in those exposed to fifty milliSieverts.[264] The argument is not over yet; a recent thoughtfully argued analysis suggests the data is still consistent with a threshold.[266]

People with highly pigmented skin suffer much less skin cancer because the pigments absorb the light. Skin cancer was quite rare a century ago amongst whites, who were generally averse to excessive exposure to the sun, but with increased enthusiasm for sunbathing, the disease has become much more common. The UV-B range of wavelengths, which is invisible to our eyes, damages skin cells by provoking the release of free radicals. The precise link between sun exposure and melanoma, the most dangerous of the skin cancers, is uncertain. Although the disease can develop in only two or three years, an episode of acute exposure in childhood may be a crucial factor.

While excessive exposure to sunlight can be dangerous, a growing body of evidence suggests that vitamin D, made in the skin during small exposures to sunlight, protects against many kinds of cancer. Several long-term studies in the United States show that the incidence of colorectal cancer in people with the highest vitamin D intake is half that of people with the lowest levels. A map of mortality from a group of common cancers in the United States shows that cancers are two and a half times more common in the cities, especially of the Northeast, than in the country and in the South. The authors suggest, somewhat controversially, that the incidence of cancer is highest in regions where people receive the lowest exposure to sunlight and lowest where people tend to be exposed to the most sunlight. Apparently, no other regional dietary factors are correlated in the same way.[156]

CHRONIC INFECTION AND INFLAMMATION CAUSES CANCER: Fifteen percent of all malignancies result indirectly from chronic infection.[267] Viral hepatitis B and C and the parasite *Schistosoma* are responsible for widespread liver cancer in Africa. In East Africa, the Epstein-Barr virus in association with malaria causes Burkitt's lymphoma. The microbe *Helicobacter* infects the stomachs of about half the people of the world by creating its own special microenvironment by neutralising acidity with ammonia. From this base, it mounts an invasion of the stomach wall to create a classic stomach ulcer and provoke chronic inflammation, which in turn damages DNA and starts the process that leads to gastric cancer. Malignancies are also associated with inflammatory bowel conditions such as Crohn's disease. The DNA damage in inflamed cells almost certainly contributes to the start of the cancer. The discovery that small daily doses of aspirin and related anti-inflammatory drugs reduce the chance of developing colon cancer is, perhaps the clearest indication of the importance of inflammation as a cause of malignancy.[267]

Asbestos is not intrinsically mutagenic but causes cancer because it provokes chronic inflammation, possibly in association with a virus. Microscopic fibres of asbestos lodge in the lungs and over thirty years cause a very distinctive cancer known as mesothelioma. Significant numbers of reports of ill health were associated with asbestos manufacture before the First World War, and the data continued to pile up until the 1960s, when the danger was more widely recognised. Throughout this period, the industry did little to protect its workers, or indeed its own long-term future, as it stubbornly denied any responsibility, and successive governments showed no reaction. Since then, the manufacturers and their insurers have paid a fortune in damages. The product was banned in the EU in 1998, but the extremely long latency period of the disease means that new and inevitably fatal cases will continue to appear for several decades into the new century.[268] A suspicion also exists that asbestos causes cancer in collaboration with a virus (SV40).[269]

UNSATISFACTORY DIET: Diet can be a cause of cancer either for what it contains or for what it lacks. Dietary deficiencies of micronutrients are probably as important a cause of cancer as smoking. Of the twenty or so vitamins necessary for human growth and development, deficiencies of at least eight can damage DNA, creating conditions that might initiate cancer.[270] Recommended dietary allowances of most individual micronutrients are the amounts required to avoid acute deficiency. Vitamin C and other antioxidants almost certainly have other benefits if taken in substantially larger amounts. Their vital role is to neutralise the free radicals that cause the oxidative damage to DNA that starts a cancer. The importance of a substantial intake of fruit and vegetables to provide sufficient micronutrients emerges from many surveys. Strikingly, the Americans who consume the least fruit and vegetables have twice the rate of cancer of those who consume the most. In a desperate and surreal attempt to reduce the risk of cancer, nutritionists urge the public to consume at least five portions of vegetables and fruit per day.

Plant chemicals distinct from vitamins (glucosinolates, iso-flavones, and, most recently, acetyl salicylic acid) are increasingly in the spotlight because of possible anticancer properties. Broccoli and cabbage contain glucosinolates that seem to reduce the risk of cancer of the lung, stomach, and colon by stimulating the detoxification system. As we have seen, the iso-flavones found in soya beans may reduce the impact of hormonal surges and lower the risk of contracting breast or prostate cancer.[201] Acetyl salicylic acid, a compound related to aspirin, found in many plants, may, like the well-known pill itself, be protective against cancer and heart disease through its anti-inflammatory effects. Some Buddhist vegetarian monks have as much acetyl salicylic acid in their blood from their diet as other individuals receiving aspirin as medication.[271]

Stomach cancer is associated with poor dietary practices. Fifty years ago, it was a major cause of lethal cancer in industrialised countries, and until recently it was very common in Japan. The precise reason for its decline is unknown, but epidemiologists suspect the crucial event was the appearance of domestic refrigerators. This probably reduced dependence on pickled, preserved, and smoked foods that may contain carcinogens or provoke inflammation. In the developing world, this cancer remains extremely common and is usually associated with food contaminated by carcinogenic fungal toxins. Malnutrition always creates opportunities for *Helicobacter*, which in turn may cause cancer (see p. 168).

Information from dietary surveys involving humans is complicated and hedged with all sorts of qualifications. Animal experiments, although not easily related to human physiology, suggest a more stark conclusion. Rats fed 20 percent more food than they need tend to develop tumours of the endocrine and mammary tissues much more frequently than control rats. While not exactly comparable, overweight humans also are more predisposed to cancer.[250]

Epidemiology has a way of throwing up gloomy hints about dietary practices. Dietary fibre may be mildly protective against colorectal cancer, possibly through the antioxidant properties of the associated phytic acid. High dietary fat intake, particularly in the form of dairy products, has the reverse effect and also seems to be associated with breast and prostate cancer. Too much red meat may increase the incidence of breast, colon, and prostate cancer. The Seventh Day Adventists and the Mormons, the American religious sects that forswear the drinking of alcohol, coffee, and tea and are often vegetarian, suffer less from cancer than other strata of society.[272]

Control and Prevention of Cancer

Anyone who has ever read Fanny Burney's unforgettable account of the excision of her breast tumour without an anaesthetic is aware that in 1811, doctors were certain that the best hope for survival was to excise a malignancy. The novelist lived for another thirty years, and more than a century elapsed before an alternative was available. As the skill of surgeons and radiographers increased, remission rates steadily improved, providing that tumours were discovered early. Without a concept of the cause of malignancies, more intellectually incisive treatments were not forthcoming. Early efforts by William Coley to provoke the body into making tumours regress was an entirely exceptional enterprise, which he claimed cured more than 10 percent of his patients with inoperable sarcomas (see p. 116).

Paul Ehrlich, whose idea of a "magic bullet" to cure infection so captivated the pioneers of modern medicine, believed chemotherapeutic agents might be found to kill tumours selectively. However, no breakthrough was made until the 1950s, when huge screening campaigns unearthed compounds with some merit, particularly in combination. Notable successes were achieved with non-Hodgkin's lymphoma and childhood leukaemia, and substantial remissions were obtained for other cancers, but progress was very slow. The sheer unpleasantness of most treatments and the spectre of drug-resistant cancers eventually left some therapies powerless. This meant chemotherapy could never be a miracle cure comparable to antibiotics. No one can look back on the early days without a rueful feeling that these were only the most preliminary solutions to the real scientific problem.

Chemotherapeutic agents were supposed to be magic bullets that would kill only malignant cells and leave normal cells unscathed. For many years, they were used in the highest doses in the hope of securing total obliteration of the cancer. It seems now that this was somewhat misguided, as the chemotherapeutic onslaught provokes suicide of the cancer cells at lower doses via the p53 protein.[273] Another sinister problem that is slowly gaining recognition is that some cancers may originate in stem cell populations, so that any cancer may contain stem cells and their differentiated daughter cells. Because stem cells are characteristically infrequent dividers, they may be less susceptible to chemotherapeutic agents than differentiated cells; they may survive treatments and may be the most effective cell type in reestablishing the tumour.[253] Further progress in chemotherapy will require better methods of distinguishing between normal and malignant cells. The tools of molecular biology are now the key weapons in devising drugs that we must hope will be as subtle and incisive as the pathology that creates the disease. We know a lot about targets specific to cancer cells that may be vulnerable to novel small compounds or monoclonal antibodies armed with a deadly warhead.

If agents in the environment or our lifestyles provoke 80 percent of cancers, one might suppose it is in our power to eliminate a large part of the risk. Unhappily, experience so far indicates this is likely to be a slow process.

Challenges, Risks, and Rewards – a Summary

Once the risk of microbial infection had receded during the twentieth century, cancer emerged as the commonest challenge to people still in their middle years. Our vulnerability to cancer increases with advancing age and originates in our bodies' lack of effective methods of eliminating the risk from mutations that initiate malignancies. A minority of cancers involve an obvious inherited genetic component, but most are provoked by diverse life experiences, including excessive exposure to hormones, deficiencies in micronutrients, environmental carcinogens, radiation, and chronic inflammation caused by infection. Risks originating in life experience are often difficult to understand in detail because these subtle and complicated diseases emerge after very long incubation periods, some even originating *in utero*. Some risks arise from industrial products that manufacturers recklessly refuse to acknowledge are dangerous, while others, such as smoking, are self-inflicted. Efforts to reduce risks originating in life experience could be enormously beneficial.

Chemotherapy as a means of controlling ongoing cancer remains a strenuous and unpleasant experience, but advances in understanding of the disease and the discovery of better chemotherapeutic agents are slowly improving prospects. As in the days of Mary Lasker's campaigns, the laity, often driven by a sense of outrage at the unfairness of a disease that strikes its victims in their prime, still believe cancer research progresses too slowly. Clifton Leaf, writing recently in *Fortune* magazine,[273] makes a persuasive case, based on candid admissions of cancer scientists, that existing schemes for clinical trials are preventing recognition of the merits of some theoretically promising drugs. He suggests that drugs would have better opportunities to show their value if trials were made, not on patients with advanced cancers, but on patients with primary malignancies that have not undergone many mutations. We will return to this issue in Chapters 10 and 11. While nothing is certain, the variety of new possibilities suggests that the tyranny of cancer could end. Whether such cures will be affordable worldwide is another matter.

GENERAL SOURCES: 250, 254, 273.

9 Fighting Infection

Infectious disease used to be the major brake on population growth. Nobody was free of risk, but children and women in childbirth were especially vulnerable. This dismal situation improved in industrial societies of the nineteenth century because of a host of social advances, including improvements in sanitation, nutrition, and housing and better understanding of hygienic principles. Although the potential of vaccination was known in the early nineteenth century, mass immunisation and antibiotics contributed importantly only after 1940. We now have weapons of enormous power that drastically reduce the incidence of infection and terminate ongoing infections, such as blood poisoning and pneumonia, that formerly took so many lives. However, the public health officials of the 1960s who complacently rejoiced in the success of preventative medicine would have been surprised to know how frequently new diseases would emerge in the subsequent forty years.

Historic Epidemics

Nobody knows exactly when infectious disease became a great threat to human existence, but the first settled communities were probably much more vulnerable to devastating epidemics than their nomadic ancestors. Demographers believe that although the birth rate was high, these communities grew slowly and spasmodically because of periodic outbreaks of disease. Without efficient sanitation, sound hygienic practices, or immunity to these diseases, human societies were always at the mercy of microbes, a susceptibility greatly increased by malnutrition. The Old Testament is, again, an indication of the perceptions of an ancient people (see Fig. 9.1).

174

"The Lord shall smite thee with a consumption and with a fever, and with an inflammation, and with an extreme burning, and with the sword, and with blasting, and with mildew; and they shall pursue thee until thou perish" (Deuteronomy 28:22). The books of Moses have a lot to say about infectious diseases. The ranting quality suggests a desperate effort to maintain discipline on the long journey to the Promised Land, but fear of epidemics was evidently very real. The explicit rules about public health suggest that the Israelites understood enough to reduce the risk of infection. The list of banned foods (pig meat, shellfish, and animals that had died of natural causes) is an entirely rational way of avoiding the danger from parasites and food-poisoning microbes (Leviticus 11). Similarly, instructions that cooked food should be eaten immediately or destroyed cannot be faulted, nor the remarkably explicit insistence that defecation should occur outside the camp. Two chapters are devoted to the need for a strict and verifiable quarantine to contain a mysterious skin disease, which the King James Version unreliably translates as leprosy (Leviticus 13-14). It is entirely possible that the Israelites vividly appreciated the dangers of sexually transmitted disease and saw strict monogamy as a solution (see also Fig. 2.1).

Figure 9.1. Microbes in the books of Moses.

Our knowledge of epidemics in the ancient world is somewhat conjectural, but we can appreciate their impact from firsthand reports of the Black Death of 1338–46 and the sixteenth-century epidemic of smallpox in the New World. The medieval plague, now unambiguously attributed to *Yersina pestis*, arrived in Europe from the steppes of Russia on black rats and then transferred to humans on fleas. Europeans had no significant resistance to the disease, which raced through Europe in just a year, killing at least one-third and possibly two-thirds of the population.[47,274] No society could remain unchanged by such an onslaught, and indeed population growth in Europe stalled for almost two hundred years, provoking social and economic changes that inexorably destroyed the feudal system. In the absence of a rational explanation, people ascribed the cataclysm to retribution by the Almighty for their wickedness, as prophesied in the Old Testament.

By the mid eighteenth century the great population centres of the world, including China, had reached a turning point. The populace was developing resistance to epidemic diseases, such as plague, through increasingly frequent exposure to infections carried along the trade routes. Epidemic diseases were essentially domesticated and increasingly confined to childhood. Even so, in 1832 the first cholera pandemic arrived in Europe from the Indian subcontinent and spread through every major city in Europe and North America within a year, exploiting the inadequacy of sewage disposal. At the margins of European exploration, however, native peoples existed with no resistance to epidemic diseases who would remain highly vulnerable to measles, tuberculosis, and influenza until modern times.

THE END OF CIVIC FILTH – THE RISE OF PUBLIC HEALTH: Epidemics in Victorian England were usually ascribed to a "miasma" – the smells emanating from decomposing organic matter of all kinds – but this odd theory suggested no positive course of action except to avoid the smell. John Snow, a surgeon and London's leading anaesthetist, thought differently. He believed that cholera was transmitted through the water supply, and indeed he was able to prove it with some rigour during the third pandemic of 1854 by famously removing the handle of the "Broad Street pump." An immediate decline in the number of cases followed, as people used water from an alternative, cleaner source.

In London, the richest city on earth in the 1850s, sewage from more than two million people was deposited into the river Thames, where it would slosh up and down with the tide. Cartoons of the period indicate that the public was perfectly aware of the dangers of "the great stink," but Parliament made no response until the smell penetrated Westminster during the hot dry summer of 1858. Confronted with an extraordinary nuisance at its foul apogee, the government agreed to finance a monumental and unprecedented project to install sewers that could pump sewage in a closed system many miles from the city. London, by good fortune, had the services of a great engineer able to rise to the formidable challenge, and within ten years the key elements of the project were in place.[275] From that moment, the cause of public health in Britain rarely looked back. In 1861, Prince Albert, Queen Victoria's consort, died of typhoid, but there was no reason at that time to relate this poignant event to the democracy of faecal contamination.

Within a decade, mortality from enteric diseases in London was falling fast, and life expectancy was also generally improving. When the fourth cholera pandemic ravaged central Europe in the decade after 1865, London was only marginally affected.[112] Although the improvements in water and sewerage eliminated the most flagrant danger, Victorian London and most of Europe were still at the mercy of a host of infectious diseases. In 1900, one in seven British babies died in infancy; one in two hundred mothers died in childbirth; and life expectancy for men and women was under fifty years. Rampant infectious disease – tuberculosis, diphtheria, whooping cough, and scarlet fever – prematurely terminated many lives. Other kinds of infection were spread through food and by the flies that thrived in Victorian cities on rubbish and horse droppings.[29] Unhygienically collected milk carried many kinds of microbe, including *Mycobacterium bovis*, the probable cause of intestinal tuberculosis and of many childhood deaths.

In spite of this, in 1900 the prevalence of infectious disease was already on a steep downward path that owed nothing to clinical science or the medications that seem so important today. Sanitary regulation had been put at the heart of British public administration in the mid nineteenth century by the appointment of medical officers of health (MoH) and a trickle of legislative measures that initiated a clean-up of the cities. As the germ theory of infection gained acceptance amongst scientists, the implications of the theory started to inform all legislation and official interventions. The role of the MoH was crucially shaped by the germ theory when, in 1889, Parliament invested them with authority to control the spread of infectious diseases by the only weapons available, notification and isolation. Householders and persons attending the sick were required to notify the MoH of every instance of communicable disease, from typhus to measles. The MoH were empowered to isolate infected individuals in fever hospitals, effectively reducing the transmission of the disease. Those who prepared food for the public were compelled by legislation to recognise their responsibilities and to ensure that food was free of infection. Municipal refuse collections and dustbins were introduced, thereby reducing the nuisance of the ubiquitous flies. Health visitors – a new profession created by legislation in 1905 – contributed hugely to the improved survival and health of children by introducing to socially disadvantaged families the skills of hygienic mothercraft.

When the germ theory arrived, John Wesley's dictum – "cleanliness is next to godliness" – acquired a focus. "Germs" entered the English language as the embodiment of a public enemy that needed to be eradicated. The emergent advertising industry, seeing the prospect for new products, began encouraging hand washing and the use of soap, disinfectants, and handkerchiefs.[8,9,29]

HYGIENE AND TUBERCULOSIS: Tuberculosis – or consumption, as it was known – may have existed since at least 20,000 B.C.[276] It became a severe epidemic in the late eighteenth century and the major cause of death in the nineteenth. In Britain, Blake's "dark satanic mills" recruited labour, including hordes of children, on an unprecedented scale to work in over-crowded, poorly ventilated factories that were the key to the spread of this airborne disease. Later, the disease became equally common in town and country, affecting all strata of society, although poverty and malnutrition were its greatest friends.[29] The severity of the epidemic gradually abated, and deaths in Britain fell by 75 percent between 1840 and 1900 – and by more than 90 percent by 1950, the year in which streptomycin became available.[51] Whether this momentous decline was due to a secular decline in the virulence of the epidemic we will probably never know. However, al-most certainly, reduced transmission of the disease occurred because victims of the disease were increasingly isolated from the community, in sanatoria and other establishments, minimising their capacity to infect their families and the public. This development gained momentum with the discovery by Robert Koch in 1882 of the organism (*Mycobacterium tuberculosis*) that causes tuberculosis. Doctors everywhere became alert to the ways in which the disease was spread; indeed, the French government, in 1886, banned spitting in public.[44] At the same time, a significant effort was being made to reduce the severe overcrowding that existed in most European cities, where ten or more people sometimes shared a room.[44] Whether nutrition was improving during the late nineteenth century is far from clear; but if so, improved immunity may have reduced the virulence of infections. However, most authorities believe that social isolation was a critical step in reducing transmission of TB.[8]

Tuberculosis of cattle is caused by *Mycobacterium bovis*, a close rela-tive of the microbe that causes the human disease. It also infects humans and causes several forms of tuberculosis that affect the intestines and other organs. Even in the 1890s, scientists were convinced that humans caught a form of tuberculosis from milk and ascribed about one in six cases of TB to *M. bovis,* but by 1907 the evidence was overwhelming. The Danish dairy industry concluded that they should slaughter infected animals to eliminate the danger and breed TB-free herds. British farmers, seeing only extra costs, marshalled their considerable political muscle to sabotage all proposed legislation intended to improve the quality of milk. Pasteurisation was belatedly introduced in Britain in the late 1940s, but in those fifty years millions of children were needlessly infected with TB.[44]

The absurdities of milk hygiene notwithstanding, the developed world of the twentieth century was a healthier place, as food and housing steadily improved and hygiene penetrated every corner. In Britain, mortality from tuberculosis, scarlet fever, diphtheria, whooping cough, and measles all fell substantially in the first four decades, a success entirely attributable to improvements in hygiene in which antibiotics and vaccination played no part.[8,51,52]

Vaccines

FIRST SUCCESS: In seventeenth- and eighteenth-century Europe, no disease was more frightening than smallpox. In a bad year, it accounted for a tenth of all deaths and killed perhaps 400,000 people every year. The disease usually erupted explosively, killing a third of those affected and horribly disfiguring the survivors. With no respect for social origin, it killed or at least infected all but one member of two generations of descendants of Charles I of England.[277] In the century after Europeans arrived in the Americas, smallpox participated in the tragic annihilation of the indigenous population – the Aztecs, the Incas, and the populous Amerindians of the Mississippi Valley. These people had no resistance to smallpox or to other infectious diseases brought from the Old World. Within a century, less than five percent of the original population remained.[30,47]

In eighteenth-century Turkey and in China many centuries before, people had discovered that an "inoculation" with a tiny speck of material from a smallpox blister conferred protection against smallpox. A slight infection followed, with none of the blisters characteristic of smallpox, and this gave immunity for life, albeit with some risk of developing the disease. The system became widely used in Britain and the United States after 1721 through the advocacy of a Lady Mary Montague,[47] and the practice spread to France after Louis XV's untimely death from smallpox.

An astute observation by the British country doctor Edward Jenner, who routinely performed "inoculations," led to the next development. Milkmaids in his neighbourhood, Jenner noticed, had no reaction to inoculation, as though they were already immune to smallpox. He also knew that the cows and the milkmaids sometimes developed pustules, which he called cowpox, that resembled the blisters of smallpox. This suggested a bold experiment: he would take material from a milkmaid's cowpox pustules, immunise a child, and then several months later test the child's immunity by the standard inoculation procedure. The first subject, eight-year-old James Phipps, showed no reaction and was therefore immune to smallpox.

Jenner's report to the Royal Society in 1796 was followed by immediate and widespread interest at home and abroad in what quickly became known as vaccination. Four years later, a survey of smallpox mortality in districts of London showed that smallpox was less prevalent in districts where vaccination was practised than in others. Napoleon was exceedingly impressed and ordered his armies to be vaccinated. Several European states also instituted compulsory vaccination (Bavaria, 1807; Denmark, 1810; Prussia, 1818), substantially reducing the incidence of the disease within ten years.[29] Britain, the apostle of laissez-faire, let vaccination spread voluntarily; by 1840, affluent people almost invariably vaccinated their children, but the poor remained indifferent. Eventually, in 1853, prompted by unnecessary deaths from the disease, Parliament made vaccination of infants compulsory, whereupon apathy changed to defiant hostility. Some believed that vaccination was a device to circumvent God's will; others doubted its safety or merely disliked compulsion. The Anti-vaccination League, founded in 1866, forged an alliance with the anti-vivisection movement and started the tradition of vociferous protest that has continued to the present day in Britain and North America. Meanwhile, in 1871, the government introduced coercion with fines in order to strengthen its policy. Forty years later, George Bernard Shaw quipped, "The reason doctors love vaccination is that it has solved the great economic problem of how to extract fees from people who have nothing the matter with them."[31]

Scandinavia and the German states implemented vaccination better than Victorian Britain. Even so, by 1900 it was manifestly working in Britain, if not in the vast imperial territories.[9] By the 1950s, mass vaccination had almost eradicated smallpox from Europe and the United States, but an estimated five hundred million lives were lost in the developing world in the twentieth century before the disease disappeared. With this disparity in mind, the WHO, in 1953, proposed the most radical initiative ever contemplated in the field of public health – the eradication of smallpox throughout the world using a vigorous vaccination policy. Unrealistic and utopian as it seemed to some scientists, the WHO could only endorse the proposed program when, five years later, China and Russia had eliminated the disease from their territories. By 1977, the very last case occurred in Somalia, and two years later the WHO assembly declared the disease eradicated. The end came not through compulsory vaccination but via a targeted scheme in which every contact was vaccinated after every outbreak.

The end of the smallpox program was to be celebrated by the destruction of all stocks of the virus except those at two WHO reference centres. These were also scheduled for destruction in 1996, but the deadline was postponed on several occasions; the anxieties this provoked we will return to in the next chapter.

RISKS AND REWARDS OF VACCINATION: Once specific disease-causing organisms could be identified, there was immense interest in making vaccines and emulating the achievements of Jenner and Pasteur. Indeed, vaccines against cholera, plague, and typhoid that protected susceptible individuals during epidemics were produced quite easily. The first attempt at mass vaccination using these new vaccines was made when the U.S. army introduced compulsory vaccination against typhoid in 1911. In a small triumph for humanity, the vaccine was subsequently used by both sides in the First World War, apparently preventing the severe epidemics that accompanied the Boer War.

Mass vaccination protects a population in excess of the immunity accorded to each individual – an effect that epidemiologists call herd immunity – because the reservoir of potential infection is greatly diminished. Mass vaccination against smallpox and polio virus was especially promising because of their inability to propagate in other animals and form a reservoir of infection. This meant that once human cases were completely suppressed, the diseases would be eradicated, providing that mutations did not by-pass the immune response. Mass immunisation, notwithstanding the success with smallpox, remained an extremely audacious notion. Not only must the vaccine be safe and effective, the public must also be convinced that it is. Small preliminary trials could be used to assess immunity and adverse reactions, but for a definitive trial of the practical efficacy of the vaccine against the naturally occurring disease, at least 100,000 subjects were needed.

The idea of mass vaccination was largely American, driven by bold and innovative initiatives in research, organisation, and financing starting early in the last century.[278] The organisers obviously assessed risks at every stage, but because they were pioneering efforts, the outcomes were always uncertain, and indeed grave setbacks foiled early attempts to make a polio vaccine. With the exception of the French tuberculosis vaccine trial (BCG), no other large trials were undertaken outside the United States during this period.

No sooner had the success of mass vaccination in the developed world reduced the severity of childhood diseases than some parents began to feel that the risks of vaccination outweighed the risks of the natural infection. The British whooping cough campaign of the 1970s encountered this problem when reports of adverse effects of the vaccine made some parents panic. Within a short time, the number of vaccinations fell substantially and the number of cases of children with the disease increased dramatically. Organisers of vaccination campaigns make colossal efforts to devise ways of mitigating risks, but the efficacy of a vaccine can be known with certainty only by completing a large-scale trial on a human population at risk.

DIPHTHERIA: Diphtheria was endemic in Europe and America throughout the nineteenth century, and although it later became less prevalent, it still terrorised poor neighbourhoods well into the twentieth century. The use of a mixture of diphtheria toxin and antitoxin to provoke short-term immunity was reported in Europe in 1913, together with a test to confirm immunity to the disease. This set the scene for the city of New York's Health Department to undertake the first bold attempt to control the disease by mass vaccination. Following a preliminary small-scale test with their own preparations of vaccine, they started a trial on 90,000 nonimmune schoolchildren in the early 1920s. Only fourteen cases of diphtheria occurred in the vaccinated group, compared to fifty-six in the control group. The organisers found this sufficiently encouraging to launch mass immunisation of all the children of New York, a campaign promoted with a novel and persuasive publicity drive. Within a few years, the death rate from diphtheria in the city was halved. The health authorities in Britain believed the trials were too daring and in need of more careful statistical analysis. However, by the 1940s, they and other European countries were convinced and the vaccine was adopted.[278] A decade later, the disease was extremely rare in the United States and Europe. Russia, too, adopted universal childhood immunisation in 1958, successfully controlling the disease until an upsurge of antivaccination sentiment reduced "herd immunity" and a serious pandemic ensued in 1993–4. The health authorities of the now-independent states, with help from the WHO, had to restore vaccination to the former standards to control the outbreak. The moral is plain – mass immunity is required to reduce the prevalence of the disease.[279]

YELLOW FEVER: For three centuries, tropical regions of the Americas and Africa were graveyards for those without immunity to yellow fever. The viral disease is transmitted to humans by the mosquito *Aedes aegypti*. With memories of epidemics that took the lives of many Americans, including scientists trying to control the disease, the Rockefeller Foundation sponsored the development of a revolutionary new vaccine. This was to be a live "attenuated" virus that could infect the subject without causing the disease; it would be made by propagating the virus through many cycles of infection of fresh batches of monkey cells. Eventually, a strain emerged that lacked virulence while retaining the capacity to generate a strong immune response. The vaccine is still in use today and has almost eradicated the disease from North and Central America. In Africa and South America, epidemics still break out. Monkeys still carry the virus, and the mosquito is still extant, with the result that, in one instance in 1962, yellow fever killed three percent of the population of Ethiopia (see also p. 213).[280]

POLIOMYELITIS: Summer during the early twentieth century in Europe and North America often brought epidemics of polio that killed or paralysed children and young people. The poignant anxiety this provoked – quite disproportionate to the numbers affected, compared to diphtheria or whooping cough – created a powerful imperative for a vaccine. An ill-fated trial in the mid-1930s in which a number of children died from polio made the case totally compelling for a well-financed, highly organised campaign to develop a vaccine. With high-profile support from President Franklin Roosevelt – himself a victim of polio at the age of forty – the burden was shouldered by a well-known American charity, the March of Dimes.

Most advisors to the charity favoured an attenuated virus, like the yellow fever vaccine, believing that it would be safest and give the best long-term immunity. The alternative, a killed virus, would carry a small but finite risk of incomplete inactivation. Either way, they would need to propagate the virus in cultured human or monkey cells. At that time, nobody suspected that killed and live vaccines conferred fundamentally different kinds of immunity. A dead virus would probably stimulate only antibody production by B-lymphocytes, because it would not enter cells; live virus could enter cells and provoke T-lymphocytes to destroy virus-infected cells. The March of Dimes favoured a killed virus because it would be speedier than waiting for an adequately tested attenuated strain. They argued this "calculated risk" was justifiable because their obligation was to proceed as quickly as possible to end the roll call of childhood deaths that poisoned every summer.[278,281] Using virus killed by the highly reactive chemical fixative formalin, Jonas Salk mounted a small trial on 161 children. This effectively and safely stimulated antibody production and convinced the March of Dimes that Salk's laboratory should be commissioned to prepare a vaccine using a mixture of the three known strains of polio.

This was done at breakneck speed, and the stage was set for what was at that time the biggest medical experiment ever conducted. Polio, although rarely out of mind, was not a common disease, and several million subjects were needed for a trial if the results were to be informative. Important choices had to be made about an appropriate control group; scientists wanted a placebo-injected control group, while the lay observers thought the public would be more impressed by an uninjected control group. To keep the two groups happy, both types of control were finally used.[281] More than two million American and Canadian children aged six to nine were approached, although half of the parents would not give their consent. The outcome of the trial, announced in 1955, was that the number of polio cases in the vaccinated group was half of that in the unvaccinated group. Moreover, the vaccine stimulated production of antibodies to all three strains of the virus and protected against the paralytic disease.

Just two weeks after the results were announced, the inherent danger of vaccine making was painfully brought to mind. Two hundred and sixty cases of polio were reported amongst a group of children vaccinated from a single batch that had been inadequately "formalinised." Production of the vaccine was immediately suspended until the problem was solved, and public enthusiasm plummeted. Later, when the usual summer toll of polio deaths occurred, a brave confidence in the vaccine returned, and by 1959 the death rate was in sharp decline.[278]

The protection conferred by Salk's vaccine was not long-lasting, just as the virologists Albert Sabin and Hilary Koprowski had expected. Independently, the two men had embarked on the longer route to a safe attenuated strain of polio virus that could induce immunity as effectively as the yellow fever vaccine. They believed that the vaccine virus would be most effective if it entered the body through the gut, where it would multiply in enteric lymph tissue and establish a focus of immunity where real infections normally start. Sabin's vaccine was eventually chosen for widespread use, and by 1968 it appeared to be manifestly more effective than Salk's, although carrying a remote but significant risk of paralytic polio. In the next phase, children were given both vaccines; the Salk vaccine for immediate protection against live virus that might be present, and the Sabin for long-term immunity; a diverse repertoire of antibodies was expected. The Dutch vaccination authorities decided to improve on Salk's killed vaccine and by 1978 had developed one that could be given in combination with diphtheria, whooping cough, and tetanus vaccines that became the cornerstone of the global eradication strategy.[282]

Twenty years after the program began, global eradication of the disease seemed feasible within a few years, as just three major foci remained, in South Asia and in West and Central Africa. Sadly, the program has stalled in parts of West Africa as religious leaders in the Muslim provinces of Nigeria have counselled their followers to refuse the vaccine on the grounds that it is a Western plot against Islam. Almost immediately, reported cases of polio in Nigeria and the neighbouring regions surged. There was no reason to suppose that the nonvirulent live vaccination strain would not persist in the environment, but nobody suspected that it could mutate back to a virulent, paralysis-inducing form that could infect unvaccinated individuals. In 2000, six years after the last case of polio was recorded on the island of Hispaniola, eight cases of paralytic polio were reported, caused by strains of polio virus in which the disease-causing properties had been restored by mutation. Clearly, the decision to stop the eradication campaign will not be taken lightly.[283]

These were daring medical adventures in which accidental outbreaks of polio were not the only risk. The possibility that viruses derived from monkeys or chimpanzees might contaminate batches of the oral polio vaccine has also provoked anxiety. Traces of a virus called SV40 were discovered in 1960 in the monkey tissue used to make the polio vaccine.[284] As this virus causes tumours in animals, the authorities reflected on the millions of Americans who might have received this potentially contaminated vaccine with some trepidation. Disconcertingly, the virus turned up recently in human tumours for the first time, but as some of these patients were never vaccinated, its significance remains unexplained.[285] Another damaging allegation suggests that HIV-1 was introduced into human populations in Koprowski's oral polio vaccine, although this seems to have been conclusively disproved (see Chapter 10, p. 207).

THE STRANGE HISTORY OF BACILLE CALMETTE ET GUERIN (BCG):
Once the causative organism of tuberculosis was known, a vaccine against
this deadly scourge was one of the most important goals of contemporary
bacteriologists, but no quick breakthrough came. The key development
began in 1908, when Albert Calmette and Camille Guerin in France decided that a weakened form of *Mycobacterium bovis*, a close relative of
M. tuberculosis, might make a useful live vaccine against human TB. This
organism causes a similar disease of cattle and also the intestinal form of
human TB. They hoped the capacity to cause the bovine disease would
be lost completely (attenuated) when the microbe was subcultured repeatedly in the laboratory. In that case, this strain might be suitable as a live
vaccine that could be taken orally to confer immunity to the disease. Thirteen years later, they had such a strain. The first "trial," made in dramatic
circumstances, was on a newborn baby who was expected to die of TB because her mother had died of the disease immediately after giving birth.
Against very considerable odds, the baby did not contract the disease.
Although spectacular in its way, a single case was hardly a ringing endorsement of the vaccine, but the authors initiated a major trial involving
more than 100,000 French children. The trial, in Calmette's estimation,
was hugely successful because good protection against TB was established,
although the doyen of British medical statistics, Major Greenwood, and
some American epidemiologists found the statistical analysis unconvincing.
Before their differences of opinion could be resolved, however, a tragedy
unfolded that unjustly and almost fatally discredited the vaccine. The incident occurred in Lübeck in 1930, when about 250 babies were mistakenly
given a virulent strain of *Mycobacterium tuberculosis* instead of the attenuated form. In the confusion that followed – an attempted cover-up and
an effort by Nazi propagandists to vilify the French – the vital momentum
was lost.

Undeterred, Swedish health authorities continued with small-scale trials
that received little recognition until after the Second World War, but they
eventually encouraged the British Medical Research Council to undertake
its own full-scale and apparently highly successful trial. Seventy percent of
the British schoolchildren in the trial were protected, and BCG was adopted
for mass immunisation. This degree of protection was sustained in later
trials in Britain, but elsewhere confusion reigned. Protection was found in
some and not in others, and the organism used to make the vaccine had
almost certainly undergone genetic changes since it was first used.[286] The
vaccine seems to be less effective against pulmonary TB (affecting the lungs)

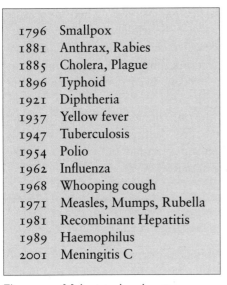

1796	Smallpox
1881	Anthrax, Rabies
1885	Cholera, Plague
1896	Typhoid
1921	Diphtheria
1937	Yellow fever
1947	Tuberculosis
1954	Polio
1962	Influenza
1968	Whooping cough
1971	Measles, Mumps, Rubella
1981	Recombinant Hepatitis
1989	Haemophilus
2001	Meningitis C

Figure 9.2. Major vaccines in use.

than against systemic forms of the disease, and it is much less effective in adults than in children who have had no exposure to *Mycobacteria*.

THE RECENT HISTORY OF VACCINES: Many other vaccines have been developed since the pioneer days (see Fig. 9.2). Efforts to immunise against whooping cough started soon after the causative organism was discovered in 1912, but convincing protective immunity eluded scientists until 1943, and mass vaccination of infants started only in the early 1950s. Vaccines against measles, mumps, and rubella (usually administered in one shot known as the MMR) were introduced in the 1970s. Gene technology has been used to improve old vaccines and create new ones. An entirely different method of provoking immunity that looks promising for vaccination against malaria and HIV is now attracting attention. DNA containing one or more genes of the organism to which immunity is sought can be injected into muscle in the traditional way; this DNA enters cell nuclei and programs synthesis of a protein that elicits the immune response. A recently invented device painlessly propels DNA attached to minute particles deep into the skin, using compressed helium.[287]

Historically, parents have usually accepted implicitly the advice of health professionals on vaccination; a newer generation has become less trusting.

The question of MMR vaccine, in particular, has recently become a cause célèbre in Britain because of a report that the triple vaccine might cause autism, based on the history of just twelve children. Epidemiologists familiar with the statistics on MMR vaccination, which come from many countries over a thirty-year period, are convinced that no basis for this claim exists.[288] Nonetheless, the British press prefers to describe the vaccine as "controversial" (rather than the one rogue report), with the result that some parents are refusing to have their children vaccinated, seriously damaging the campaign. In the United States, a small but vociferous lobby reminds the public that not everyone agrees with vaccination.

Magic Bullets – Antibiotics

The sudden devastating advance of septicaemia or pneumonia was a feared and familiar accompaniment to family life in the nineteenth century. Nothing could halt its progress or prevent the deaths of children and women in childbirth, whose memorial is so conspicuous in the dismal statistics of those times (Fig 2.2). A vision of how this could change emerged in late nineteenth-century Germany, championed by Paul Ehrlich. In the new science of organic chemistry, Ehrlich saw the means of generating an infinite variety of chemicals from which a "magic bullet" could be selected that would kill only microbes, through a specific target, and leave the patient unscathed. He died in 1915 having discovered drugs that could cure syphilis and sleeping sickness, but with his major dream unfulfilled. However, his idea of screening for magic bullets was eventually adopted on a massive scale by the pharmaceutical industry.

A methodical search organised by Gerhard Domagk, starting in 1927, finally vindicated Ehrlich's vision with the discovery of a compound that could cure streptococcal septicaemia. This microbe has always been one of humanity's most deadly adversaries; in 1936, at least two out of every thousand women giving birth in the developed world died from puerperal fever caused by *Streptococci*. Scarlet fever was another extremely serious disease caused by *Streptococci*. Domagk's direct approach was to inject candidate compounds into mice infected with a lethal *Streptococcus* in the hope of finding one that could cure the infection. Amongst the thousands tested, the most effective was a red dye called prontosil, which Domagk used in desperation to cure his own daughter of septicaemia. The first clinical trial was carried out in London on patients suffering from puerperal fever who were very likely to die. More than eighty percent of the subjects were cured, and within a year the risk of death in childbirth began to decline in many centres.[289]

Chemists at the Institut Pasteur looked at the structure of prontosil and immediately recognised the active part as sulphanilamide. This simple compound – one of the reactants in preparing prontosil that Domagk had not tested! – had been known for thirty years and was not protected by patents. The pharmaceutical industry, always appreciating an historic opportunity for a quick profit, sprang into frenzied activity and in quick time had made and tested hundreds of related molecules. Wherever they were used, they revolutionised medicine – a singular triumph for Domagk, who was awarded the Nobel Prize for Medicine in 1939. Hitler barred Domagk from accepting the prize and even imprisoned him briefly to ensure his compliance, outraged at the committee who had awarded the Nobel Peace Prize to the German pacifist Carl von Ossietsky in 1936. Happily, the Führer never realised that sulphapyridine had cured Winston Churchill of pneumonia on a visit to North Africa in 1943.[289]

The accidental discovery of penicillin by Alexander Fleming was the next great milestone in the quest for magic bullets. Fleming was unable to prepare the drug himself, but Ernst Chain and Howard Florey at Oxford University obtained enough to demonstrate its unprecedented efficacy against severe septicaemia. With the aid of the American pharmaceutical industry, penicillin became available on a massive scale during the last stages of the Second World War; it saved countless lives and completely overshadowed sulphonamides (see p. 223).

RESISTANCE TO ANTIBIOTICS CHALLENGES THE NEW ORDER: Before long, a diverse range of antibacterial substances became available that could challenge almost every microbial disease, including tuberculosis. For the first time, physicians had a variety of potent weapons that could arrest infection and save lives dramatically, but very soon they made a disquieting discovery; disease-causing organisms were becoming resistant to the new drugs. Dysentery outbreaks in Japan were occasionally resistant to antibiotics in the late 1950s, but by 1964 about half of all isolated dysentery bacteria were simultaneously resistant to streptomycin, chloramphenicol, tetracycline, and sulphonamides. Similar observations were soon made with other bacteria in many centres. Infections that once yielded to antibiotics became untreatable in less than twenty years – an acute and unforeseen disappointment that steadily gained momentum. Forty years later, some disease-causing organisms are resistant to all available antibiotics, and organisms as rare as plague sometimes carry antibiotic-resistance genes picked up in the wild.

The cause of this perverse development is a previously unknown and remarkable genetic exchange between bacterial strains, or even between different species. Resistance to antibiotics is carried on small circular pieces of DNA – known as plasmids – that can "infect" other bacteria by inserting themselves into chromosomes. Gradually, this facility for genetic exchange – an obscure phenomenon first discovered in bacteria by Fred Griffith in 1929 – gained recognition as the invisible hand that drives microbial evolution. Antibiotic therapy has dramatically contributed to this process by encouraging antibiotic resistance genes to be drawn into this invisible web of microbial genetic transactions, favouring the survival of resistant bacteria and their cargo of acquired genetic information.

Animal breeders, pioneering factory farms, were quick to see that antibiotics could improve the health of intensively farmed animals in an otherwise very unhealthy situation. By the mid-1960s, farm animals were a major reservoir of antibiotic-resistant organisms containing plasmids that could transfer to other microbes and enter the human food chain. Eminent epidemiologists of the day were infuriated to discover that outbreaks of infant gastrointestinal disease were caused by organisms with multiple drug resistances, traceable to farm animals. Chloramphenicol, at that time the main weapon against typhoid, figured in one of the most serious losses of security. Mere bacteriologists, no matter how loud their protests, made no impact on commercial interests, and the practice continued unabated for more than four decades.[58,290] Happily, a sensible end is now in sight. Denmark, a country where pigs outnumber people six to one, stopped mass medication of its pigs in 1997. No loss of meat production ensued, but the amount of antibiotic-resistant bacteria recovered from the pigs fell dramatically. The World Health Organisation is sufficiently encouraged to pronounce that the routine use of antibiotics on farm animals can be abandoned without affecting farm incomes or endangering the animals. The European Union will ban the practice throughout the EU in 2006,[291] but the American authorities are still holding back in deference to their farm lobby.

Blood poisoning caused by *Staphylococcus aureus* was frequently fatal before 1940, but with the discovery of sulphonamides and erythromycin in the 1950s, the danger lifted. An ominous warning that this menace might recur turned up in Britain around 1960 in the form of bacteria resistant to a new type of antibiotic called methicillin. This strain – now usually known as MRSA or methicillin-resistant *Staphylococcus aureus* – also carried resistance to penicillin, streptomycin, tetracycline, and erythromycin.[292] These antibiotics were introduced in 1945, 1948, 1950, and 1960, respectively, and the resistance genes were added sequentially to the plasmid during the 1950s, presumably in hospitals and their effluents. Even in its early stages, the strain was a serious challenge because it spread readily to other patients in hospital. It has spread for unknown reasons more widely than related strains and now contaminates many European and American hospitals, where health care personnel carry the organism without any sign of disease.[292] A few years ago, the antibiotic vancomycin was held in reserve for infections of MRSA, but this last weapon in the armoury may already be compromised, as rare vancomycin-resistant strains of MRSA already exist. Bacterial strains as dangerous as our old enemies of the preantibiotic era now cast a gloomy shadow.[290]

The future of antibiotic therapy is hard to predict. The Cassandra position, that antibiotic resistance is inevitable and alarming, may be too extreme. Antibiotics could be used more intelligently in order to reduce the reservoir of drug resistance entering hospitals from the farmyard and the environment generally.[290,293] We hear optimistic talk about the design of novel antibiotics, but few new ones emerge. Major drug companies seem to be abandoning research because the costs outweigh the rewards, in spite of the gathering anxiety about MRSA and other antibiotic-resistant organisms.[294]

Why Microbes Will Always Be a Threat

Hundreds and possibly thousands of existing disease-causing microbes bear witness to the rich potential of the mammalian body as a venue for microbial evolution. Their capacity for rapid multiplication and their ability to maximise opportunities through genetic change has driven the process relentlessly. It was a necessary condition for mammalian survival to develop appropriate defences, against every microbial assault, and indeed most people in the developed world effortlessly avoid infection most of the time. However, complacency is inappropriate; the greatest dangers occur when well-established equilibria are broken by famine, earthquake, upheavals in land use, or new contacts with wild animals.

The arrival of Europeans and African slaves with their distinctive burdens of disease in Central and South America in the seventeenth century brought a catastrophe of almost unimaginable ferocity on a completely defenceless Amerindian population. Animals had never been domesticated in the New World as they were in the Old (with the exception of the guinea pig and the llama). In consequence, New World populations had no history of exposure to diseases derived from animals, least of all the epidemic diseases that contributed so fatefully to Old World history. The deadly cargo of smallpox, measles, influenza, tuberculosis, yellow fever, and malaria gradually gripped and eventually destroyed their society; intruding Europeans remained virtually unscathed. The only exception, perhaps, was an extremely virulent epidemic of syphilis that ravaged Europe in the two decades after 1492 that may have been caused by a hybrid between varieties of the New and Old World *Spirochaete*.[47] The disease eventually calmed down, but later became widespread as a slow-burning nemesis of Victorian rakes.

We can still be surprised by our vulnerability to microbes. Since 1960, more than ten new infectious diseases of varying severity have caused medical emergencies under unusual circumstances. Each disease-causing microbe successfully exploited a novel opportunity or flared up when a well-established biological equilibrium involving wild animals was disturbed (see Figure 9.3; Chapter 10). A group of severe viral diseases originating in Africa has come to prominence, associated with radical disruption of the primeval forest; another set has emerged in Southeast Asia. The abundance of new ecological niches and the unparalleled capacity for epidemiological investigation available in America was responsible for the identification of another set of microbes that probably are ubiquitous. The evidence is strong that microbes will always be a threat and even lethal in our first encounters, but our systems for managing new diseases seem to be remarkably effective in the longer term. A few classes of microbe that in the past three decades have caused alarming moments are considered in the next few pages.

GASTROENTERITIS: The human intestine is a key battlefield in the long history of our relationships with microbes, and enteric microbes have evolved a highly successful strategy to exploit this ecological niche. By provoking diarrhoea, the organism guarantees dispersal into the environment. Starting with a few hundred cells, infections grow to billions in the gut, with

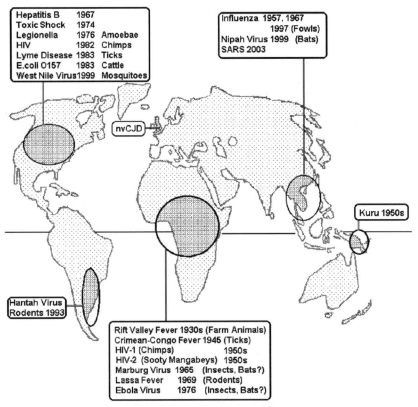

Figure 9.3. Global distribution of diseases and epidemics that have emerged in recent times, with names of suspected animal sources.

enormous potential for initiating epidemics when dispersed effectively. One might guess that enteric infections became a problem the moment our ancestors started to eat meat; this was perhaps alleviated to some extent by the invention of cooking but remained a problem until the invention of safe sewage disposal.

Communities with a reliable clean water supply are not threatened by typhoid, dysentery, or cholera, but many enteric infections, such as *Salmonella enteritidis* and *Campylobacter,* can be acquired from contaminated animal carcasses and routinely create public health problems. Food, inadequately cooked or contaminated after cooking, can have the same effect. *Salmonella enteriditis* has discovered a novel niche in chicken eggs, presumably having found a way of entering the egg by intracellular invasion within the hen.

The food-poisoning organism classified as *Escherichia coli* O157: H7 causes a potentially fatal infection characterised by bloody diarrhoea and kidney damage. This was linked to human disease for the first time in 1983 in the United States and then discovered elsewhere. Infections usually originate from ground beef, milk, or even vegetables contaminated by manure. Clearly, immense care must be taken in slaughtering animals to avoid contamination of the carcass by gut contents, an especially important consideration in a land where ground beef is a dietary staple.

The DNA sequence of the genome of O157 is moderately similar to its harmless cousin *E.coli* that inhabits the gut. The genome is interrupted by literally hundreds of pieces of DNA derived from other organisms, their viruses or plasmids, some carrying genes for nasty toxins. The two strains started to diverge around 4.5 million years ago and have continued to diverge by the insertion of new pieces of DNA.[295] Present-day O157 has a defective DNA-repair system that permits mutation rates far higher than those in its stable, friendly cousin, and this makes the emergence of new and more dangerous forms a distinct possibility.[296] The dysentery organism *Shigella*, although apparently of a different genus, is more closely related to *E. coli* than to O157 and probably evolved from a common ancestor on four separate occasions between 27 and 35 million years ago.[297]

INCIDENT AT A CONFERENCE: At a meeting of the American Legion in 1976, more than two hundred people contracted a form of pneumonia, of which thirty-four died. No well-known culprit seemed to be responsible, and several years elapsed before the outbreak was traced to a previously unknown microbe, now called *Legionella* in honour of its peculiar provenance. The fatal infections occurred because *Legionella* was disseminated by the air conditioning system of the hotel. The microbe could grow happily in condensed water in the system but had also learnt to live in the intracellular vacuoles of amoebae found in the same place. Learning this trick was an important preparation for infecting macrophages of the human lung, where, once established, it could provoke the unpleasant pathology of pneumonia. The first outbreak and others were alarming and tragic, but in reality *Legionella* is not very virulent and cannot spread easily from person to person. It seems to be an opportunist waiting for people whose lung function is compromised by smoking or other problems. The disease is now avoidable by routine disinfection of air conditioning facilities.[297]

VIRAL HEPATITIS: Historically, epidemics of jaundice (or hepatitis A) were relatively mild, originating in poor hygiene of the oral-faecal kind usually found in orphanages and military establishments. A more serious problem surfaced in 1942 when many American military personnel (28,000 cases and 62 deaths) who had been immunised against yellow fever became ill. This was eventually attributed to an unknown infectious agent carried in human serum used to stabilise the vaccine. The problem was not easy to understand or to eliminate, because the agent seemed to remain dormant for long periods. The same syndrome recurred throughout the Second World War and later in the Korean War, overcoming victims with fever, tiredness, vomiting attacks, and in some cases fatal liver damage.[298] A dreadful conclusion eventually became inescapable: the hepatitis agent was being transmitted by blood transfusion from apparently healthy individuals to the patients receiving blood.

How could carriers be identified? What was the agent? Increasingly frantic efforts spanning twenty-five years led eventually to a crucial clue in the blood of an Australian Aborigine in 1967. Normal control blood precipitated when mixed with this serum. The reactive component was a microscopic particle that looked like a virus under the electron microscope. When injected into a brave human volunteer, the particle – the so-called Australia antigen – caused hepatitis. A test could then be devised that quickly established who was a carrier. The outcome of these investigations was extremely disturbing: symptomless hepatitis B carriers were evidently extremely numerous, transmitting the infection to others, including health care personnel, and importantly would contaminate blood transfusion services if they were donors. The problem of hepatitis B was largely solved once a test became available that could be used to check for the presence of virus in every unit of blood. But as the hepatitis B problem was coming under control, a new and serious threat emerged from another virus, hepatitis C. As with hepatitis B, there are carriers, and the disease can be transferred in blood through the usual routes. Although the treatment is quite severe, the virus can now be eliminated by a course of alpha-interferon and ribivarin.

The need for perpetual vigilance against novel viruses that might compromise the safety of blood transfusion is the paramount consideration of the services. One of the first objectives of the gene technology industry was to make vaccines against all known varieties of hepatitis and to prepare blood components in a definitively virus-free state (Chapter 12).

OLD ENEMY, NEW NICHE: The sudden death of thirty-eight healthy young women, and the severe sickness of thousands more, dramatically alerted the public to toxic (or septic) shock syndrome in the 1970s. The condition was associated with a tampon designed to be hyperabsorbable during menstruation, in which the bacterium *Staphylococcus* flourished during prolonged use. A so-called superantigen made by the microbe was released into the bloodstream and provoked a severe reaction in the victim. A dramatic loss of blood pressure, cardiovascular collapse, and the failure of other critical organs, with death often ensuing within seven days, was the consequence.[189] The circumstances of these outbreaks were novel, but *Staphylococcus* and *Streptococcus* species are notorious causes of septicaemia and will always be a potent danger whenever hygienic safeguards fail.

BAD NEWS THAT INSECTS BRING: The immense threat posed by malaria is well known (see Chapter 10), but insects also transmit many other kinds of microbial disease. Historically, insect-borne diseases such as plague and typhus were associated with violent upheavals such as wars and earthquakes, but even in untroubled times, insects bring microbial diseases to new locations. West Nile virus, once confined to the northeast quadrant of Africa, appeared in a New York suburb in the late summer of 1999, infecting sixty people, of whom seven died of encephalitis. The virus was carried by mosquitoes that arrived unexpectedly with migratory birds and then survived the cold winter and the energetic efforts to eradicate them. In three years, the virus spread to the West Coast, transmitted by many species of mosquito and occasionally by blood transfusion. Although about 80 percent of infections are asymptomatic, some individuals suffer serious neurological damage. The scale of the problem, in the absence of a vaccine, is becoming more serious, with at least one hundred deaths in 2003 and thousands of cases.[299]

Lyme disease is named after the Connecticut town where, in 1983, the bacterium *Borrelia burgdorfia* was definitively associated with a new disease. This largely American disease transfers to humans by means of a deer tick and sometimes causes arthritis. Fortunately, the disease is successfully cured with antibiotics. The disease was probably seen earlier in the century, but the recent upsurge is related to reforestation close to homes.

SARS – SEVERE ACUTE RESPIRATORY SYNDROME: The recent emergence of a new and dangerous disease in the Far East provoked the traditional ritual of apocalyptic speculation. The outbreak was serious and tragic, but international infectious disease surveillance performed impressively and set new standards for quickly identifying the source of the problem. The disease spreads through the air in droplets of fluid and kills about one in six of those infected – for people over sixty-five, perhaps one in two. News of an acute and potentially lethal respiratory disease in the Guangdong province of China (a territory adjoining Hong Kong) started to circulate in January 2003. This aroused the interest of Hong Kong's influenza surveillance unit, which was investigating what could have been a new flu outbreak. Outbreaks occurred in other countries – including Canada, Singapore, and Taiwan – and on March 12, the WHO announced a global alert for a new disease syndrome that was definitely not the flu. In the absence of drug treatment, the only course of action was to impose quarantine and trace all known contacts. Within ten days, several labs were convinced that the disease was caused by an infectious agent belonging to the coronavirus family of viruses. By the beginning of May, there were more than 6,000 cases known, with at least 400 deaths. Later in the month, there was experimental proof that this virus could cause the disease in monkeys. Meanwhile, epidemiologists were beginning to think that the disease had jumped from animals to humans and were particularly suspicious of the masked palm civet, whose meat is a gastronomic delicacy in China. The complete DNA sequence of a number of isolated viruses was determined and a diagnostic test developed. At the same time, comparison of the human and civet isolates showed that a twenty-nine-base deletion had occurred in a single gene from the human form.[300] By mid-July, the outbreak was completely under control, but a vaccine or drug treatment is still some way off. The outcome of this crisis was reassuring, but the potential for harm inherent in an airborne disease is evident, particularly if it remains undiscovered for some time in a populous and poor tropical country. The SARS crisis was also another timely reminder of the importance of quarantine and contact tracing when no cure is available, just as in the time of Moses (Fig. 9.1).

Challenges, Risks, and Rewards – a Summary

Once humans started living in cities, the greatest challenge to their survival and the principle causes of premature death was microbial infection. Compromised water supplies, inefficient sanitation, close proximity to animals, malnutrition and the consequent lack of immunity, and acute overcrowding were the factors that favoured the spread of infectious disease. The Victorian commitment to closed sewers and piped water, better hygienic practices, and improvements in every aspect of public health reduced the menace of infection and gradually increased life expectancy. The introduction of mass immunisation and antibiotics in the 1950s made the developed world unprecedentedly safe. Dread childhood diseases of the nineteenth century are now uncommon and rarely serious; smallpox is eradicated, and polio is likely to follow quite soon. The great immunisation campaigns of the last century were the culmination of very determined efforts by public health authorities over many years. The move from small-scale trials to mass vaccination required breath-taking confidence that no potential tragedy hidden in the figures would not be multiplied many times in a full-scale campaign. At least until the 1990s, the public trustingly accepted the professional judgement of the designers of these schemes. By accepting a remote individual chance of an adverse effect from vaccination, members of the public who agree to mass vaccination recognise that they are protecting themselves and their entire community from the much bigger risk from the disease. Now that gene technology can provide the means to make even better vaccines, we may see determined efforts to eradicate more disease-causing organisms in the future.

Antibiotics still retain their ascendancy over microbes, but the emergence of resistance to all classes of antibiotics makes us apprehensive for the future. Our microbial adversaries remain as treacherous as ever, even in the industrial world, as new infectious diseases – more than ten in the last four decades – discover chinks in our hygienic armour. Reported cases of food poisoning and sexually transmitted disease rise inexorably as hygienic principles are neglected.

Infections on medical devices implanted in the human body are relatively new threats to human health that arise in perhaps one in a thousand medical interventions. They are caused by organisms that grow on surfaces as close-knit communities known as biofilms, enveloped in a sticky sheath that makes them difficult to culture and peculiarly resistant to antibiotics and disinfectants. This lifestyle is a vexing problem that today accounts for 65 percent of all bacterial infections reported, including organisms such as MRSA. Under the best circumstances they are difficult to avoid, and important scientific breakthroughs will be needed to comprehensively eliminate the risk.

GENERAL SOURCES: 8, 47, 278, 298.

10 Are Devastating Epidemics Still Possible?

Great epidemics of the past occurred following a collision between a rampant microbe and a totally defenceless population living in circumstances in which the disease could spread freely. The enormous menace of HIV and the prion diseases originate in the novelty of their disease-causing capability and their ability to exploit weaknesses in our hygienic arrangements. Other formerly dangerous diseases still have an undiminished ability to kill humans on a large scale. Malaria – less menacing outside of Africa than in former times – remains one of the world's greatest health problems. Tuberculosis is, in principle, curable, but the difficulty of implementing effective treatment means that it persists. Indeed, it infects one-third of the world's population, and in HIV-infected people it is now a severe danger. The influenza virus has a history of creating unpleasant surprises, but it is now under control because of a remarkable surveillance and vaccination program. Bioterrorism makes health officials nervous. Complacent in our belief that smallpox has been eradicated, and with our immunity to smallpox steadily diminishing, we could be vulnerable to bioterrorists. The important lesson of the last thirty years is that we should never assume that microbial enemies are vanquished and should recognise that microbes exploit chinks in our hygienic armour.

Should We Worry about Unknown Infectious Enemies?

Industrial countries with well-developed sanitation and public health systems are no longer at risk from the old villains such as plague, cholera, and typhus, but new infectious diseases can still surprise us. Most of these have proved manageable by antibiotic therapy, by provoking immunity, by adopting better hygienic practices, or by breaking the link with animal

sources of infection. However, we may still be vulnerable to a highly contagious agent that spreads quickly or to one that spreads unseen as a latent
infection with a long incubation period.

The peculiar horror of the diseases caused by the rare African viruses
Ebola, Lassa, and Marburg – the massive and lethal internal bleeding, their
apparent high local infectivity, and our lack of immunity – suggest the
potential for a serious epidemic. Ebola virus, first encountered in the Congo
in 1976, causes substantial sporadic epidemics in central Africa from time to
time – the most recent was in Uganda in 2000 – and is apparently decimating
populations of great apes.[301] However, in more than two decades Ebola
virus has never spread beyond central Africa and seems unlikely to start
a global epidemic. Although the Ebola and Marburg viruses can infect
monkeys, the primary source of infections is unknown and may be an
insect. Lassa fever is carried by a West African rodent. For all three diseases,
most cases of human infection originate in transfers of bodily fluids from
bleeding patients to inadequately protected health care workers. Although
the likelihood of these viruses spreading globally is not high, they are RNA
viruses with potentially high rates of mutation that could evolve into more
efficiently transmitted forms.

New disease-causing viruses surface frequently in the developing world,
very often spreading from animals as a result of a change in human-animal
interactions (see Fig. 9.3).[284] In an era of high-speed international travel
and high-density living, the threat of a new viral disease spread by coughs
and sneezes is an obvious and dangerous scenario. The monkeypox virus,
for example, which has some similarities to the airborne disease smallpox
and which occasionally kills humans in Africa, would be a serious worry if it
became more proficient at infecting humans.[302] In the sections that follow,
we shall consider five situations with a major potential for creating a public
health disaster. Missing from the list is tuberculosis, a disease we considered
in its historical context in the last chapter, although this in no way belittles
its potent menace. As approximately one-third of the world's population
carries this organism in a latent form, the disease could become active under
nutritionally challenging circumstances, in old age, or in people infected
with HIV who have lost an effective immune response. The special tragedy
of the disease is that it is usually curable with existing drugs, providing
treatment is continued for at least six months. If treatment stops before
the organism is eliminated, however, drug-resistant strains usually emerge
to become a major health threat, as in the former Soviet territories. After
a period during which TB was coming under control, the disease is now

advancing in many parts of the Third World and in the former Soviet
territories. Refugees and displaced persons play an important part in the
resurgence of the disease, because it spreads quickly in makeshift camps,
and the victims unwittingly spread the disease to industrial countries from
which it had almost been eliminated. About two million people die of the
disease every year.

Influenza

The influenza pandemic of 1918 may have killed forty million people (more
than the death toll of the First World War).[303,304] Highly contagious epidemics were circulating in the United States and in both armies on the
Western Front in the spring of 1918. By August, a second wave of infection, of a far more deadly kind, sprang up in France, Boston, and Sierra
Leone and then spread to every point on the planet within four months.
This second wave was particularly devastating to young adults and differed
from any form of flu described before or since in the fierce inflammatory
response it provoked in the lungs. The end frequently came with acute respiratory distress and haemorrhage, the lungs filled with fluid in a manner
that seems unique to the 1918 flu.

The devastation that could be inflicted on communities with no previous
experience of flu was extreme in remote settlements in the Pacific and Arctic,
where death rates of 70 percent were recorded. In one almost unique instance, a strict quarantine imposed by a strong-minded official in American
Samoa protected the inhabitants from the rampaging disease; in former
German Samoa, one-fifth of the population perished because no quarantine
was in force. By 1919, the pandemic had nowhere else to go and gradually
subsided. From today's perspective, the 1918 flu seems uniquely terrifying,
but serious pandemics also occurred in the nineteenth century, and there are
indications of a particularly devastating one in the seventeenth century.[47]
In 1957 and 1968, pandemic flu would return on a massive scale, but this
time considerably less dangerous to the young than to the elderly.

Without knowledge of the organism that caused the disease, contemporary medical scientists could do little to help. Once the outbreak subsided,
the most urgent priority was to find the causative organism, a task that took
fifteen years. The key step was finding material from a human flu victim
could produce a flulike disease in ferrets. The infectious material was evidently a virus and not a bacterium, as it could pass through a special filter

designed to retain bacteria. Virus recovered from infected animals was used to reinfect more animals. In fact, it was, as the pathologists say, "passaged" through 196 ferrets, after which a scientist was infected who developed the classic symptoms of flu. The experiments completed in London around 1933 proved beyond any doubt that this particular virus caused flu, but whether it was exactly the same as the 1918 flu was a more difficult question. People who had survived a bout of the 1918 flu had antibodies that recognised the virus, and this was the best and only evidence that the two were related. However, the character of the 1918 flu could not be defined more precisely without greater knowledge of the viruses.

Once the flu virus was conclusively identified, a monumental international investigation of the natural history of the virus started; it has continued to the present day. As in all viruses, the genetic material is wrapped in a tight-fitting membranous coat that helps the virus to enter cells. The viral membrane sticks to the membranes of lung cells, which then engulf the virus. Once inside, the virus reproduces, and the progeny are packaged into new coats ready to infect more of the victim's cells or to spread the infection to new victims in a sneeze.

The genetic material of the flu virus is present in eight separate strands of RNA. This arrangement makes flu peculiarly adept at evolving into new forms that might be more virulent. If two unrelated strains of flu infect the same cell, the two sets of RNA "reassort" at random and can make a hybrid virus that might combine the worst features of both strains. This kind of event is known as a "genetic shift" and involves the introduction of a novel RNA molecule into a virus, frequently from a strain that normally infects a different animal species.[305] The evolution of flu strains is crucially affected by an episode of selection that occurs in every infection and favours viruses with enhanced virulence, viruses that reproduce more efficiently, and those that are invisible to our immune system. This contributes to a steady accumulation of mutations, called "genetic drift," that gradually changes the "appearance" of the virus through modifications to viral coat proteins. Crucially, these changes enable viruses to evade antibodies that might have formed in response to an earlier infection.

The epicentre of these sudden genetic shifts has usually been the Far East. Although less vicious than the 1918 pandemic, the Asian and Hong Kong flu pandemics of 1957 and 1968 infected huge numbers of people worldwide, and mortality amongst the elderly was severe. These strains were hybrids between the common strain of flu and viruses present in ducks – a regular occurrence in the Far East, where poultry are frequently reared close to human habitation.

The start of what could have been a major epidemic was intimately observed in 1997 after the international surveillance system was alerted by a number of deaths in Hong Kong. These deaths were attributed to a new strain of flu that seemed to originate directly from chicken flu, at the same time that local flocks of chickens were being decimated by a serious outbreak of flu. The experts immediately saw a parallel with an American outbreak of an extremely virulent chicken flu that had occurred in 1982, which was caused by a single mutation in the H5 RNA. The new virus had the same mutation in H5, and a swiftly conducted screening of local people identified some individuals who already had antibodies to the virus but no recollection of feeling ill. The warning was very clear: a new strain had emerged with the potential to start a dangerous epidemic if it formed a hybrid with a highly infectious form of human flu. When a highly virulent flu infects an American chicken farm, the practice is to destroy the entire flock in order to prevent the evolution of a new strain of virus that combines virulence to birds and the ability to infect humans. The Hong Kong authorities responded in exactly the same way and destroyed all the chickens on the affected farms.

The crisis was not averted. In December, chicken flu erupted in other farms, and more Hong Kong residents died from the H5-containing flu. The situation looked extremely grave. Until then, humans had never been infected directly by a wholly chicken flu. Recognising that the entire chicken population of Hong Kong was a reservoir of potential infection, the authorities decided that they must all be destroyed, to widespread derision from the international nonscientific press. However, a pandemic may have been averted (who can tell!), and only six deaths and eighteen cases were recorded from the novel flu. At the start of 2004, reports began to reach the West that huge epidemics of bird flu were occurring in flocks in a number of Southeast Asian countries. A few humans also died of bird flu, but there was no evidence of the disease spreading between humans.

The mysterious and extreme virulence of the 1918 flu has never been forgotten by virologists. Inevitably, when appropriate technology became available they looked for an explanation and made efforts to recover RNA of the 1918 virus from the remains of victims. One source was the body of an Inuit woman exhumed from the permafrost in Brevig Mission on Alaska's Seward Peninsular, a remote village where 85 percent of the inhabitants had perished in 1918. Two more sources were pathology samples of lung preserved in formalin, obtained from American soldiers who had died in separate military camps in the United States during September 1918. Miraculously, RNA could be extracted from this material and used as a template to make a DNA copy – using a technique whose details need not concern us – from which a nucleic acid sequence was unambiguously reconstructed. This revealed two flu genes that were nearly identical in the three subjects, which could have undergone little change as the virus rampaged from the Atlantic coast to the wildest reaches of Alaska. The 1918 flu seems to have evolved from a bird flu just prior to 1918; it was also quite similar to a pig flu first identified in the 1930s that probably originated from humans.[304] Two slender nuggets of information suggest that the provenance of an ancestor of the 1918 flu might have been south China. H1-like virus was present in blood samples taken in the decade before 1917 in south China, and thousands of Chinese labourers were employed in France during the First World War to dig trenches.[306]

As we saw in Chapter 9, the most elegant way of preventing epidemics is vaccination. Unfortunately, because of the ceaseless genetic changes of the flu virus, no vaccine could confer lasting immunity, and consequently a sensitive global surveillance campaign is needed to identify new strains that might be a threat.[305] Since 1948, this has been organised under the auspices of the WHO, with large numbers of collaborating laboratories in many countries engaged in monitoring local influenza viruses and in plotting the spread of strains and emergence of new viruses. New vaccines have been made for many years using the natural tendency of flu viruses to undergo "reassortment" to create a hybrid. These are composed of a harmless laboratory strain that propagates in high yield together with a new epidemic strain,[307] but better methods are constantly being considered.[305]

What did we learn from the 1918 flu, and will something similar happen again? At a biochemical level, virologists know our enemy in ever-greater detail. We have a system for flu vaccination and antibiotics to treat secondary infections. A well-rehearsed contingency plan exists to deal with possible pandemics, which performed well in practice after the 1997 Hong Kong outbreak. But is this enough? Everybody hopes the answer is yes, but it does depend on the singular lottery that generates new strains of flu from animals and the accretion of mutations. Reducing the ancient practice of living cheek-by-jowl with livestock would diminish one risk factor, but the virus could evolve to make proteins that force an entry into cells more efficiently or that provoke inflammation more effectively. As vulnerable elderly people make up a greater proportion of our population and the great cities of the developing world become even more populous, more opportunities for the virus may be appearing.

AIDS

AIDS (acquired immune deficiency syndrome), the greatest medical emergency of our time, emerged in June 1980. Four previously healthy young men were hospitalised with a rare form of pneumonia and a bizarre set of symptoms, including multiple infections with viruses and fungi.[298,308] Their immune systems were evidently defective, because the level of one class of T-lymphocytes was greatly reduced. One patient had a rare but distinctive tumour, known as Kaposi's sarcoma, and an extreme wasting of muscle. The press quickly labelled the syndrome a "gay plague," but the profile of the group soon expanded to include haemophiliacs and women. From the start, the chief suspect was an infectious agent. Some believed the syndrome originated in an overload of infections, but in the winter of 1982 this was quickly forgotten, when the first baby died of the disease two years after a blood transfusion from a gay donor. The world was suddenly facing the alarming prospect of an extremely serious infective agent transmitted in blood. Infections started with a mild fever that was followed by a symptomless period during which T-lymphocytes were progressively destroyed until the immune system could no longer fight "infectious opportunists," such as pneumonia and tuberculosis. This marked the last dreadful stage of the disease, known as the full-blown AIDS syndrome. The victim's muscles began to waste severely; hitherto rare cancers such as Kaposi's sarcoma and lymphomas developed; and the victim's cognitive skills began to dwindle away.

Epidemiologists from the Centers for Disease Control (CDC) in Atlanta wanted immediate action to prevent further contamination of the blood supply by people known to be at risk. This was resisted, fatefully, by a coalition of the gay lobby and the blood industry; the former objected to perceived stigmatisation and the latter to what they supposed would be prohibitive costs (see Chapter 13).[298] Both would regret their rashness when the immensity of the tragedy became apparent.

While the eyes of the world were focussed on American anxiety, AIDS was soon diagnosed in Europe. Amongst Belgian and French victims, there were significant numbers of people of both sexes from central Africa. As soon as a blood test for HIV became available, sub-Saharan Africa was recognised as the real centre of the epidemic, with very large numbers of HIV-positive individuals and many already suffering from full-blown AIDS. Blood samples stored for many years in African hospitals showed that HIV was present in Kinshasa as early as 1959 and could have been introduced into East and West Africa in the 1960s by promiscuous truckers and refugees. As the epidemic gathered momentum in Africa, sexual contact and blood transfusion played an important part in transmitting HIV to both sexes. Reciprocal contacts between Haiti and central Africa in the 1960s, and then between Haiti and the United States, probably account for the spread to America. By 2003, over fifty-eight million people had been infected by HIV worldwide, of whom twenty-three million were already dead. Two-thirds of HIV subjects alive today reside in sub-Saharan Africa, but six million Asians are also infected, and the disease is advancing rapidly in Russia, India, and China.[284]

The virus was eventually propagated on a large scale in human lymphocytes using novel methods until enough material was obtained for devising a blood test for the virus. The virus evidently entered the bloodstream during sexual activity (presumably through lesions), in blood transfusions, or during drug taking. Insect bites and exposure to other bodily fluids, such as urine and saliva, were not sources of infection. In the early 1980s, many haemophiliacs were victims of a tragic accident – in Japan, 50 percent; in Britain, 20 percent – when they were infected after receiving contaminated blood or blood products.[298]

By 1984, there was no doubt that AIDS was caused by an unusual retrovirus that could kill T-lymphocytes.[308] At a time when retroviruses (an RNA virus that is transcribed into DNA) were thought to be interesting academically but of uncertain significance, HIV revealed what a deadly enemy this class of viruses could be. The enormous firepower of molecular biology was now trained on HIV, and the virus quickly became the most completely studied self-replicating entity in existence. The virus enters T-lymphocytes by attaching to a surface molecule known in lab shorthand as CD4, which takes the virus into the cell. The enzyme, reverse transcriptase, makes a DNA copy of the viral RNA and the product is inserted into the chromosome, where it becomes a template for making more RNA copies of the virus to be packaged into new infectious units. HIV-positive individuals develop AIDS over a period of between two and fifteen years. In the early days, most patients died, but in recent years, patients receiving the best treatment are surviving quite well.

Where did HIV come from? HIV strains seem to have transferred to humans from primates on a number of occasions. HIV-1 is a far more common cause of AIDS than the related strain, HIV-2, which is about 60 percent similar and confined to West Africa. The closest known relative of HIV infects nonhuman primates (simian immunodeficiency virus, or SIV).

HIV-1 originated from chimpanzees; HIV-2 came from sooty mangabeys. A calculation based on how many base changes are required to generate HIV from SIV suggests that the strains diverged between 1910 and 1930.[309] The most favoured explanation of how this occurred is that chimpanzee blood entered cuts or wounds when chimpanzee meat was butchered for human consumption. This was a common practice, and indeed still may be, during logging operations in the forests of central Africa. But apart from being untestable, this theory fails to explain why a number of HIV strains emerged within one short period and not, as far as we can tell, during any previous era. If HIV evolved from SIV, a succession of mutational events would probably be required to develop virulence against humans. One plausible suggestion is that when injectable antibiotics arrived in the Congo during the 1950s, medical personnel reused unsterile syringes and needles out of dire economic necessity. This would have transferred the virus from person to person, possibly selecting mutant viruses with greater virulence towards humans.[310] About ten varieties of HIV-1 (the M group) were extant by about 1950 and disseminated rapidly, propelled by migration, sexual promiscuity, and intravenous drug taking.

A sensational alternative theory, proposed initially by Louis Pascal in 1989 and further developed by Edward Hooper,[311] claimed that Hilary Koprowski's oral polio vaccine (see p. 185) was made using virus grown in HIV-contaminated chimpanzee kidney cells. Allegedly, the first victims of AIDS were infected during the field trials of the vaccine performed in the Congo, where the HIV pandemic started. No evidence supporting this scenario was uncovered by recent investigations. These established conclusively that the kidney cells used came from Rhesus monkeys and that existing vaccine samples contain not a trace of HIV or of chimpanzee DNA.[231,312] Estimates of the age of divergence of HIV and SIV suggest an origin substantially before the field trials.[309]

Until the 1980s, no drug treatment could control viral diseases, but the AIDS emergency provoked a search for drugs that blocked virus replication. One set prevents reverse transcriptase functioning (azidothymidine [AZT] and dideoxyinosine [DDI]); another blockades an enzyme that cleaves into separate units a large viral protein required to package the virus. Used in combination, these drugs extend the latent period of the disease for many years, but in the end, viral mutations may make the virus resistant to these drugs.[313,314] The immediate problem is that the drugs are too expensive to be affordable by health care providers in sub-Saharan Africa. Consequently, the momentum of the disease is unlikely to be checked in the foreseeable future, except through improved efforts to prevent new infections.

When the potential seriousness of the emergency in the developed world became apparent, most countries decided on bold and rather frightening publicity campaigns to encourage more circumspect sexuality. The explicitness of the message was a landmark in this kind of public relations, but nothing that has occurred since suggests this strategy was misjudged. Indeed, the pity is the campaign was not sustained and targeted more effectively at the groups at risk.

Prion Diseases

Prions are the strangest infectious agents we know. The word "prion," introduced in 1981 by Stanley Prusiner, defines a group of infectious agents ostensibly made only of protein that might have remained obscure academic curiosities had their role in lethal incurable diseases of humans not surfaced. They cause a catastrophic brain damage characterised by gaping holes and strange-looking fibrils – called spongiform encephalopathies by pathologists – never seen in uninfected animals.[78] The prion disease known to science for the longest time is scrapie, a disease of sheep that occurs sporadically throughout the British Isles. Material from the brains of these animals can transmit the disease to other sheep, as though it were a virus. To widespread consternation, when the first biochemical analysis of scrapie particles was made, they seemed to be composed almost entirely of one protein. There was too little nucleic acid for them to be a credible virus, and their viability was almost unaffected by ionising radiation and ultraviolet light – treatments that would probably destroy all viruses. At a time when DNA was being recognised as a universal genetic material, these were highly paradoxical observations, but a novel and ingenious proposal formally reconciled them with molecular biology.[77] A currently widely accepted version of this scheme is that all sheep cells contain a protein with an amino acid sequence identical to the scrapie agent but with a different three-dimensional structure. The scrapie agent, the prion, rather like an evangelical preacher, enters the cell and converts cellular forms of the prion into the infectious form, thereby increasing the total amount of infectious material present. This peculiar process then hijacks these cells, causing them to accumulate masses of fibrils that kill the neurones, creating the weird trademark pathology.[78]

Meanwhile, the telltale pathology of spongiform encephalopathy turned up in extraordinary circumstances associated with an invariably fatal disease known as kuru (the "laughing sickness").[30] This epidemic afflicted the South Fore people of Papua New Guinea in the 1950s, killing women and children within a year of the first appearance of symptoms, but not adult men. Suspicion fell almost immediately on cannibalism, which had been practised there since about 1910, because only women and children ate the cooked brains of dead males in the ritual feasts. Once Papua New Guinea came under Australian administration in 1954, the authorities vigorously suppressed cannibalism, and new cases of kuru in children declined immediately. Definitive proof that kuru was an infectious disease was finally obtained by showing that chimpanzees developed a spongiform encephalopathy when brain material from a kuru victim was implanted in their brains. The peculiar pathology of kuru and scrapie is present in several other extremely rare human diseases never previously thought of as infectious – such as Creutzfeldt-Jakob disease (CJD), which occurs in one person per million each year. The most likely scenario is that the first victims of kuru caught the disease when they ate the brains of an individual who was developing the sporadic form of CJD. Cooking does not inactivate the agent, and tragically, the disease usually remains dormant for as long as fifteen years.[315]

Ominous intimations of another spongiform encephalopathy affecting humans surfaced under extraordinary circumstances just as the kuru investigation was being concluded. In the late 1970s, two people contracted CJD and died after undergoing a neurosurgical intervention with electrodes that had been used on the brain of a patient with unrecognised CJD. The procedures used to clean and sterilise the instruments after each treatment – organic solvents and formalin – would certainly have killed bacteria or viruses extremely effectively. The disturbing implication was that prions were bound to the electrodes incredibly tightly and yet could start an infection. To extend the investigation, the same instruments were implanted in a chimpanzee brain, and the animal succumbed to a CJD-like disease. Not only had the prions remained tightly bound to the instruments, but they could also activate the cellular version of the prion without letting go of the electrode. Continuing studies using only stainless steel wires confirmed this conclusion and demonstrated beyond any doubt the extreme and extraordinary durability of prions and their remarkable power to remain in an active form when stuck to steel. The message was crystal clear: prion disease could be transmitted on surgical instruments.[316]

More instances of iatrogenic (i.e., caused by medical intervention) prion diseases have surfaced. Growth hormone obtained from pituitary glands removed from cadavers was used for many years to overcome growth hormone deficiencies. In the mid-1980s, three Americans who had received human growth hormone therapy died of a disease similar to CJD. The FDA, knowing the origin of the hormone, concluded that the material must have been contaminated with prions derived from a victim of CJD. Use of the hormone was immediately banned in the United States and elsewhere. Unhappily, by 1999 about a hundred cases were known worldwide, each of which had started before the ban and had taken more than twelve years to develop.[315] Fortunately, for children deficient in growth hormone, a genetically engineered form, definitively free of prions, became available in the same year.[317] Transplanted corneas have also been a cause of iatrogenic prion disease.

Just as these sinister developments were becoming common knowledge, the notorious "mad cow" disease of cattle, known as BSE or bovine spongiform encephalopathy, appeared in British herds. In 250 years, there had been no indication that cattle were ever infected by scrapie. Nonetheless, an infectious particle isolated from affected cattle was similar to the particle that transmitted scrapie in every respect except that scrapie did not readily infect cattle. The origin of the epidemic was eventually traced to a nutritional supplement made from meat and bone meal. Although every child knows that cattle are wholly herbivorous animals, modern farmers – since 1926 – had defied the wisdom of millennia by feeding a meat-derived supplement to the animals to stimulate their growth, without apparent undesirable consequences. In the 1970s, the prion probably underwent a critical change that allowed it to infect cattle. The modified form became more prevalent through recycling of the infected meat, possibly exacerbated by changes in the rendering process, so that a decade later the BSE epidemic was probably inevitable. Whether the BSE agent originated in a mutation of a cattle prion or from scrapie is still not known with certainty. Sadly, farmers continued to use compromised feed even after suspicions had been aroused. The fear that "scare-mongering" might damage the beef export business made the government reluctant to specify the risks, inevitably damaging public confidence. By the late 1980s, the gloomy implications were very clear: almost the entire national herd of cattle was compromised by prion-infected dietary supplements. Although these materials were banned forthwith, the tragically mistaken assumption that humans would not be infected persisted for a little longer.[315]

The risk to humans from the meat of BSE-infected but symptomless cattle was publicly acknowledged in 1989, when bovine offal for human consumption (i.e., nervous tissue that might be enriched with prion particles, kidneys, liver) was banned. The evidence that prions could jump a species barrier strengthened in the following year, when encephalopathies probably originating from BSE-affected meat were diagnosed in cats and other animals. Finally, the worst fears of the experts were realised in 1994, when twenty-four deaths were reported that were attributable to a prion disease probably contracted before the "specified offal ban." This prion is now called new-variant CJD (nvCJD).[315]

Predicting the scale of the nvCJD epidemic is a very difficult exercise. The number of cases is small, and analysis of diet or occupational exposure (e.g., in farmers or abattoir workers) does not explain the incidence of nvCJD. At least one nvCJD victim had been a vegetarian since 1985. The length of the incubation period is uncertain, although the age of the youngest victim (sixteen) puts a limit on the maximum incubation period.[315] The disease afflicts younger people principally (the average age is twenty-seven), with no gender bias. Early estimates of mortality were very high, but the best predictions now suggest several hundred deaths.[318] The tragedy will take many years to play out, but the culling of infected animals has ensured that further infections are unlikely. Although no treatment is in prospect at present, research will continue, as prions will remain a constant if shadowy danger.

In the next section, we will see how human exposure to the malaria parasite is etched indelibly in the genomes of people of African descent. A similar process of selection may have conferred resistance to kurulike prions in prehistoric times, when cannibalism may have been widely practised. Survivors of the cannibalistic feasts associated with the twentieth-century outbreak of kuru have a higher frequency of one particular version of the cellular prion gene, which may indicate that it confers resistance. As the gene is present in many different ethnic groups, it may have conferred a selective advantage (protection against prions) in the dark days of prehistory when cannibalism was widespread.[319]

Mosquito-borne Diseases

Just one hundred years ago, half the population of the world was at risk of contracting malaria and perhaps one-tenth of those were destined to die from it. Today, the global death toll is just one percent of that figure; but even so, 400 million people catch malaria every year, and more than a million children die of it, predominantly in Africa.[320]

Mosquitoes and mosquito-borne diseases have probably tormented humans of the Old World since agriculture began, and they continue to create huge and growing problems throughout the developing world.[320] Using exquisitely engineered mouthparts to suck blood from mammals and birds, these tiny self-propelled hypodermic syringes transmit, with extraordinary efficiency, the parasites and viruses they encounter to other individuals at the next bite. Malaria, the most notorious of these, is caused by *Plasmodium*, a single-celled organism with a complicated life cycle. However, the same fiendish efficiency disseminates many other curses of the tropical world, such as yellow fever, dengue fever, West Nile virus, and filariasis (the parasitic disease that causes elephantiasis).[321]

The intensity of the struggle with malaria is burnt into the genetic constitution of present-day inhabitants of the tropics. Mutant genes, widespread in the tropics, that subtly disrupt red blood cell function confer some resistance to malaria. The selective value of the sickle-cell trait, the best-known of these, is so powerful that one in three people in some regions of central Africa carry it. The key oxygen-carrying protein of red blood cells in these individuals contains a single amino acid substitution that confers resistance to malaria in the heterozygous state at the expense of a significant burden of ill health. At least seven genetic diseases that confer some resistance to malaria exist in different parts of the tropical Old World, causing thalassaemia, glucose-6-phosphate dehydrogenase deficiency, ovalocytosis, and other conditions. All of them impair the health of the carriers, and together they mark the lives of perhaps half a million new babies each year.[320]

The *Anopheles* mosquito disseminates malaria, but other mosquitoes spread different diseases. The yellow fever virus, the mosquito *Aedes aegypti*, and malaria probably all arrived together in the Caribbean from West Africa in the mid seventeenth century on slave ships and then spread throughout the subequatorial Americas.[47] As we saw in the last chapter, the menace of yellow fever has not disappeared, but thanks to the vaccine, it no longer threatens as in former times. Dengue fever, a sometimes lethal and agonisingly painful disease, marginally less dangerous than yellow fever, is also carried by *Aedes aegypti*. The disease was only sporadic while mosquito eradication programmes were active in the 1950s and 1960s, but without vaccine or prophylactic treatment, reported cases have increased twentyfold in forty years. Epidemics of new strains start at three- to five-year intervals in the cities of Southeast Asia, spreading into the country and frequently into new territories.[322] The potentially lethal West Nile virus that is widely distributed in Africa has recently appeared in the United States, spread by migrating birds and mosquitoes.

Hopes for breaking the grip of malaria have been invested in efforts to eradicate the mosquito and to control infections with drugs. European penetration of the African interior was possible in colonial times only with the antimalarial drug quinine. Better synthetic drugs, such as Atabrine and chloroquine, became available in the 1940s, and many more are available today. However, *Plasmodium* develops resistance to drugs, and different kinds will be needed if a revival of malaria is to be avoided.

A very effective vaccine against yellow fever, available for more than sixty years, keeps this disease under control. Epidemics still occur in remote equatorial forests, where *Aedes aegypti* is still rife, and the virus still infects monkeys. Vaccines against other mosquito-borne infections have been more difficult to develop. *Plasmodium* provokes significant immunity, but the parasite has a perverse facility to evolve into strains that evade the immune response by changing its cell-surface proteins. More than four decades of research have not given us a vaccine against malaria, and even the availability of the complete genome sequence of *Plasmodium* does not suggest this will be achieved soon. A vaccine against dengue virus is also problematic, because at least four varieties of the virus are known, although a live attenuated vaccine that gives immunity to all of them is just entering clinical trials.

ERADICATION OF MOSQUITOES: There would be few mourners if mosquitoes disappeared from the planet. Eradication of the insect was the main thrust of malaria control since the connection between malaria and the mosquito was first made in 1897. Very quickly, it became common practice to drain swampy regions and spread oil on standing water to prevent larvae from hatching. By 1914, malaria had almost disappeared from the United States, northern Europe, and many subtropical cities where once it had flourished.[321] Amongst the pretexts for the American invasion of Cuba in 1898 was a wish to eliminate the sources of yellow fever, widely believed to seed the outbreaks that periodically wreaked havoc in the United States. With no knowledge of how the disease was transmitted, the American military began reforming sanitary arrangements on the island, but that did not prevent a large proportion of the soldiery from catching the dread disease. Meanwhile, under the leadership of Walter Reed, a team started to investigate whether mosquitoes transmitted yellow fever, and by 1900 – and following the deaths of several colleagues – they could prove that *Aedes aegypti* was the culprit. The army of occupation was now in a strong position to control the disease by quarantining patients in buildings with screened windows, by oiling ponds, and by imposing draconian punishments upon property owners who permitted larvae on their land. In five months, the curse of more than three centuries had been completely exorcised.

French-led efforts to build a canal through the Panamanian isthmus in the 1880s petered out because malaria and yellow fever killed a third of the European contingent. Twenty years later, the Americans tried again with a real conviction, after their experiences in Cuba, that they could make the region free of disease-carrying mosquitoes. With yellow fever eliminated from the canal zone and malaria greatly reduced, the remarkable project was completed between 1906 and 1914.

After 1918, with international communications set to bring remote regions into ever-closer contact, the Rockefeller Foundation pondered the dangers of mosquito-borne disease. Even if mosquitoes could not fly the oceans, they could stowaway on planes or ships. Was it possible that Africa's most virulent malaria could spread to Brazil and perhaps throughout the Americas? Yellow fever, too – hitherto absent from Southeast Asia, where *Aedes aegypti* was exceedingly common – could devastate the region. Biologists familiar with history understood the catastrophic impact of yellow fever and malaria in the New World in the seventeenth century.

Anopheles gambiae, a mosquito that carried highly virulent African malaria, arrived in Brazil in the 1930s on the new airline connecting West Africa to Rio de Janeiro. By 1937, *A. gambiae* was established two hundred miles inland, and in the following year an epidemic exploded, killing fifty thousand people, catastrophically disrupting the local economy. With consternation growing in the United States (for their own safety), the Rockefeller Foundation organised a campaign to prevent further incursions, to be led by a now-legendary fighter of mosquitoes, Fred Soper. The outbreak was to be tamed by disinfecting all transport leaving affected districts, spraying houses with the insecticide Paris Green, and treating every victim of malaria with quinine to prevent *Plasmodia* from being passed on. By 1940, the emergency was over, and everyone believed the Western Hemisphere was saved from a terrible fate.[323]

Soper's campaign strategy was imitated extensively. In Egypt in 1942, an epidemic that killed 135,000 people was brought under control by Paris

Green and a new insecticide, DDT (dichloro-diphenyl-trichloroethane). The latter, patented in Switzerland in 1940 by Paul Müller, killed most kinds of insects efficiently on contact, was long-lasting, and seemed to be harmless to vertebrates. The allies used it with startling success to suppress an epidemic of lice-borne typhus fever in Naples in 1944. After the war, Soper, supported by the Rockefeller Foundation and the governments of Sardinia and Greece, used DDT in hundreds of tonnes to eradicate mosquitoes and the malaria that had troubled those regions since Roman times. Müller's contribution was recognised with the Nobel Prize for Medicine in 1948, but the victory proved only partial when mosquitoes reappeared, ominously resistant to DDT, albeit free of malaria, in just a few years.

Although DDT resistance was a developing problem, the American government was prepared to finance an eradication campaign mounted by the WHO in every country tyrannised by the mosquito. If these countries were drenched in DDT as quickly as possible, the argument went, malaria would be vanquished and the inhabitants would be forever indebted to the United States. In the 1950s, the incidence of malaria was spectacularly reduced in many countries, with help from the WHO. Perhaps the most remarkable campaigns were fought in India and Sri Lanka, where, in the early twentieth century, endemic and epidemic malaria had inflicted huge mortality on a population always on the edge of famine. Many islands – such as Japan, Taiwan, Zanzibar, Cyprus, Singapore, and Jamaica – were completely freed of malaria. China, too, took extraordinary steps to control malaria, using its own resources and methods.[320]

Even as malaria was being successfully eradicated from large tracts of the world, a big challenge was developing. Experts were recording the advance of mosquitoes resistant to DDT and malaria parasites resistant to chloroquine. Malaria was staging a revival in Sri Lanka and elsewhere, and a new question was being asked: if the populace had no immunity in regions from which malaria had been eradicated, would the disease return more dangerous than ever? If malaria ceased to be endemic in tropical Africa, the loss in immunity amongst the older generation could make the entire population extremely vulnerable to epidemic malaria.[320] An indication that this might be the case came from Madagascar, which had been free from malaria for twenty years, when an exceptionally severe epidemic broke out in the highland region between 1986 and 1988. The WHO has now abandoned the global battle for eradication of the parasite or the mosquito and concentrates on fighting them using locally available drugs and public health measures.

Publication of Rachel Carson's *Silent Spring* alarmed the American public. Her passionate denunciation of DDT and all agricultural chemicals provoked people into wondering whether the chemicals that were damaging wildlife were threatening humans, too. The euphoria driving the mosquito eradication campaigns began to fade, and Congress, taking the book's claim at face value, banned DDT in 1972 because of fears that it might be a carcinogen. Most entomologists came to believe the ban was misguided, because cheap effective insecticides were still needed for public health purposes, and DDT was still one of the best.[321]

Although the worst excesses of malaria came under control in the twentieth century, the immediate future is less promising, as the disease regains its former dominion in Africa. The parasitology community is engaged in a race to develop novel solutions before it is too late. The complete sequence of the genomes of the malaria parasite and of *Anopheles gambiae* were reported in 2002 and will underpin efforts to find new drug targets and proteins suitable for making vaccines. On another front, scientists have devised a method for making transgenic mosquitoes that they hope will infiltrate wild mosquito populations and interfere with their capacity to spread *Plasmodium*.[322]

The Threat of Bioterrorism

Although biological warfare was used only minimally during the last century, a number of countries mounted development programs. This phase seemed to be ending in 1972, when more than seventy countries (including the Soviet Union and Iraq) signed the Biological Weapons Convention. This prohibited production, stockpiling, or development of biological weapons and required destruction of existing stockpiles. The once-secret world of biological warfare is candidly and surreally anatomised in a major textbook and associated internet site,[324] where we can learn how certain organisms were "weaponised." Every obscure corner of the microbial kingdom was trawled for nasty incapacitating fevers with potential as offensive weapons. *Bacillus anthracis* and *Clostridium botulinum* are described as "suitable" for germ warfare because they produce deadly toxins and cause a lethal disease if inhaled. Moreover, they can be cultivated easily and delivered to their target as spores – dormant forms of the bacteria that are extraordinarily resistant to sunlight, heat, and disinfectants. The idea was to deliver bacteria by missile into enemy lines, where they would be released as an aerosol.

The treaty signed, with all parties apparently in good faith, the unsavoury business of biowarfare seemed to be over. Complacency settled on the West until Iraq invaded Kuwait in 1989, when we were suddenly told that Iraq had a substantial arsenal of botulinum toxin and anthrax spores and the means to distribute them. Three years later, President Yeltsin also confessed, on behalf of the recently dissolved Soviet Union, that the old regime had continued to make biological weapons secretly after the 1972 treaty. Even smallpox virus, an organism believed to have been eradicated in 1979, had been prepared in a form suitable for delivery by intercontinental ballistic missile. Ebola and Marburg virus were also under study for the same purpose.

An alarm bell should have rung in 1979, when an outbreak of anthrax in the Siberian City of Sverdlosk killed sixty-six people. The Soviet authorities attributed this to consumption of contaminated meat, but later investigations showed that the victims contracted anthrax from spores released from a military establishment.[325] The collapse of the Soviet regime awakened interest in the fate of skilled microbial warfare personnel, which remains unknown. The fear is that they may be working for a clandestine organisation with destructive intentions.[326]

Even before the attack on the World Trade Center on September 11, 2001, terrorist groups with obscure ambitions for world domination were lurking in the shadows. A Japanese religious cult, Aum Shinrikyo, indicated their intentions with sarin, the deadly nerve gas invented by the Nazis, by deliberately releasing it on several occasions in public places, causing deaths and injuries to many people. Botulinum and anthrax were also released on eight occasions, but the group has never shown any competence at bioterrorism.[326] When "weapons-grade" *Bacillus anthracis* spores, sent through the American mail service by an unknown small-time terrorist, killed or injured several individuals in the autumn of 2001, bioterrorist weapons came to mind once again. Nobody believed the incident was an international conspiracy for very long, and the public quickly accepted that the material must have originated in a U.S. government facility, even though such stockpiling was illegal under international law.

With mass vaccination now a distant memory, and immunity to smallpox fading fast, Donald Henderson, a former director of the global smallpox eradication campaign, warned that serious preparations should be made against a bioterrorist threat. If the virus is ever released, nobody is likely to recognise it until a second wave of infection has started. Quarantine and vaccination could then be effective in controlling the outbreak. If not, a catastrophic fast-moving epidemic would ensue that might kill one-third of the victims. The American government took this warning seriously and signed a contract for the manufacture of forty million doses of smallpox vaccine in September 2000. As a solution, that too has dangers, because a substantial number of adverse reactions – particularly amongst immuno-compromised people – would be inevitable.[302] Smallpox and anthrax, in Henderson's estimation, are more serious prospects as bioterrorist weapons than botulinum toxin and plague because the skills and equipment required to manufacture them are within the capabilities of a determined, well-financed foe.

The science profession attaches great value to unrestricted publication, but they and governments are becoming increasingly conscious that published research could help terrorists. One recent example showed how a genetically modified mousepox virus was lethal to vaccinated mice. Similar treatment of monkeypox virus might create an agent as dangerous as smallpox. Another report, condemned by some biologists, described how an infectious polio virus could be constructed using the published DNA sequence of the virus and commercial enzyme kits.[327]

Challenges, Risks, and Rewards – a Summary

Pandemics as severe as the Black Death or the great epidemics that annihilated the Amerindian civilisations of Central and South America are amongst the worst disasters our species has ever suffered. Such extreme events can occur only when highly virulent infections, uncontrollable by public health measures, make contact with a population that has no vestige of immunity. Could we ever face such a challenge again? Sophisticated international surveillance systems and an extensive armoury of anti-infective drugs and vaccines mean we are well prepared, but nonetheless the arrival of prion diseases and HIV were clearly beyond the imagination of our best talents. Those experiences tell us we are most vulnerable to previously unknown infectious agents, particularly ones with long asymptomatic latency periods during which the disease spreads. Novel epidemics could conceivably emerge from developing countries that contain reservoirs of hitherto unknown viruses in wild animals. The emergence of AIDS in sub-Saharan Africa is the grimmest illustration of the disruption that such an epidemic can unleash. In the absence of a realistic cure, the disease threatens to kill one-third of the inhabitants of some regions. Expensive drugs capable of controlling the disease exist, but the political will to make them affordable is lacking, and an economic dislocation is imminent as severe as that in the aftermath of the Black Death. Better understood diseases, such as malaria and flu, can still surprise us by evolving into forms that are more infectious, that cause more acute disease symptoms, or that evade the immune response and drug treatments more effectively. In the future, if population trends continue and the public health infrastructure does not improve, huge Third World cities could become the epicentres of serious epidemics.

The possibility that bioterrorist groups might ignite an outbreak of smallpox when the last vestiges of resistance are disappearing from human populations would be the worst and most stupid kind of inhumanity our species has ever contemplated.

GENERAL SOURCES: 207, 298, 304, 315, 321.

11 Discovering Medicines: Infinite Variety through Chemistry

The strangely audacious notion that human sickness could be cured with concoctions made from natural products existed in most ancient civilisations. This fantastic dream began to come true in the twentieth century, as chemistry and pharmacology were harnessed to make reproducibly effective versions of folk medicines. Prospects for discovering useful drugs greatly improved with the discovery of the first antibiotics and the realisation that these were representatives of a chemical universe of hitherto unimagined richness containing molecules of extraordinary biological potency. With inspiration from nature and growing sophistication, synthetic chemists have learnt how to generate vast libraries of chemicals, which the drug discovery industry screens for useful medicines. We will consider the prospects of protein medicines in the following chapter.

The Power of Chemistry

Experimentation with the physiological effects of natural products must be as old as human civilisation. Alcohol, narcotics, mind-affecting drugs, poisons, and many other natural products with pharmacological effects were widely known in ancient times. More puzzling, though, is why the ancients thought these products might cure human sickness. The ancient Egyptian Ebers' papyrus lists many medicines; whether they were efficacious we shall never know, but the betting must surely be not. One offering contains human excrement![328] The puritanical Hebrews, who would have known the Egyptian predilection for medicine, seemed to reject this kind of hocus-pocus along with alcohol. References to "medicines" and alcohol in the Old Testament are rare; they appear only in the later books, with the exception of Noah becoming disgracefully drunk (Genesis 10: 21–24) and

Jacob's wife, Rachel, dabbling with mandrakes to make her fertile (Genesis 30: 14–17).

If the idea of "medicine" before the dawn of scientific understanding of pharmacology is puzzling, nineteenth-century indulgence in the apothecary's art seems downright mysterious. Potions like Morison's Pills were used on a massive scale. Their weird ingredients suggest they were both laxatives (aloes and colocynth) and emetics ("cream of tartar," a salt of antimony). Confusingly, they were said to cure virtually everything from nameless fevers to measles, smallpox, consumption, and the afflictions of old age. Deaths from overdose were common, and although the medical profession clearly knew these nostrums were utterly worthless, no coroner ever found fault with them. James Morison grew phenomenally rich selling them throughout the British Empire and Europe.[29] Why Morison's Pills and late Victorian patent medicines were so widely used is not known. Their dramatic pharmacological effects presumably gave an illusion of "doing something," and their addictive character presumably encouraged repeat prescriptions. Several medical historians believe that what we now call the placebo effect may have contributed to their success.[8] The first of a number of remarkable studies by clinicians in 1954 showed very clearly how 30 to 40 percent of surgical patients reported satisfactory relief of pain when injected with a placebo – in this case, one millilitre of saline.[27] Why this should occur is mysterious and may be a subconscious learned response to medication. The phenomenon raises a serious question about how we test drugs, because if a significant proportion of patients report feeling better after receiving a placebo, we must expect some false positives from experimental drugs.

Medicines from plants that conferred undeniable benefits for specific conditions were known before the modern period. Opium is the best known, but the bark of the South American cinchona was widely used in seventeenth century Europe to cure tertian malaria.[328] Inevitably, medicines of this kind attracted the pioneer organic chemists, who learnt to purify the active principles and identify the chemical features critical for their pharmacological activity.

As natural-product chemistry was beginning to make advances, Paul Ehrlich, a director of one of the great nineteenth-century German scientific institutes, conceived a sharply focussed view of how chemistry could be used. His idea was to create compounds that would be "magic bullets," capable of killing bacteria, parasites, or even cancer cells by attacking particular targets but leaving the patient unscathed. Ehrlich's inspiration came from chemical dyes that stained infectious agents in pathology specimens, but not the surrounding cells, with extraordinary specificity. Arsphenamine (salvarsan) and suramin were two of the first fruits of this search for selective poisons that Ehrlich used to call "chemotherapy," although today the term usually refers to anticancer drugs. Around 1900, a severe epidemic of trypanosomiasis (sleeping sickness) was threatening to make the new colony of German East Africa (Tanzania) and neighbouring areas uninhabitable. Over a twenty-year period, starting with a dye that stained trypanosomes, Ehrlich and his industrial colleagues explored a succession of chemical derivatives, looking for drugs that were more effective at killing trypanosomes. After Ehrlich's death, this search culminated in suramin, a drug that is still in use today. Arsphenamine emerged in 1910, after incremental improvements of an arsenic compound with low activity, as a treatment for trypanosomiasis and later for syphilis. Arsphenamine was compound number 606 in this series, made by a process that foreshadowed the mammoth screens that would dominate pharmaceutical chemistry fifty years later. This, the first chemotherapeutic agent, was a revolutionary discovery that remained the only effective medication for syphilis until penicillin appeared.[259]

In Ehrlich's time, organic chemists were already immensely skilful in constructing new compounds, and by 1900, the extraordinary possibilities inherent in the chemistry of carbon were well understood. If every carbon atom could form four different bonds, fantastically complex carbon skeletons could be created in the manner of a wonderful child's construction toy. Chemists had grasped that chemistry had a kaleidoscopelike potential to create an infinite variety of products, limited only by their imagination, that could range from dye-stuffs to artificial fabrics. As the study of natural products advanced, novel chemistry of a kind never conceived by the human mind came into view, indicating that nature had already harnessed a vast repertoire of chemical architectures for the most surprising purposes. After a century of drug discoveries, new natural products with important pharmacological effects are still emerging. Indeed, half of the five hundred new chemical entities approved by regulatory authorities during the 1990s came from natural sources.[329]

Chemists believe that they can accelerate the discovery of new compounds with pharmaceutically interesting properties using a scheme they call combinatorial chemistry. This procedure employs robots to perform the steps of a generic synthesis in many parallel reactions in which individual components are varied systematically to generate every permutation

of the possible variables. Early versions of the procedure made short peptides (linear chains of amino acids) of random sequence that have been immensely valuable. Using the same principle, apparently unlimited varieties
of chemicals are already being generated in every other branch of chemistry.
Amongst these libraries of compounds, entities can be found that recognise virtually any novel target, and some are already in clinical trials.[330]

Lessons from Nature

If natural-product chemistry opened a new frontier, the discovery of antibiotics revealed an undreamt-of chemical universe populated by remarkable
structures with extraordinary biological properties. The story started on a
bacteriologist's culture dish in 1928, with a legendary encounter between
a mould and colonies of *Staphylococcus aureus* at St Mary's Hospital in
London. Alexander Fleming famously realised that the mould was secreting
a microbe-killing chemical that would become the best-known medicine of
all time, penicillin.[16] The formative experience that prepared Fleming's
mind for this crucial discovery was the First World War. As a doctor,
he tended countless young men dying from infected wounds that nobody
could save and wondered about entirely new ways of controlling infection.
Viruses that destroy disease-causing organisms, and the antibacterial effects
of tears were two of the avenues he explored before his great insight.

The preparation of penicillin for a clinical trial was a difficult project
for a clinician with no chemical training, but with the help of some colleagues, Fleming succeeded in preparing a concentrated extract of the substance. This was remarkably untoxic, extraordinarily effective at killing
Staphylococci, and even cured a few cases of conjunctivitis. Although these
findings were published, Fleming made no further attempt to develop the
drug. Ronald Hare, a one-time colleague,[16] believes that his enthusiasm
evaporated in the face of its instability; he was never sufficiently motivated
because he could not foresee its potential against septicaemia. Penicillin remained in the shadows for ten more years, until Ernst Chain and Howard
Florey got interested. Initially, their ambitions were limited to finding out
how it worked, but with the growing threat of war, a more important idea
came to dominate their thinking – the possibility of a powerful antibacterial
substance that could prevent septicaemia. By 1940, they were certain that
penicillin was remarkably effective in preventing systemic streptococcal
infections of mice, without being toxic. A year later, their faith was

rewarded when a number of dangerously ill patients were cured. Early trials with real wounds confirmed Florey's view, that if the most dreadful of wounds were flooded with penicillin, infection was preventable. They were now certain that penicillin could prevent sepsis in heavily infected battlefield wounds in a way that sulphonamides could never match. One problem remained: how could it be made on an industrial scale?

Sadly, the British government and Britain's leading chemical companies (Boots and ICI) could provide no assistance, but, undaunted, Florey and Chain decided on an American collaboration. This began at the U.S. Department of Agriculture Research Station in Peoria, Illinois, in 1942. There, the fungus could be grown on a huge scale on cheap waste products from the cornstarch industry and higher-yielding strains selected. From then on, events moved quickly. The first clinical trial on badly burnt survivors of a devastating nightclub fire in November 1942 was a resounding success. By then, the enormous resources of leading American companies had been mobilised, and factories in Britain and Canada were making penicillin on a colossal scale. When allied troops landed in Normandy, sufficient penicillin was available for every military eventuality and without doubt spared thousands of lives. Even so, administration of penicillin was not simple; it had to be injected or, better still, delivered by a drip to ensure that the concentration present in the body was as high as possible. Much later, acid-stable forms of penicillin that could be taken orally would become available.[328]

Fleming was not alone in seeking microbial products to control disease-causing microbes. Selman Waxman of Rutgers University, one of the leading soil microbiologists of his day, had a vision of the soil as a battlefield in which human pathogens were destroyed by the natural flora of the soil. Waxman's immediate reaction on hearing the electrifying news from Chain and Florey in Oxford was to start screening soil organisms for antibacterial substances that he christened "antibiotics." Using quick tests that demonstrated whether an isolated organism could prevent growth of another microbe, he focussed on the *Actinomycetes*, a group of soil microbes, obscure at the time, on which he was the greatest if not the sole authority. Antibiotics were easily found, but they were usually too toxic to be useful clinically. Waxman astutely directed his enterprise toward seeking an antibiotic against *Mycobacterium tuberculosis*, an organism unaffected by penicillin whose menace was still very potent. The search was sharply focussed by testing against a fast-growing relative of the TB organism. His perceptiveness was quickly rewarded when in 1942 he and his colleagues found a substance to which tuberculosis was susceptible – streptomycin, from *Streptomyces griseus*. Soon it was in clinical trials at the Mayo Clinic, at first indicating enormous promise; but the disease eventually resumed its progress as the bacteria became resistant to the drug. The conundrum was quickly solved, however, by treating patients with streptomycin and a less effective second drug (initially aminosalicyclic acid and later iso-niazid). Thereafter, infections came under control without the appearance of drug-resistant strains, and patients were cured after several months of treatment. In 1950, TB was still a dangerous disease, although mortality was only 10 percent of the figure recorded in 1840 in Britain; almost immediately TB began to drop out of mortality statistics. Hospital beds were no longer allocated in significant numbers to treat the disease. Streptomycin, though never as safe as penicillin, was used to cure a number of serious diseases that penicillin could not touch, including bacterial dysentery, brucellosis, and plague.

Within a few years, with the massive revenues from successful drugs beckoning, most major American pharmaceutical companies were engaged in a fierce competition to discover their own antibiotics by screening soil microbes from every point on the planet. The soil, and indeed the entire living world, was evidently full of these extraordinary treasures, but each had to be tested and marketed as quickly as possible. Typical of the Klondyke spirit prevailing was the introduction of chloramphenicol, an antibiotic produced by *Streptomyces venezuelae* discovered at Yale University in 1947 that was effective against a broad spectrum of infectious diseases, including typhus (caused by *Rickettsia*). In record time, the active component was prepared by the pharmaceutical company Parke-Davis, and full-scale trials were launched immediately against an epidemic of typhus in Bolivia and against scrub typhus in Malaya. The drug proved extremely effective against both outbreaks, without apparent adverse consequences. By accident, typhoid patients were also treated during the trials, and these too were cured. Although the drug was an immediate commercial success, it was superceded within a few years by broad-spectrum penicillins, but it remained the last-resort treatment for typhoid, meningitis, and rickettsial infections.

American companies did not monopolise the field entirely. Beecham's, in Britain, discovered how to make a substance that was easily converted chemically into a family of novel penicillin-like antibiotics, some of which were sufficiently stable in the acid conditions of the stomach to be taken orally. The Oxford Group had one more success with cephalosporin C, an antibiotic made by another serendipitously discovered fungus, which became the starting point for useful derivatives.

The high-water mark of antibiotic discovery was reached in the early 1960s, and although many modifications of known antibiotics were made

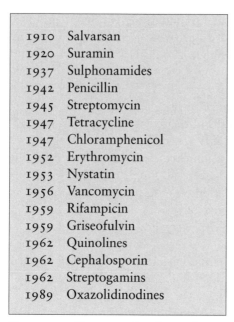

1910	Salvarsan
1920	Suramin
1937	Sulphonamides
1942	Penicillin
1945	Streptomycin
1947	Tetracycline
1947	Chloramphenicol
1952	Erythromycin
1953	Nystatin
1956	Vancomycin
1959	Rifampicin
1959	Griseofulvin
1962	Quinolines
1962	Cephalosporin
1962	Streptogamins
1989	Oxazolidinodines

Figure 11.1. First examples of new classes of antibiotic.

after this time, very few represented new chemicals classes (Fig. 11.1). As multiple drug-resistant bacteria steadily advance, this is beginning to look like an ominous warning. However, insights into the genomes of infectious organisms and of potential antibiotic producers may improve our chances of finding new targets and new antibiotics.[331]

The great screens for antibiotics, conducted with such single-minded zeal in so many companies, also turned up substances with pharmacological effects of an entirely different kind that became the basis for novel clinical uses. Notable examples include cyclosporin, a fungal product later used for suppressing the immune response for organ transplantation, and lovastatin, the founding member of the family of statins that are used to reduce cholesterol levels in the body.

Sulphonamides have a special importance in the annals of bacteriology as the first drugs that could control ongoing infections of bacteria (see p. 189). Sulphonamides dramatically brought to an end the era of "Eve's curse." The chemist's art, in making derivatives, turned up new treasures with unrelated uses, and over more than thirty years, new family members were discovered regularly, with uses as diverse as lowering blood pressure and controlling type II diabetes.

In bacteria, sulphonamides compete with a chemical needed to make the vitamin folic acid (para-aminobenzoic acid), which is in turn required to make nucleic acids. By contrast, mammals obtain folic acid from their diet and do not make it. Consequently, bacteria are killed selectively by sulphonamides, and the patient remains unscathed.

No antiviral drugs emerged from the pioneer antibiotic screens, for the simple reason that assays for antiviral activity were not very practical. There was no reason to suppose that antibacterial drugs could prevent viral infection, as with few exceptions the targets available in bacteria were not needed for viral propagation. The first antiviral drug, marketed as Acyclovir® in 1977, was discovered by screening for compounds that prevent replication of herpes virus in cultured human cells. The drug was incorporated into the precursors of DNA by the virus and then into viral DNA, and this blocked further viral replication.[259] Later, more drugs were found to prevent viral DNA synthesis, with no apparent serious consequences for human cells.

The drug companies responded with enthusiasm and cunning to the AIDS crisis developing in the early 1980s. Borroughs-Wellcome successfully applied for a patent to use an old drug (AZT – first made by the National Cancer Institute [NCI] in 1964) for the new disease, even before any clinical trials. Three years later, they marketed the drug as Zidovudine®, at $8,000 a year for a course of treatment. AZT and several other drugs already in existence antagonised reverse transcriptase, but they were all far from perfect, causing severe anaemia and permitting the virus to develop resistance to the drugs by mutation. Other drugs, such as Kaletra®, that could antagonise a specialist protein-cleaving enzyme required for the making of viral coats were also helpful in reducing the viral onslaught, but with significant side effects.[332] A therapy based on three phenomenally expensive drugs has reduced mortality from HIV sharply in recent years in the developed world. What the future holds, though, is far from certain, as drug-resistant varieties of HIV are likely to emerge. In sub-Saharan countries, most victims of HIV will probably die, as the cost of the triple-therapy and the additional medical care required for monitoring their use is beyond their resources.

Making New Drugs

ANTICANCER DRUGS: For many years, cancer scientists had hoped to find drugs that would selectively kill cancer cells and leave normal cells unscathed – just as penicillin kills bacteria without harming the patient. As the chief difference between normal cells and cancer cells is the latter's capacity for rampant cell division, the objective of research was to find drugs that targeted dividing cells. From the start, cancer specialists realised that they would have to test potential chemotherapy agents on terminally ill patients and these treatments would cause severe and unpleasant side effects, with perhaps just a small remission. The campaign would be long and hard and require extensive medical and scientific support to evaluate the trials and to undertake the improvements necessary to make the treatments bearable. Frustratingly, new leads emerged more often by accident than by design, and successful drugs were rarely useful against many kinds of cancer.

The first glimmering of progress in chemotherapy emerged from the nightmares of chemical warfare. Foul First World War weapons, such as nitrogen mustards, dramatically eliminated white blood cells. In the Second World War, doctors of the American chemical warfare service began to explore the possibility of humane applications in chemotherapy. One compound that successfully made mouse tumours disappear also dramatically reduced an advanced lymphosarcoma in a patient, although within a month it recurred and the patient died. The known nitrogen mustards were evidently far too toxic, but in the following decade better derivatives, such as cyclophosphamide (1957), were devised and are still in use today.

Even in the late 1940s, scientists were looking for specific targets in cancer cells for which an antagonist might be found that would kill cancer cells preferentially. They were not obvious, but one emerged from the discovery that certain leukaemia cells grew much better if they were provided with more of the vitamin folic acid. This suggested similar molecules might interfere with the role of folic acid and damage cells undergoing DNA replication in the malignant cells. Obligingly, the pharmaceutical industry made a set of similar molecules, and in the clinical trials conducted in 1948, two of these – aminopterin and methotrexate – looked promising for controlling childhood leukaemia.[328] Meanwhile, George Hitchings and Gertrude Elion at Borroughs-Wellcome in the United States – who would be awarded the Nobel Prize for Medicine in 1988 – began a more comprehensive search for drugs that could antagonise cancer cells in the same way. Their idea was to look for compounds that interfered with nucleic acid synthesis in a microbe – a remarkably prescient idea at a time when the importance of nucleic acid was not firmly established. From this emerged the most effective and safest anti-leukaemia agent of those times (6-mercaptopurine), and later 5-fluorouracil and a naturally occurring nucleotide (cytosine arabinoside), all of which are still in routine use. Another serendipitous discovery, azathioprine, was an immuno-suppressant that was to play a crucial role in the development of organ transplantation.

New chemotherapy agents emerged from many sources. Plants have contributed extraordinary chemicals – vinblastine, vincristine, and taxol – that in various ways disrupt the apparatus for separating chromosomes at cell division (the mitotic spindle). The accidental discovery of compounds of platinum – an element with famously limited powers of chemical reaction – led to several new chemo agents, of which cis-platin is the best known. The launch of the Sloane-Kettering Institute in the United States by two eponymous philanthropists marked a crucial phase in mounting a massive screening program for new anticancer drugs, which was not a realistic undertaking for industry. Later, the centre expanded, augmented by the vast resources of the NCI, and by 1975 more than half a million compounds had been screened, of which just a handful were useful.

The story of the development of chemotherapeutic treatments is a remarkable tribute to the fortitude of patients and physicians. James Le Fanu tells the poignant history of the evolution of treatment for childhood leukaemia at St. Jude's Hospital in Memphis, Tennessee, between 1962 and 1987.[165] Starting when fewer than one in a thousand children with acute lymphoblastic lymphoma survived for more than five years, they raised survival in small steps to 70 percent by patient exploration of combinations of drugs and other treatments. The most effective treatment to emerge used three separate drugs – methotrexate, 6-mercaptopurine, and cortisone – together with irradiation of the child's cranium to eliminate leukaemic cells that might enter the brain and evade the drugs. Similarly, notable successes were achieved in treating both Hodgkin's and non-Hodgkin's lymphoma, but the optimism for blood cell cancers did not transfer easily to solid tumours. Progress in developing chemotherapy has always depended on unsentimental and determined experimentation that at times must have seemed questionable, as tiny improvements of remission were bought at the expense of opportunities for kindly palliative care.[165] Most societies commemorate altruism of another kind when they remember "those who died that we may live," and it is appropriate that we should not forget the countless subjects of chemotherapy trials, a unique experiment in improving the human condition.

The quest for better anticancer drugs is now focussed on the so-called oncogene proteins – the mysterious proteins that activate during the cell's descent into malignancy. Foremost among these are enzymes that modify essential proteins by adding phosphate groups to growth factor receptors found on cell surfaces. This complicated area of biochemistry is far beyond the scope of this book, but phosphorylation, as a chemical phenomenon, is of enormous significance because it radically and dynamically changes the shape and activity of proteins. Phosphorylation was considered a target for potential drugs for two decades, but FDA-approved drugs that antagonise the process have emerged only recently. The first was Glivec (Imatinib), an inhibitor of a phosphorylation associated with an oncogene, which is now in use for patients with certain kinds of leukaemia and solid tumours.[333] Iressa, another inhibitor of phosphorylation, may prove more versatile because it targets epidermal growth factor, a protein active in many kinds of tumours. Both treatments are in their infancy, but they point the way to a time when solid tumours may be annihilated by better-targeted drugs.

STEROIDS: Scientific interest in steroids started when a rather disreputable idea began to circulate amongst gentlemen of a certain age, that extracts of gonads and endocrine organs had "powers of rejuvenation" (see p. 150). Academic research never confirmed this claim, but these early adventures in the gland trade led to the vast family of steroids. The chemical core of steroids, the four contiguous six-membered rings, is modified in nature in countless ways as hormones, in reproduction and control of inflammatory responses, and as vitamins, cholesterol, bile, and a host of pharmacologically active plant products. In the middle years of the twentieth century, steroids were one of the most competitive and important frontiers of industrial chemistry. Male and female steroid hormones were identified during the 1930s. The adrenal gland was recognised as a source of an important steroid-rich secretion necessary for the life of a mammal, and although many steroids were identified, their individual roles remained tantalisingly mysterious. Endocrinologists hoped to cure with an appropriate biological chemical Addison's disease, an adrenal insufficiency disease known for a hundred years. Making steroids chemically on this scale was an heroic undertaking, but it was a vital breakthrough in establishing their importance in biology. The breakthrough came with a thirty-six-step synthesis, devised by the American biochemist Edward C. Kendall, that converted a component of ox bile into a substance he called Compound E, later known as cortisone. The objective was to see whether Compound E relieved rheumatic arthritis, a disease that in those days compelled its victims to adopt pathetically distorted postures to cope with severe pain. Kendall's colleague Philip Hench had long suspected that adrenal secretions could reduce the inflammation of rheumatic arthritis under stressful circumstances, such as in pregnancy or during attacks of jaundice. This inference proved brilliantly correct when a single dose of Compound E transformed an inert cripple into someone who could walk without difficulty within three days. Such a drama had not been anticipated; the press pronounced it a miracle. Confusingly, they wanted to call Kendall's compound vitamin E, but he preempted such nonsense with a better name, cortisone. With incredible speed, the arduous synthesis was undertaken commercially, and the drug was soon prescribed widely for a host of chronic inflammatory problems. Cortisone spectacularly relieved symptoms of rheumatism in the short term, but it failed to cure the underlying disease permanently and, disquietingly, upset the salt balance of the body. However, the way to finding better anti-inflammatory drugs was clearly visible. Within a few years, the chemical kaleidoscope had generated many derivatives of cortisone, culminating in 1958 in dexamethazone – a far superior anti-inflammatory agent that left the salt balance unchanged.[328] Soon, most serious inflammatory illnesses were manageable to some degree using a steroid.

The cost of the thirty-six-step synthesis from ox bile, initially two hundred dollars a gram in 1949, sparked a headlong competition among four companies. The winner was twenty-seven-year-old Carl Djerassi, a rank outsider from an obscure little Mexican chemical company (Syntex) staffed by apparently inexperienced youngsters. Starting with diosgenin, a steroid present in tubers of a wild Mexican yam, they made cortisone through an intermediate, the sex hormone progesterone, which was already attracting attention as a potential component of the birth control pill. However, cheap cortisone finally came from the Upjohn Company of Kalamazoo, which used the fungus *Rhizopus* to convert progesterone into cortisone in a single step. By 1965, cortisone cost just fifty cents a gram.

Steroids used in reproductive medicine – some of the safest medications ever invented – are an application of chemistry that has affected more lives than almost all the chemical classes described in this chapter (see p. 32). Research has continued to find better formulations of the Pill ingredients, although academic experts suspect the industry is half-hearted. Drospirenone, said by the press to be "the Pill that makes you lose weight," is reported to prevent fluid retention. Other ideas include suppressing the number of menstrual cycles women have in a lifetime and developing pills that reduce the risk of breast cancer. The development of a male contraceptive pill seems long overdue, although the suspicion that "they will never use it" is always in the air.

In the late 1960s, doctors began to explore the possibility of correcting the unpleasant consequences of the menopause, which historically have always troubled women, by restoring the level of oestrogen to premenopausal levels. This so-called hormone replacement therapy (HRT) won a reputation for solving the problem, and women began taking HRT in steadily increasing numbers over a twenty-year period. In the last few years, this confidence was abruptly shattered by the results of surveys that seemed to indicate a small but significantly increased risk of breast and endometrial cancer in HRT users, as well as other adverse effects. These surveys were extensively publicised before the data were peer reviewed and provoked many women into abandoning the treatment unnecessarily. Now it is clear that many medical researchers believe the risks attributed to HRT were too high and that taking the treatment is still probably the best policy for some women. Physicians, generally, are anxious to make sure HRT is not seen as a lifestyle drug with miraculous abilities to delay ageing but only as a medication to treat serious problems. Attempts are being made to design steroids that reduce osteoporosis – loss of calcium from bone, which sometimes occurs in older women – without affecting reproductive tissues. The artificial steroid derivative clomiphene became widely used in the 1960s to stimulate ovulation of involuntarily childless women; indeed, it remains the first line of treatment before considering IVF.[334]

An alliance of a few disreputable chemists and athletic coaches has created a clandestine trade in so-called anabolic steroids that are used for body building, ultimately to obtain an illicit advantage for athletes in competitive situations. These steroids stimulate the growth of muscle but also produce adverse effects including loss of fertility, stress on joints, and, in high doses, extremely aggressive behaviour. They were commonly inflicted on hapless teenagers in the former communist countries by unscrupulous coaches looking for national glory. More recently, the ingenuity of chemists has created new steroids that evade standard doping tests. Happily, methods of analysis are improving all the time, so that the culprits can be caught and disgraced.

MOOD-AFFECTING DRUGS: Irrational moods, hallucinations, self-destructive impulses, and other kinds of serious mental instability trouble the lives of surprisingly large numbers of people. The discovery that irrational moods originate in the chemistry of the nervous system and can be dispelled by medication is a major triumph of twentieth-century medicine. The modern approach started with an appreciation that nervous activity reflects the activity of special chemicals within the brain. As with every other effort of drug discovery, the magic of the chemical kaleidoscope was the leading influence in finding potential mood-affecting drugs. The effects of these drugs at a biochemical level and on patients gradually illuminated neurochemical pathways on which aspects of mood were critically dependent. Broadly speaking, the new drugs that emerged in the 1950s fell into two categories: those that raised the spirits (antidepressants) and those that calmed irrationally troubled minds (tranquillisers).

Clinical and commercial success in the treatment of depression came in a rush during the 1950s from chance discoveries. Later, it was evident that most of these new drugs affected processes that maintained levels of a class of chemicals called monoamines (nor-adrenaline and serotonin) in the region of nerve endings where synaptic transmission occurs. The importance of this idea was recognised in 1970 by the award of the Nobel Prize for Medicine to Julius Axelrod.

Typical of the serendipity that prevailed in drug discovery in the 1950s was the discovery in the United States, by an observant clinician, that the anti-TB drug iproniazid profoundly improved the mood of some very gloomy patients. A proper clinical trial with chronically depressed psychotic subjects in 1957 confirmed this unequivocally, and within a year more than 400,000 patients were being treated. Later, it became apparent that the drug antagonised an enzyme called monoamine oxidase and consequently prevented the monoamines from dissipating. Within a few years, starting from this lead, safer and more specific antagonists of the same type were discovered.

At about the same time, a similar episode of serendipity led to the discovery of the so-called tricyclic antidepressants. These also slow the re-uptake of monoamines at nerve endings and proved hugely lucrative to Swiss and American companies of the 1950s. Later, Prozac® emerged as a selective inhibitor of the re-uptake of serotonin, a chemical messenger of the monoamine class, and also broke many records for earnings. Nobody supposes the problem of clinical depression is completely solved, but the discovery of inexpensive, safe medications that support individuals who would otherwise be unaccountably blighted by depression is a substantial triumph.

Schizophrenia is the twentieth-century name for a malady people have found frightening throughout recorded history. The lives of about one per-cent of the adult population are ruined by it, usually starting in adolescence; the disease causes irrational fear of persecution, emotional unpredictability, hallucinations, and the hearing of voices. In the first half of the twentieth century, the madness of psychoanalysis or the savagery of lobotomy were used to treat the condition, but no evidence was found that suggested these treatments were of any use. A breakthrough in the management of this troubling condition emerged from an investigation during the 1930s of an Indian medicine used for lowering blood pressure, made from the plant *Rauwolfia*. Two decades later, American investigators found the main com-ponent made subjects relaxed rather than drowsy – an outcome they called "tranquillized" – which they proposed could calm schizophrenics. The drug was marketed as Reserpine® but was soon eclipsed by a better class of tran-quilliser – the phenothiazines – that emerged by chance from a search for antihistamine drugs that could overcome the shock associated with surgery. In clinical trials with the first of these – chlorpromazine – patients suffering from severe manic behaviour were dramatically calmed, and their halluci-nations, irrational fears, and self-destructive impulses were greatly reduced. Sadly, this drug too proved unsatisfactory when serious side effects like Parkinsonism appeared in some patients. However, for the management of patients with mental disorders, the introduction of these drugs marked an important transition from the need for containment in institutions to reha-bilitation in the community. Better antipsychotic and tranquilliser drugs were quickly discovered – such as the benzodiazepams Librium® and Valium® – which reduced anxiety and restored normal sleeping habits re-markably. Eventually, they became the most commercially successful drugs of their era, although they proved addictive to certain patients.

A completely different treatment for manic-depressive illness – which affects perhaps one in a hundred people and drives more than a tenth of these to suicide – emerged accidentally from a perceptive observation made by an Australian doctor, John McCade, in 1949. Guinea pigs treated with salts of the element lithium became unusually lethargic and unresponsive to handling. Wondering if they had any potential for controlling mania, McCade first tested the salts on himself to establish their lack of toxicity and then on a patient who had been in a state of acute manic excitement for five years. The excitable and distracted manner was dispelled sufficiently for him to return to his old occupation. Because of its potential toxicity, and the obscurity of its pharmacology, twenty years elapsed before lithium received the support it deserved for treatment of manic depression; even in 2003, it is still not adequately understood. Sadly, although academic research indicates some potential for lithium in the treatment of Alzheimer's disease and possibly schizophrenia, pharmaceutical companies are not taking it further, as it lacks any potential for serious profit.[328,335]

The discovery of mood-affecting drugs transformed the situation of people with serious depression and mental illness, but these drugs were easy to prescribe excessively and without proper consideration of environmental causes of behavioural problems. The pharmaceutical industry benefited enormously from this regime and skilfully distracted attention from the serious adverse effects on small numbers of individuals.[336] The latter allegations relate particularly to drugs of the Prozac family that intensify suicidal feelings and precipitate a significant number of suicides. At the same time, drug companies seem to find ways of restricting the freedom of independent academics to criticise the industry.[337,338]

Mood-affecting compounds that are used for controlling the symptoms of attention-deficit disorder (ADHD) are also being overpromoted. The drug methylphenidate, better known as the best-selling medication Ritalin®, has been in use since the 1950s for calming down hyperactive children. It seems to increase the powers of concentration and acts by stimulating re-uptake of dopamine by neurones. In the 1990s, the drug became the first-call treatment in the United States for wayward behaviour by children, rather than any effort to investigate the environmental origins of the problem or to devise methods of stimulating good behaviour. Perhaps one in ten children (predominantly boys) are taking the drug, and students are using it to improve their concentration while studying. Children under the age of five are also being given the drug, although no trial has been conducted to determine whether brain development will be affected. No tangible adverse medical effects are reported, but concern is mounting that accepted traditions of child rearing are being abandoned in favour of a "convenient" pharmacological solution. Parents and teachers have become willing victims of a concerted sales campaign that has enormously enriched the drug companies and given credibility to a dubious diagnosis.[339]

Drugs affecting memory have immense commercial potential, and a substantial number of companies looking for compounds that affect plausible targets are emerging from academic neuroscience. Clearly, drugs that could enhance the memories of people with degenerative diseases or even of the "worried well" are likely to be a bonanza. The converse, drugs that erase painful memories, are a potential humane solution to post-traumatic shock disorder, and indeed beta-blocker drugs may already serve this purpose. However, there is the potential danger that such drugs could be abused if the memories of witnesses to shocking incidents were erased, or if they were used to suppress appropriate feelings of shame or guilt.

ANTI-INFLAMMATORY AGENTS: The capacity of extracts of willow bark (and other trees) to reduce fevers has long been known. The active ingredient was probably salicylic acid, and it was acetyl salicylic acid that became the world's most widely used drug early in the last century. This was the eponymous aspirin – whose name derives from "a" for acetyl and the plant source, *Spirea ulmaria* (meadow sweet) – that was synthesised chemically by Bayer, a German pharmaceutical company, and sold in tablet form for headaches and reducing fevers. Aspirin was probably the first medicine that underwent trials for safety and efficacy as an anti-inflammatory and pain-killing agent; it remains notably useful after more than a century. Aspirin is not entirely perfect, as it can, in unusual cases, irritate the stomach lining and cause kidney damage.

How aspirin relieves pain remained a mystery until the 1970s, when John Vane found a novel explanation that marvellously illuminated an important area of biology. The key insight was that aspirin and several other well-known painkillers, such as ibuprofen and indomethacin, inactivate the recently discovered cyclo-oxygenase enzymes. One of these (known as COX-1) is always active in normal circumstances – particularly in blood platelets (the key cell type in blood clotting), the stomach, and the kidney – and makes a class of chemicals called prostaglandins that have important roles in physiology. COX-2 only becomes active when tissues are inflamed in response to injury.

Two effects of aspirin underlie the pharmacological value of the drug. By antagonising COX-2, aspirin reduces inflammation, while very small daily doses of aspirin gently control the activity of COX-1 in platelets and reduce the risk of blood clot formation and consequently of heart attacks and strokes.[340] For obscure reasons, aspirin also seems to be protective against colonic cancer, too, probably in its role as an anti-inflammatory agent or COX-2 inhibitor. Aspirin, together with statins, are examples of medications being recommended to healthy people on the grounds that they will reduce their risk of heart disease or cancer.

A conclusion from John Vane's work was that a drug that antagonises COX-2 only and not COX-1 might be a better painkiller than aspirin and free of side effects. One recently discovered drug (Celecoxib®) seems to fit this description, but whether it is more effective than aspirin is not yet entirely certain – it will certainly not be cheaper! Paracetamol is chemically related to aspirin but seems to be less anti-inflammatory and more effective against pain and fever, perhaps acting on a third type of COX found only in the brain.[340]

Anaphylactic shock and severe allergy are the alarming face of the inflammatory response, conditions in which blood pressure suddenly plummets, threatening a fatal circulatory collapse because of the abrupt release of histamine. The search for drugs to prevent this reaction has preoccupied pharmacologists since histamine was discovered almost a century ago. Antihistamine drugs have evolved through many generations, yielding new products suitable for controlling severe reactions and others appropriate for milder allergy and hay fever and even less obviously connected symptoms such as travel sickness. The discovery that histamine controls acid secretion by the stomach triggered a race between the pharmaceutical companies, which lasted more than a decade, to find the best compound to control gastritis and protect against stomach ulcers. Zantac® and Tagamet® emerged victorious at first, but a rival that controlled acid secretion directly was soon in competition (Losec®). The revenue from such products is massive; five percent of all drugs sold in the world in 1999 were for gastritis and related problems.[259]

Meanwhile, two Australian physicians identified the real cause of stomach ulcers as the spiral-shaped bacterium Helicobacter. This organism inhabits the stomach lining of perhaps one out of every two people, where it overcomes the hostility of the environment by neutralising stomach acidity locally with ammonia. Infections cease to be harmless if the bacterium invades the stomach wall, initiating chronic gastritis and ulcers and creating the conditions for stomach cancer to develop. Eradicating the organism with antibiotics makes stomach ulcers disappear, so that "wonder drugs" are no longer needed.

CARDIOVASCULAR DRUGS: As the threat of infectious disease receded in the early twentieth century, diseases of the cardiovascular system emerged as the most common cause of death. Narrowing of the arteries, caused by accumulation of atherosclerotic plaque, creates the crucial conditions that make an interruption of blood supply to the heart muscle or the brain possible, with a consequent risk of coronary heart disease or stroke. Raised blood pressure is a key physiological warning of this development; it may have many causes arising from lifestyle. For many years, the pharmaceutical industry has been trying to find better drugs to reduce blood pressure, prevent accumulation of atherosclerotic plaque, and manage degenerating cardiovascular systems. The vast range of drugs now available that partially solve these problems has contributed notably to the extension of life expectancy of the last fifty years.

The most effective route to reducing blood pressure, recognised many years ago, is to promote urine production and the excretion of sodium ions using diuretics. The potential of drugs derived from sulphanilamide for this process was known in the 1930s, but three decades of research and many twists of the chemical kaleidoscope have generated much better drugs. In the 1960s, James Black formulated a different approach. His idea was to look for drugs to prevent the rapid heartbeat provoked by a surge of adrenaline, as in the well-known fight-or-flight reaction. The drugs he discovered, called beta-blockers (because they affected beta-receptors), are now a mainstay of clinical medicine because of their capacity to reduce blood pressure and the amount of work done by the heart. They are especially useful in calming performers of all kinds who are significantly incapacitated by feelings of nervous anticipation. Another class of drug promotes relaxation of the muscles lining blood vessels by blocking the channels through which calcium moves and can be used to manage heart conditions.[259]

The tendency of arteries to constrict is controlled by angiotensin – a short chain of amino acids called a peptide – which is cleaved from a bigger protein by the angiotensin-converting enzyme or ACE. This peptide also stops the kidneys from making urine and creates the sensation of thirst via the brain. The possible benefits of controlling production of angiotensin using drugs came sharply into focus when a snake venom was discovered that dramatically reduced the blood pressure of animals by antagonising angiotensin. Scores of drugs, now usually called ACE inhibitors, are currently used to protect the failing heart by blocking this process and reducing blood pressure.[259]

Warfarin is well known to people of a certain age as a drug for reducing the chance of blood clotting and consequently the chance of suffering strokes. It originated, in the 1920s, in a natural product found in spoiled clover that caused severe haemorrhage in cattle through interference with vitamin K and the synthesis of blood-clotting factors. After undergoing the usual pharmaceutical evolution, a safe and highly effective drug emerged.[328]

The single greatest risk factor for chronic ill health in late middle age is atherosclerosis. An excessive level of blood cholesterol, the majority of which originates in the liver, is an important factor contributing to atherosclerosis and heart disease. A class of chemical called statins can reduce the level of cholesterol in blood by as much as forty percent by slowing the synthesis of one of its precursors. The first member of the family originated from a mould, but synthetic versions are now available, with proven value in reducing the risk of coronary heart disease. For the first time, large numbers of essentially healthy people are taking a drug as a precautionary measure – perhaps for the rest of their lives – to reduce the risk of heart disease.

Quacks, Carelessness, and Misadventure

Paradoxically, the Age of Enlightenment was also the golden age of quackery. Just as science and rational thought were starting to improve the human condition, the ignorant and desperate were enchanted by extraordinary concoctions sold by charlatans on a mammoth scale.[112] By the nineteenth century, in Britain and probably throughout the developed world, self-medication was institutionalised, and the purveyors of bogus medicines, such as James Morison (whom we have already met), were becoming notably professional. Soon the likes of Thomas Holloway, Thomas Beecham, and Jesse Boot joined the fun, amassing huge fortunes from the sale of similar nostrums but eventually leaving important endowments for the public good. Their success was based on "patent medicines," a concept introduced in the British Food and Drugs Act of 1875, ostensibly to improve the quality of medicines by disclosing their ingredients in a trade journal in return for duty-free tax status. We have little idea of what the public thought of these recipes, which contained unlikely-sounding ingredients such as antimony salts, capsicum, cannabis, methylated spirits, turpentine, soap, acetanilide, and opium.[29] Later in the century, medications laced with narcotics or alcohol were common in most countries, but law eventually controlled the practice, so that the more outrageous patent medicines disappeared as legal redress became a tangible possibility. Twentieth-century drug regulation concentrated on sealing the loopholes that allowed the unscrupulous to sell inadequately tested or downright fraudulent drugs.

In the United States, lack of regulation came to a head in the 1930s after the "Elixir of Suphanilamide" scandal, in which seventy-six people died from severe kidney damage caused by a solvent, diethylene glycol, used in preparing the drug. At that time there was no legal framework governing the way manufacturers prepared for sale a drug that was not governed by a patent. Congress responded to the disaster with the Food and Drug Act of 1936 to establish standards for proprietary medicines and prevent any repetitions. The provisions were sufficiently robust to spare the United States the miseries of the thalidomide tragedy that unfolded twenty years later in Europe.[328]

Chemie Grunenthal, a German drug manufacturer that understandably no longer exists, promoted thalidomide as a highly effective and completely safe treatment for morning sickness in the 1950s. The drug was distributed worldwide by major pharmaceutical companies, except in the United States, where one redoubtable figure in the FDA withheld a licence because of the flimsiness of the safety data. The public was alerted to the disaster by a letter to *The Lancet* from the Australian obstetrician William McBride, who reported serious defects of limb development associated with taking the drug. By the time the drug was withdrawn, about 12,000 babies had been born with limb defects. A single tablet taken between the sixteenth and twenty-sixth day of the pregnancy, when the limb buds are starting to form, could cause the damage. In Britain, the distributors of the drug, Distillers Ltd., refused to accept any responsibility, although they eventually compensated the victims when the European Court threatened to make a ruling. Later, when the lack of realistic trials on pregnant mammals became apparent, legislation was introduced to make exhaustive animal tests for teratogenicity obligatory for any medicine to be used by pregnant women. A further advance came in 1962, when Congress compelled drug manufacturers to demonstrate *the efficacy and safety* of their products in a new Food, Drug and Cosmetic Act.

Thalidomide was not the only instance of an unsafe drug being used on pregnant women during the 1950s. For more than two decades, a synthetic oestrogen (diethylstilbestrol or DES) was prescribed to prevent miscarriage – until 1970, when the FDA withdrew its approval of the treatment. An observant oncologist found a common factor that linked seven cases of an extremely rare vaginal cancer in young women: their mothers had taken DES in the third month of pregnancy. A more detailed investigation showed the situation to be even blacker; developmental damage to the reproductive organs was found to be widespread amongst the daughters of the women who had received DES. The typical outcome was ectopic pregnancy or miscarriage, the very conditions the medication was supposed to prevent. The drug should undoubtedly have been abandoned in the early 1950s, as clinical trials indicated that DES was ineffective in preventing miscarriage. Moreover, there were reports from 1938 indicating DES could cause cancer in animals. Why doctors continued to prescribe DES, when no evidence for any benefit existed, is a mystery. As the drug was cheap, the companies were not greatly concerned about the falsity of their claims, and physicians felt compelled to prescribe the drug because of pressure from their patients.[171]

Thalidomide and DES were scandals of the most serious kind, but drug developers often face difficult situations in which huge immediate benefits must be weighed against remote risks of serious harm to a few individuals. Because of penicillin's astonishing capacity to save lives, the drug was widely used before there was any indication of the allergic reactions that afflict one in ten people. Chloramphenicol was in use for several years before a strange lethal anaemia was found in one in every 100,000 people treated. The risk was small, but the FDA ruled that the risk must be clearly stated on the label. Lawsuits followed, and the drug was superceded by other antibiotics.

Whenever drugs are introduced clinically on a large scale for the first time, there is always a risk of small numbers of serious immediate adverse effects or longer-term risks not observed in a small short-term trial. The contraceptive pill was widely prescribed after successful, relatively small-scale trials. When a million subjects were included in the statistics, an elevated risk of blood clots was evident. The risk was no greater than the risk attributed to childbirth, except for smokers over the age of thirty. Nonetheless, the dose of oestrogen was subsequently reduced, with no loss of efficacy.

Unexpected adverse effects from a widely prescribed drug surfaced again recently when one statin-type drug (used for reducing cholesterol levels) was alleged to damage seriously the health of some patients. If the thousands of complaints now being considered by the courts are successful, Bayer, the venerable German pharmaceutical company, may collapse.[341]

Governments are expected to regulate the evaluation of drugs developed by pharmaceutical companies, but the public has an oddly schizophrenic attitude toward so-called alternative medicine. Dispassionate observers are at a loss to understand the success of alternative remedies, because clinical trials rarely indicate any efficacy by the criteria of mainstream medicine. Nonetheless, about a third of adults in the industrial world use the products of this largely unregulated industry. One can suppose either the old placebo effect is working its magic, or that users are gambling on unproven claims. The problem has a special poignancy when individuals with terminal diseases place unrealistic hopes in unscrupulously marketed alternative medicines. Notable bogus remedies of the last fifty years include laetrile (the cyanide-containing glycoside found in almond kernels) and kerboisin.

Challenges, Risks, and Rewards – a Summary

The idea that ingested chemicals might have beneficial consequences was one of humanity's more daring experiments, one that has evolved into an abiding feature of modern life, central to both human welfare and to commerce. A tradition that for millennia was based on hard-to-substantiate claims can now deliver products with precisely defined characteristics and effects that are both safe and efficacious. Paradoxically, while the humanitarian benefits of medicinal chemistry are recognised as one of the triumphs of the age, the first duty of the pharmaceutical companies is to their shareholders. This means the industry focuses on that part of humanity that can pay for the biological benefits it confers. This generally excludes the diseases of people in the Third World with low spending power, or indeed intractable degenerative diseases that affect everyone. Drugs are perceived as expensive, which the industry claims is understandable because of the expense of development and the wasteful late-stage failures that are commonplace. However, critics note that promotional costs often exceed the costs of research; competition fails to bring down prices; and company mergers seem designed to prevent competition.[342] Other models of drug discovery do exist, such as a Harvard-based not-for-profit biotech company that hopes to get the support of philanthropists and patient groups to mobilise academic research to develop drugs for neglected areas.[343]

Powerful generic methods of drug discovery, methods that contrast strongly with the famous serendipity that characterised the early days, are now available. Tens of thousands of compounds are being generated every year and tested against thousands of novel target proteins whose existence has been established from the human genome sequence. Many more successes, with clear uses against organic disease, are expected from the well-travelled routes of modern medicinal chemistry. Scientists already anticipate that important proteins will be identified that play a part in more subtle human attributes such as personality, moods, libido, perceptions, memory, maintenance of body mass, and ageing. With their talk of "lifestyle" or "designer" drugs, the media has already put the idea into the public's mind that special "pills" will give a new twist to old problems. Examples of such drugs already exist – hormone replacement therapy, Viagra, Prozac, Ritalin – that demonstrate in practice our dangerous ambivalence toward drugs. They can be used to solve pressing clinical problems but also win credibility through a subliminal feeling of fashionability. The pharmaceutical industry welcomes our interest in novelties of this kind, and the economic imperative that drives such developments is likely to pose more serious ethical problems than those raised by genetics.

GENERAL SOURCES: 165, 259, 328.

12 Protein Medicines from Gene Technology

Diabetes was the first example of a human disease that could be attributed to a deficiency of a specific protein (insulin) and that could be corrected by an equivalent derived from animals. That startling success was the inspiration for replacement therapies for other diseases. With the advent of gene technology in the 1970s, a generic method for making human proteins on an industrial scale became available. This provoked a gold rush to start the process of making a host of potentially useful proteins on a commercial scale. The discovery came at a critical moment, when infectious agents had seriously compromised the safety of clinically used blood proteins and hormones of human origin. Using gene technology, blood-clotting factor VIII and growth hormone could be made conclusively free of viruses. Gene technology has now revealed many proteins that have potentially important roles to play in controlling pathological symptoms of hundreds of clinical disorders.

Genetically Engineered Insulin

Insulin has a special place in the annals of biotechnology. In the 1920s, insulin transformed the short and sickly lives of diabetics and was arguably one of the most significant achievements in any field in those dismal times. The astonishingly bold idea was that a human physiological function lost in diabetics – the capacity to regulate the level of glucose in blood – could be restored by an injected protein obtained from cattle. No concept of the nature of the protein existed in 1922, and it was 1953 before a precise chemical structure was revealed. Nonetheless, insulin was made on a large scale and quickly entered clinical practice, bringing an immediate improvement in the quality of life of diabetics. However, even with the most careful management, insulin and glucose levels could not be adjusted to match exactly the pattern found in a nondiabetic after a meal. In particular, unregulated surges in blood sugar level cause many of the complications of diabetes, such as damage to the retina, that typically undermine the health of diabetics.

Insulin was made commercially from cows and pigs for the first time in 1922 by a long-established medical supplies company called Eli Lilly, under license from the University of Toronto, where Banting and Best's famous work was done. We know now that beef and pork insulin differ from human insulin by a single amino acid at the end of one of the two peptide chains, chemical differences that are responsible for minor immunological reactions to animal insulin. That problem was finally solved during the 1970s, when the Danish pharmaceutical company Novo-Nordisk found a chemical method of modifying pig insulin to make it identical to human insulin.

The idea of using gene technology for commercial purposes dates only from 1976, when the twenty-eight-year-old Robert Swanson made his ground-breaking proposal. After just four years in venture capitalism with an American bank, Swanson persuaded the molecular biologist Herbert Boyer to join him in an enterprise to make pharmaceutically interesting proteins using the techniques of gene technology.[344] Boyer was already well known for important contributions to gene technology. Their company, to be called Genentech, founded on an initial stake of $1,000 each, would become a $100-billion-dollar business before it was sold to the Swiss chemical giant Roche in 1989.

Routes for making proteins in bacteria were taking shape in the minds of academic biochemists in the mid-1970s. The idea was that genes could be inserted into a naturally occurring self-replicating plasmid that would make many copies in bacteria (one of the meanings of the word "cloning" – see p. 80). The protein in question would be made providing it was spliced next to a sequence that could control protein synthesis (a *promoter* – a DNA sequence that organises mRNA synthesis). Swanson saw the potential and initiated Genentech's first efforts by financing the efforts of Boyer's lab and another university lab in California to make insulin in bacteria. Very quickly, a race developed between a number of American establishments to be the first to make a protein from cloned DNA in bacteria; Genentech, with the crucial help of Eli Lilly, won easily. Genetically engineered insulin was approved by the FDA in 1982 and immediately put on sale.[345]

Once genetically engineered insulin was established as an exact physiological equivalent of natural human insulin, some long-standing clinical problems could be overcome. Variants of insulin could be made in which certain amino acids were changed by redesigning the template – an unattainable goal with animal insulin. More effective control of blood sugar levels could be obtained using genetically engineered insulin modified to be either fast-acting or able to make a long-lasting response.[346]

In the regulatory climate of the 1970s, Genentech's strategy proved crucial to the outcome of the race. A group of prominent pioneers of molecular biology, led by the Nobel laureate Paul Berg, wrote a letter to *Science* magazine in 1974 drawing attention to potential hazards that might arise from cloning of DNA sequences. Their argument was that a novel plasmid – a self-replicating piece of DNA – loaded with a foreign gene, if free in nature, could possibly infect other cells and spread like a weed. The worst scenario they envisaged was a plasmid containing a cancer-causing gene or an entire virus evolving into a disease-causing organism in the human gut. The signatories suggested there should be a moratorium on all genetic engineering work until a regulatory framework was in place. An international conference was convened at Asilomar in California in 1975 to discuss the containment problems arising from this hypothetical problem. The key proposal was that DNA sequences from mammals or viruses that were potential hazards had to be contained in such a way that the material could not escape. The cloning of DNA from cancer-causing viruses and dangerous toxin-producing organisms was banned completely until the risks involved could be assessed. Such a public outburst of introspection inevitably surprised and alarmed the public.[347]

Boyer and colleagues quickly identified a way to move forward rapidly and legitimately, using containment measures that were convincingly watertight. The idea was to chemically synthesise DNA templates corresponding to the two chains of insulin and then clone them in two separate bacterial cultures. This would have no ecological impact, because active insulin could not be made even if bacteria escaped into the environment. The proteins recovered from bacteria would then be processed in the laboratory by an entirely chemical procedure in order to establish the correct interchain links and reconstitute physiologically active insulin.

The decision to seek the support of Eli Lilly, whose pharmaceutical experience with insulin was unrivalled, made the final stage of preparation and quality control completely painless. The product was subjected to searching physiological evaluations that showed it to be indistinguishable from pig insulin. The success of the project rested on a number of very fortunate practical decisions that we need not consider here. Each hurdle was cleared comfortably, and the product was manufactured on a huge scale – a resounding triumph that illustrated unequivocally the robustness of the molecular biology that had started with Watson and Crick's model of DNA twenty-five years earlier. Lilly never sold the new insulin at a lower price than pig insulin, and by 1985 the revenue from the new form was $100 million a year, with a steady stream of royalties for Genentech.

It should be said that the reputation of recombinant insulin has not always been immaculate. A small group of diabetics has complained that when they switched from the tried and tested animal insulin to the genetically engineered form, they suffered hypoglycaemic comas (too little blood glucose) and other severe adverse effects. Most clinical studies indicate no excessive incidence or increased severity of hypoglycaemia in users of the new insulin, but the situation has generated considerable publicity for those who ardently believe that genetically engineered products are inherently untrustworthy.

Swanson's conception of business management at Genentech was an unorthodox blend of flexibility and drive, without the hierarchical features of the major pharmaceutical companies.[344,345,348] Skilled scientists abandoned their careers in academia for the company, seeing there the best of two worlds: the exceptionally deep pockets of Genentech and an opportunity to publish their scientific findings in a personally satisfying fashion. Crucially, the early recruits were often motivated by a certain animosity toward their erstwhile academic superiors who had denied them credit for their contributions to gene technology.

The critical moment for the commercial future of gene technology came when Boyer and Cohen filed a patent for making a "cellular factory" based on their gene-splicing technology. The idea of patenting the process had not been in their minds earlier; they believed their ideas were mere extensions of other work in the public domain. This perspective changed under the expert guidance of a patent lawyer, who could see that the process was in principle applicable to every kind of cell. The patent was assigned to Stanford University, and major licensees quickly emerged – Amgen, Eli Lilly, Schering Plough, and many others. Ultimately, the patent realised at least $300 million for the university and established a virtuous circle that justified the investment of two decades by the NIH.

In an atmosphere of intense intellectual excitement, the Genentech staff quickly compiled a register of all the proteins for which there was the remotest clinical demand that could be met by making a genetically engineered protein. The roll call included human growth hormone, blood factor VIII, interferons, proteins for dispersing blood clots, monoclonal antibodies against many kinds of targets, and a host of newly discovered and very important cell growth factors. The objective was to clone the genes and prepare the proteins to a pharmaceutically appropriate standard.

In 1980, Genentech was the first gene technology company to sell shares to the public. The stock market debut became a legend amongst equity traders; the shares, initially priced at $35, rose to $89 within an hour, and the company remained a stock market darling for some years. Genentech's early strategic objectives were obvious to many other well-educated biochemists; inevitably, intense competition began. Later, stock market sentiment towards Genentech dwindled for several reasons. The races to clone α-interferon and erythropoietin were lost to vigorous newcomers to the field. Then, when the gigantic earnings anticipated from a novel blood-clot-dispersing product failed to materialise, mammon took its traditionally cruel revenge with a massive sell-off of Genentech's shares. This forced the company into the hands of the Swiss pharmaceutical giant Roche that for fifteen years had sat on the sidelines, deciding whether gene technology had a future. Genentech's original science-led culture still survives. The company is still inventing truly novel products to treat life-threatening conditions for which no cure exists and in 2004 is evidently ascendant again, with many new protein-based drugs in the pipeline.

A sordid footnote to the meteoric career of Genentech, and testimony to the *zeitgeist* of the gold rush, emerged from a Californian court in 1999. The court ruled that the cloned DNA used to make growth hormone had been wrongfully removed from the University of California at San Francisco twenty years earlier. Consequently, the company had to pay $200 million to the university for infringing its patent and was lucky not to have incurred draconian damages.[349]

Gene Technology Becomes an Industry

The emergence of gene technology as an industry dedicated to solving important health-related questions by completely novel routes was the start of an adventure as daring and as far-reaching as any humans have ever undertaken. In the race to establish the industry, new companies soon contested Genentech's commanding lead. In the early days, venture capitalists button-holed prominent academics, who, with a few like-minded friends, drew up a business plan and recruited a core of young scientists to carry it through. Later, the impetus to start new companies came from academics who set out to find venture capitalists. The major pharmaceutical houses took little overt interest for some years, only later emerging as "corporate backers" of new companies. Non-American capitalists generally lacked the necessary nerve or understanding, even when a scientific establishment existed in their own countries. British government money was needed to underwrite Celltech, an organisation set up belatedly in 1980 to exploit monoclonal antibodies; it eventually became a public company and was bought by a Belgian company in 2004.

The development of entirely novel medicines using gene technology, in spite of its theoretical attractions, was likely to be an expensive and pro-tracted process. There was no certainty that jockeys would remain mounted at the final fence in the clinical trials. The crucial step was the enthusiastic welcome given to initial stock market offerings by the American public. As in stock market booms of former times, the crucial chemistry that launched the rocket lay in a mysterious formula composed of intellectual excitement, faith in early successes, and the public's nose for a fast buck. The famous biotech offerings of the eighties were intended to raise capital for the long term, but the high turnover of shares in the immediate aftermath indi-cates unequivocally how financial speculation was the chief element in the process. The share prices of ventures that started in such circumstances al-ways reached a point where a collapse in confidence was inevitable and left the companies managing on whatever money was left in the bank, until the next cash-raising exercise. The fact is: gene technology projects usually take longer to complete than a major engineering project, and they are subject to late-stage setbacks in clinical trials that can be very damaging to business. In such circumstances, stock market sentiment is inevitably a rollercoaster ride, an evolutionary process that only the fittest survive. Members of the financial community who initially encourage the new companies to exag-gerate their prospects quickly turn against the fledglings when setbacks occur.

The early biotech companies were favoured by one other circumstance, the U.S. Orphan Drug Act. This was introduced in 1983 to encourage private industry to develop treatments of any kind for rare diseases; it provided companies a seven-year period of exclusivity if they could sponsor drugs through the FDA approval process. In the case of protein medicines, most of the early products qualified for this kind of support, and, apparently to everybody's surprise, drugs such as erythropoietin, growth hormone, and others became extremely profitable under this regime.[350]

By the mid-1990s, biotechnology was well established, and some products were making substantial earnings. Employment in the American industry was already approaching a million people. By 2001, 389 stock market–quoted biotechnology companies existed worldwide, with a total value of $24.9 billion. Three-quarters of these were in the United States, and many more abroad were financed by American money.[351] In addition, there are thousands of start-up companies, engaged in extremely diverse activities, ranging from efforts to find new medicines to development of novel technologies to facilitate biotechnology. Characteristically, they are small-scale enterprises originating in academic work that are destined to be subsumed into bigger organisations. In this no man's land between academia and industry, commercially useful research into the key molecules that regulate the behaviour of cells is developing quickly, accelerated now by the availability of the human genome sequence. Life science patents (mostly involving gene technology) held by North American organisations are now generating an income of more than a billion dollars a year from licences to business, of which over half come from just ten universities. The income of European universities from this source is just a tiny fraction of that figure.[352]

LIVING FACTORIES: The idea at the heart of gene technology is the use of living organisms to make proteins on an industrial scale. Bacteria were the first choice, because they were manifestly less trouble than higher organisms, faster growing, and more amenable genetically. However, after the initial success of insulin, scientists were disappointed to discover that many human proteins were not easily made in bacteria. The technical solutions to these problems need not concern us here, but if two unlike protein chains were involved, or if carbohydrate additions to proteins were necessary for the function or stability of a protein, serious difficulties lay ahead. Mammalian or even human cells are logical possibilities because they can make the carbohydrate parts, but they are substantially more expensive to grow than bacteria. Yeast was an early favourite because of its versatile genetics and its long association with human food and drink. Plants have their attractions, too, because products targeted to a plant organ can be harvested effectively and are likely to be exceedingly cheap.

Mammals that make valuable proteins in their mammary glands, or chickens that make proteins in their eggs, are a more radical idea that is very appealing for very high-value pharmaceuticals. For this to be possible, the animals must be "transgenic" – that is, the DNA sequence that encodes the protein of interest must be inserted into the animal's genome. In addition, this sequence must be under the control of DNA sequences (i.e., the promoter) that normally drive formation of proteins. This could be specific for milk proteins from cows, goats, or sheep, and in chicken eggs specific for egg white protein (albumen). Once the goats or chickens are bred, the proteins would be recovered from milk or egg whites.[353] Pioneering steps were taken in the late 1980s to obtain sheep that produced a human protein called α1 anti-trypsin that has potential as a cure for cystic fibrosis and congenital emphysema. One of these sheep produced milk in which 50 percent of the protein content was the product. A variety of other proteins are made in high yield by the same method. The technology is still in its infancy, and the long gestation period and the rather unpredictable expression levels of different transgenic animals remain a problem. The industry believes that this problem will be solved using the more radical process of nuclear transfer to create identical animals such as Dolly the sheep.

Several American companies are breeding roosters with human genes under the control of the promoter for the egg white (albumen) gene inserted into their chromosomes. The idea is for the rooster to inseminate hens so that the fertilised egg makes the human protein in the white of the eggs. Only a few transgenic roosters would be required to keep a flock of chickens laying golden eggs.

Making Novel Protein Medicines

Gene technology is constantly instrumental in writing new chapters in every branch of biology because it provides the means of characterising any interesting gene and for manufacturing protein products in cellular factories for biomedical use. This section is about a few of the subjects that gene technology has illuminated and the important proteins that can be made.[354] Protein medicines can be used to add something to the human body, unlike chemical medicines, which generally interfere with a process. Leaders of the industry like to remind their shareholders that their products (i.e., protein medicines) have an intrinsic exclusivity that can be established by legally comprehensive patents, which cover all substantially homologous versions of the prototype. By contrast, patents on small molecules can rarely be written so comprehensively. Some milestones in the development of the industry are shown in Figure 12.1.

1982	Insulin
1985	Human growth hormone
1986	Interferon α-2a for hairy cell leukaemia Muromonab mAb against CD3 to prevent kidney transplant rejection Hepatitis B vaccine
1987	Plasminogen activator for use in blood clot dispersion
1989	Erythropoietin promotes red blood cell formation Interferon α-1b management of certain cancers
1990	Adenosine deaminase for replacement therapy for deficiency disease Interferon γ-1b management of infections associated with granulomatous disease
1991	G-CSF (Neupogen) stimulates white blood cell formation after chemotherapy
1992	Interleukin-2 helps prevent rejection of kidney transplants Factor VIII corrects blood clotting disorder
1993	Interferon β-1b treatment of relapsing multiple sclerosis Pulmozyme clears viscous material from lungs of cystic fibrosis patients
1994	Asparaginase used to reduce levels of asparagine in blood of leukaemia patients. (Leukaemia cells have a high requirement of asparagine.) Cerezyme glucocerebrosidase for Gaucher's disease Abciximab prevention of blood clotting mAb against platelet receptors
1997	Factor IX corrects blood clotting disorder Rituximab mAb against pre-B cell protein (CD20). Non-Hodgkin's lymphoma. Platelet derived growth factor (PDGF) used for treating diabetic ulcers that will not heal. Interleukin 11 stimulates platelet formation to make good damage caused by chemotherapy
1998	Tumour necrosis factor receptor treatment of rheumatic arthritis Herceptin mAb against metastatic breast cancer. HER-2 receptor is the oestrogen receptor. Glucagen recombinant glucagon. Treatment of hypoglycaemia.
1999	Factor VII corrects blood clotting disorder
2000	Mylotarg humanised mAb for treating acute myeloid leukaemia. Recognises CD33. Follicle stimulating hormone male and female infertility and IVF Human chorionic gonadotrophin IVF Luteinising hormone female infertility
2001	Alemtuzumab mAb against CD52. Use in chronic lymphocytic leukaemia.

Figure 12.1. Milestones in establishing gene technology medicines by date of approval for clinical use.

HUMANISED MONOCLONAL ANTIBODIES (mAbs): The remarkable ability of antibodies to recognise specific chemical entities has seemed to offer important applications in medicine for many years. The prospect of using antibodies as pharmaceuticals advanced dramatically in 1975 with the discovery of a novel, artificially created cell type known as a hybridoma.[355] These cells are literally hybrids, combining the capacity to make a single antibody – hence the term "monoclonal antibody" or mAb – with the capacity to propagate freely like a cancer cell. Georg Kohler and Cesare Milstein – a Swiss and an Argentinian, respectively – working in Cambridge, UK, discovered that these cells could be made from B-lymphocytes from the spleen of an immunised mouse and an immortal cancer cell known as a myeloma. Scientists use the verb *fuse* as a convenient word to convey any process of integration – in this case, of two cells. Hybrid cell types may exist rarely in the mammalian body, but they have been made in the laboratory since the 1960s, when the phenomenon found an important use in studying the genetics of human cells. The two cell types are cultured together and made to fuse using the Sendai virus – the cause of a flulike disease of mice. The virus attaches to two unlike cells and pulls them into such intimate contact that their outer membranes fuse. Kohler and Milstein found that some hybrid cells integrated into an unusual kind of joint cell that had no tendency to separate into its component parts. It continued to divide and grow indefinitely, making the antibody expected from the immunisation procedure. The resulting cell is of immense practical value, because it makes only one type of antibody in unlimited quantities. The immense potential of mAbs as therapeutic agents was soon recognized in thousands of laboratories throughout the world. The idea of using antibodies to block physiological processes was common practice in research laboratories. Now it was possible to devise a mAb that could suppress inflammation, blood clotting, or organ rejection; alternatively, it could kill malignant or virus-infected cells.

The variety of potential uses of mAbs was never in doubt, but a great challenge remained. How would mAbs fare when injected into a patient? Early efforts were quite disappointing; the antibody was destroyed by the liver, and it was ten years before clinically useful versions emerged. The primary problem posed by mouse mAbs was their tendency to provoke a strong immunological response in the human body. Before the advent of gene technology, this problem seemed almost insoluble. Using gene technology, scientists could contemplate inserting just the recognition sequence from a mouse mAb into predominantly human antibody sequence so that the new entity provoked only a minimal immunological reaction in a human subject. Today, most mAbs intended for clinical applications are what antibody engineers call "fully humanised."

The first FDA-approved mAb was introduced in 1994. This product was a partially humanised mAb known as ReoPro, intended for use in vascular surgery and designed to suppress platelet aggregation and consequently blood clotting. As the patient required only one or two treatments, there was little immune response; consequently, complete humanisation was unimportant. This pioneering product was evidently safe and effective, because by 2000 sales registered $418 million. In the same year, one-quarter of all gene technology products under development were mAbs, and thirty had already been approved for clinical evaluation.

Diseases such as rheumatoid arthritis and Crohn's disease are inflammatory diseases caused by an abnormal attack by T-lymphocytes on joints and gut tissues, respectively. Two proteins, tumour necrosis factor (TNF) and interleukin-1, are out of control and are responsible for the pathology of these diseases. A mAb that could inactivate or neutralise TNF looked like a promising treatment for both conditions if it would calm down the pathology and make the patient more comfortable. The same pathology is involved in the fierce inflammatory disease that occasionally accompanies infections caused by *Staphylococcus* and *Streptococcus*, including toxic shock syndrome.[356] Treatments of this kind are still experimental, and physicians have to be alert to adverse effects, watching particularly for evidence of infections or activation of tuberculosis because molecules such as TNF are part of the natural immunological defences.

Designers of mAbs want to improve the immunosuppression necessary for organ transplantation. Interleukin-2 and certain T-lymphocyte proteins involved in organ rejection are targets for mAbs already in use or in clinical trials and are intrinsically more attractive than more toxic drugs such as cyclosporin.

Malignant cells are another set of ambitious targets for mAb designers. Genentech secured a notable success in cancer therapy with a genetically engineered mAb that recognises the extracellular part of a growth factor receptor known as HER-2 that is excessively abundant in certain types of breast cancer. Five years spent developing the antibody and another six taking it through clinical trials culminated in a treatment regime that grants substantial remission to patients in which the cancer has spread to other organs.

Designing mAbs to recognise cell surface proteins of other cancers has generated a number of promising products effective against leukaemia and non-Hodgkin's lymphoma, products that are now in use or in clinical trials. The lethality of these mAbs for malignant cells can also be improved by attaching radioactive sources or powerful poisons. They are used, usually in conjunction with chemotherapy, to improve the chances of a cure. Although experimental, these treatments are already extending and improving the quality of life of patients. Now that the chief hurdles in preparing effective mAbs have been cleared, their use in clinical practice is set to grow rapidly, although they will remain expensive for some time.

UNDERSTANDING AND EXPLOITING MESSENGER MOLECULES: The idea of chemical messengers emanating from one organ and controlling the function of another tissue at a distance dates from the discovery of insulin. That discovery alerted physician-scientists to the possibility of other medical conditions that may represent deficiencies of particular messengers. There exists a vast class of protein factors, in addition to the well-known endocrine hormones, that govern the growth and fate of cells at every stage of development of the mammalian body. Indeed, the human genome encodes at least 3,000 secreted proteins that are probably important messengers. Some stimulate proliferation of particular target cells (growth factors); others, called cytokines (from the Greek for "cell" and "motion"), change the character of cells.

GROWTH HORMONE: Growth hormone, a protein hormone made by the pituitary gland, is necessary for children to grow to a normal size. It was the second product to emerge from the gene technology industry – also from Genentech – and one that poignantly illustrates both the value of gene technology and a less attractive side of commercial biochemistry. In the late 1950s, enterprising doctors realised that the condition of a small number of children with stunted growth could be corrected by injections of growth hormone. The only available source was the pituitary glands of cadavers, from which the hormone was prepared at great expense; the demand frequently outstripped supply. Inevitably, a genetically engineered replacement was one of the first objectives of Genentech; the new drug reached clinical trials by 1981. Tragically, as this project was nearing completion three people who had been treated with cadaver growth hormone contracted Creutzfeldt-Jakob disease. Today, we know this was caused by prion contamination of the stock of growth hormone, perhaps from just one or a few individuals with the disease. The FDA, in 1985, immediately banned use of the hormone from this source, but fortunately they were quickly able to approve the genetically engineered form.[317] The new product, although extremely expensive, could be supplied in sufficient quantity to ensure that children who needed treatment could approach normal height. By judicious management, these children could also undergo an appropriate pubertal growth spurt – and was a remarkable triumph for recombinant proteins. A group of adult patients who no longer had a functioning pituitary gland after surgical removal of a brain tumour also benefited enormously from growth hormone treatment. Growth hormone was soon attracting attention because of its apparent ability to rejuvenate the elderly. Athletes looking for an unfair advantage are also allegedly finding growth hormone useful. At least one untimely death from heart disease – that of an Olympic sprint champion at the age of thirty-eight – is attributed to growth hormone abuse.

ERYTHROPOIETIN: The idea that the kidney makes a substance that promotes red blood cell formation was proposed early in the last century by medical physiologists who had noticed that some patients suffering from kidney failure became severely anaemic. In the three decades after 1950, research gave strong support to this theory, culminating in the purification of erythropoietin from the urine of anaemic patients in sufficient amounts to demonstrate its clinical value. The clinical demand for erythropoietin was not likely to be met, so a genetically engineered version was an inevitable goal for gene technology; a new drug was first made in 1984 by the biotech company Amgen. The carbohydrate parts of the natural molecule, which were necessary to make a stable product, could be made only in mammalian cells. The drug's approval by the FDA in 1989 and introduction into medical practice in record time was a technical and commercial triumph for the new industry. It was easily administered, completely safe, universally effective, and is now a routine treatment for the anaemia associated with kidney failure, cancer treatments, HIV infection, and rheumatoid arthritis.[357]

Illegitimate uses of genetically engineered erythropoietin quickly emerged. Because erythropoietin stimulates formation of blood cells under conditions of relative scarcity of oxygen, it was logical to expect it would improve stamina. Some competitors in endurance sports, such as long-distance cycling, have seen the opportunity and use erythropoietin to enhance performance. Whatever success it brings, the cost to health is high, because the increased blood viscosity brings an increased risk of blood clots, strokes, and heart attacks.

INTERFERON, INTERLEUKINS, AND OTHER CYTOKINES: Since its discovery by Alick Isaacs and Jean Lindemann in 1957, interferon has often been in the public eye as its promise to cure viral diseases, cancer, and multiple sclerosis has waxed and waned. It was discovered as a biological activity that conferred resistance to viral infection and that seemed to promise a great new antiviral therapy. Many years of arduous work and the advent of gene technology were necessary before the chemical and biological character of the interferons was understood. Interferon was the first of a class of secreted messenger molecules (cytokines) that control subtle and important functions in immunity. Type-1 interferons (α, produced by white blood cells, and β, produced by widely distributed fibroblasts) are anti-inflammatory; type-2 interferons (γ), made by white blood cells, are pro-inflammatory.[358]

The image of interferon as a wonder drug was established in the 1970s, when reports of its ability to suppress the growth of cancers in some laboratory animals emerged from Sweden and the United States. The first FDA-approved clinical use of interferon-α made by gene technology, in 1986, was for treatment of an incurable and very rare condition known as hairy cell leukaemia, which we know is caused by a retrovirus called HTLV-1. This disease prevents formation of red blood cells and causes an accumulation of the "hairy cells" that give the disease its curious name. Low doses of injected interferon overcame this apparent block, permitting a rapid increase in the red blood cell count, a resumption of interferon-α production, and the disappearance of the hairy cells.

When the treatment is started early, interferon-α also cures another kind of leukaemia (chronic myelogenous) associated with the Philadelphia chromosome (see Chapter 3). Blood cells carrying this abnormal chromosome become malignant, but interferon-α markedly suppresses them. Interferon therapy for solid tumours was too erratic to warrant FDA approval, but there were indications that it could be effective against some late-stage cancers by preventing the formation of blood vessels.[359]

The promise of interferon as a defence against viral infection, which caused such excitement more than four decades ago, is at last finding some use. Mice lacking the genes for interferon are extremely susceptible to infection. Conversely, interferon-α is remarkably effective in curing chronic human infections of hepatitis B and C. This treatment was approved by the FDA in 1992 and rapidly generated a very substantial income from sales.

Interferon-β was at one time considered a promising anticancer drug, but it failed in crucial clinical trials. It languished for some years but then emerged as an FDA-approved treatment (betaseron) for the inflammatory effects of multiple sclerosis – the muscle spasms and paralysis. The extraordinarily high price of the drug (£7,259 per year in 2003) periodically causes uproar in Britain; the Health Service is accused of rationing it according to a so-called post-code lottery. The problem is a delicate one, because it is the *only* drug available that gives any relief for the terrible symptoms of the disease, and yet, mysteriously, only one-third of subjects respond to the treatment.

Scores of cytokines are known, of which the large family of interleukins are representative. These control the formation of many types of lymphocytes, particularly the inflammation cascade. From a clinical point of view, decreasing their activities using monoclonal antibodies can be a useful way of preventing the inflammation that contributes so frequently to chronic health problems. Other cytokines stimulate the bone marrow to make antibacterial leukocytes; they now have important FDA-approved uses in stimulating recovery of the bone marrow following chemotherapy.

The first of these, marketed in 1991 as Neupogen®, is a cytokine called granulocyte-stimulating factor.

REPRODUCTIVE HORMONES: The practical use of protein hormones to stimulate ovulation started in the 1950s when they were used to assist a Swedish woman, previously unable to produce eggs, to become pregnant with twins.[334] The first preparations were extracted from pituitary glands obtained from cadavers, but these were soon superceded by hormones recovered from human female urine. An Italian pharmaceutical house, Farmalogico Serona, supplied the most widely used hormones as a very expensive product called Pergonal, obtained from the urine of postmenopausal nuns living in Italian religious communities.[360] Establishing the exact dose was problematic, as subjects varied enormously in response and there was evidently a narrow range of doses separating no effect from pregnancies involving multiple births. Gynaecologists using these preparations often gave the impression that childless women were clamouring for the treatment in spite of warnings that multiple pregnancies – sometimes quintuplets or sextuplets – were a possibility. The hormones – follicle stimulating hormone (FSH) and chorionic gonadotrophin (hCGH) – were also essential for the success of the world's first *in vitro* fertilisation. Inevitably, gene technology was needed to meet the growing demand for hormones. The company changed its name to Serono, moved to Switzerland, and set out to prepare genetically engineered versions of all the reproductive hormones. The company – now one of the most profitable using gene technology – sells

FDA-approved FSH and hCGH and several other hormones with important applications in reproductive-medicine products, including an FSH treatment for improving male fertility. Seven years after the birth of Louise Brown, Serono bought the famous Bourn Hall fertility clinic founded by Edwards and Steptoe (see p. 37) and later financed an international network of fertility clinics at which these costly treatments are now routinely used.[334]

BLOOD CLOTS AND VASCULAR MEDICINE: The uneventful circulation of the blood in everyday life conceals a delicately balanced system of amazing ingenuity for protecting and maintaining the bloodstream. Our bodies survive trauma because we have systems to seal and repair blood vessels. If these process fail, as in the bleeding disease haemophilia – or if they are activated inappropriately, as in a thrombosis – serious consequences are likely to follow. Understanding these processes has always been a crucial clinical issue, but the penetrating spotlight of gene technology now seems to be revealing some ingenious solutions.

More than a score of proteins are involved in the blood coagulation network. There is a fine balance between a set of proteins that favours clotting and another group that opposes the process, so that clots form only in response to injury.[361] Genetic diseases are known in which the victim has defective pro-coagulation proteins or defective anticoagulation mechanisms, which cause a failure to clot (bleeding diseases) or a tendency to develop life-threatening blood clots. Clotting factors VIII and IX, respectively, are defective in haemophilia A and B; a defect in the factor VIII–carrying protein causes von Willebrand's disease. Haemophilia A, a defect encoded on the X chromosome, was the cause of the notorious genetic disease that affected male descendants of Queen Victoria and about one in 10,000 nonaristocratic males. The childhood of victims of haemophilia is spent in recurring crises of spontaneous bleeding in any organ of the body, with periods of excruciating pain. Christmas disease is clinically indistinguishable from haemophilia A but is caused by a deficiency in factor IX. The less serious von Willebrand's disease affects perhaps 2 or 3 percent of Europeans.

Several decades before gene technology made its mark, blood-clotting proteins recovered from normal human blood were used to treat haemophiliacs.[298] At the time, the risk from contaminating viruses was not apparent, but within two decades an immense tragedy unfolded when patients were infected with hepatitis and HIV from these products. Gene technology was the obvious alternative route to making blood proteins definitively free of contaminating viruses, and today, most blood-clotting factors are prepared using gene technology. Factor VIII was prepared in the late 1980s, for treatment of haemophilia A and von Willebrand's disease; it was followed by a second-generation product in 1993. An FDA-approved form of factor IX became available in 1997.

More people are at risk from unregulated blood clots than from bleeding diseases. Blood clots are the principle cause of heart attacks and strokes, but they also can contribute importantly to the development of serious septic conditions when clots in small blood vessels prevent antibacterial or anti-inflammatory molecules from reaching a site of infection. About half of first-time thromboses occur in people with thrombophilia – a set of genetic defects that favours clot formation.[361] These mutations may be in any of three anticoagulant factors or in a pro-coagulation protein (factor V Leiden) that interferes with a process that normally dissolves blood clots. Approximately 5 percent of Caucasians carry this mutation; they are three to eight times more susceptible to forming blood clots, even in the heterozygotic state (i.e., one copy of the mutation). Individuals who are homozygous (two copies) for this trait are seriously at risk of deep vein thrombosis (30– 140 times greater risk). Thrombophilia is a risk factor for strokes and heart attacks that is quite independent of the risk involved in smoking or high cholesterol. Rare individuals exist who *lack* factor V completely and have a bleeding disorder that must be treated with injections of factor VIII.

Pioneering efforts to disperse blood clots using injected preparations of a bacterial protein known as streptokinase were made in the late 1940s and later formed the basis of a highly successful FDA-approved procedure. An early objective of the new gene technology industry was to find a replacement for streptokinase that lasted longer and that caused no uncontrolled bleeding. The system that naturally disperses blood clots in normal human blood provided an obvious starting point. This protein is plasmin, whose role is to dissolve fibrin, the fibrous protein that makes up the mass of a blood clot. Plasmin is made from plasminogen, under the influence of a protein called plasminogen activator. Genentech was the first, in 1987, of a number of companies to make an FDA-approved version of this protein, initially for treatment of coronary thrombosis and then for treatment of pulmonary embolism (clots in the lungs) (1990). Unhappily for Genentech, the great revenues anticipated never materialised because clinical trials failed to demonstrate any superiority over the old medicine, streptokinase.

Certain cases of sepsis – severe bacterial infection of the blood – occur when small blood vessels near an infected wound are sealed by tiny blood clots. These form in response to proteins released by bacteria, and they often prevent the body's infection response team from reaching the site. Sepsis of this kind is a serious problem that kills more than a third of its victims – perhaps 225,000 in the United States in 2001 – and is the leading cause of death in intensive care units everywhere. A genetically engineered version of activated protein C (called Zovant), whose normal role is to dissolve blood clots by attacking fibrin, has proved extremely effective in clinical trials and is awaiting FDA approval.[362]

New blood vessels form in a healthy adult (a process called angiogenesis) only as part of a wound-healing process, the result of a biochemical prompt that makes capillaries sprout from preexisting vessels. Important exceptions to this occur when the placenta forms during pregnancy and when the uterine lining reestablishes after each menstruation. Pathological formation of blood vessels occurs in the later stages of cancer – an event that signals the start of a more serious phase of the disease. The nascent tumour emits a biochemical signal that stimulates neighbouring blood vessels to sprout and send out capillaries that eventually feed the tumour.[251]

Using gene technology, scientists have identified a complex network of biochemical signals that control angiogenesis and are actively seeking ways of exploiting this knowledge. Physicians know that, in principle, coronary heart disease and stroke would benefit from a boost in blood vessel forma-tion, and that some restriction of blood vessel formation would probably retard the development of cancer, diabetic blindness, rheumatoid arthritis, and psoriasis. Gene technology has uncovered a score of protein factors that favour angiogenesis, of which the best known are vascular endothelial growth factor (VEGF) and platelet-derived growth factor (PDGF). VEGF, first cloned at Genentech in 1989,[363] is undergoing trials as a treatment to stimulate formation of blood vessels in heart muscle damaged by blood clots. PDGF is already used routinely to treat the chronic foot ulcers that plague diabetics. In their attempts to stop pathological proliferation of blood vessels in particular tissues, scientists have explored the possibilities of using monoclonal antibodies to VEGF. One product that is in clinical trials is a treatment to prevent macular degeneration of the eye and dia-betic retinopathy, a condition in which blood vessels form excessively in the retina, causing a progressive loss of sight.[364]

Gene technology has brought to cancer research detailed insight into the peculiar biology of blood vessels near tumours. Many proteins stimulate or restrain blood vessel formation in these regions and can even affect the be-haviour of cancerous cells that have spread to sites far removed from the original tumour. Cancer researchers are exploring the possibility of preventing formation of blood vessels to tumours using the natural inhibitors of blood vessel formation, angiostatin and endostatin. Angiostatin, discovered in 1994, is yet another protein derived from plasminogen (from which the blood-clot-dissolving protein is also made); endostatin is made from a secreted protein called collagen XVIII. They both have the unusual facility of activating programmed cell death in cells attempting to form blood vessels; when used together in experimental animals, they can make tumours regress efficiently. They are now in clinical trials, together with other antiangiogenic proteins; they have a special appeal because cancers do not seem to develop resistance to them.[364]

REPAIRING THE SKELETON USING GENE TECHNOLOGY: Bones that seem so solid in adults are constantly being broken down and rebuilt throughout human life by means of two kinds of bone marrow cells, known as osteoclasts and osteoblasts, respectively. After the age of forty, the rebuilding process slows down and bone starts to lose mass and strength, leading sometimes to a condition known as osteoporosis that is often a serious nuisance to older women. Understanding the capacity to repair accidental damage and the insidious age-related decline in bone was an important objective of biologists for many years. When naturally occurring proteins that stimulate bone formation were discovered in the 1970s, they raised the tantalising possibility that they might be used to repair damaged bone that failed to repair naturally. Using gene technology, scientists have prepared these proteins (bone morphogenetic protein) on an industrial scale, permitting serious investigation of their usefulness.

Bone morphogenetic proteins stimulate bone reformation in experimental situations. They are now being tested for their capacity to stimulate repair of seriously damaged bones or slow-healing fractures by introducing them into a lesion in a biodegradable material.[365] Gene therapists have the more radical idea of introducing genes into the patients' bone-making cells, transplanting them into the bone marrow and hoping they will correct osteoporosis by stimulating bone formation.

Biotechnology companies in the United States and Japan independently discovered an important element controlling bone resorption in 1997, by screening various novel secreted proteins for the ability to correct osteoporosis in experimental mice. This novel protein, now known as osteoprotegerin, boosts bone accumulation by preventing the breakdown process carried out by osteoclasts. The protein is already in clinical trials to explore its potential for preventing osteoporosis and for supporting bone growth as a way to prevent the spread of bone cancer.[366]

BIOCHEMICALS THAT KEEP US THE RIGHT SIZE: An epidemic of obesity is spreading through the developed world that will have well-known and serious consequences for the health of its victims. Although the role of too much food and too little exercise is well known, the biological reason why some individuals become obese and others do not remains mysterious. With the identification of genes that affect obesity, this area of physiology is becoming easier to understand. The discovery in 1994 of a protein (now called leptin) that could transform a strain of mutant obese mice into healthy animals with a normal body weight provided one striking insight.[367] Moreover, several seriously overweight children have been identified whose condition originates in a rare mutation in the leptin gene. By providing them with genetically engineered leptin, their disability has been successfully controlled. The leptin protein is a chemical signal that circulates in the bloodstream and reports the size of fat stores to a region of the brain called the hypothalamus. When stores are too big, they provoke a loss of appetite and stimulate energy expenditure. We should be clear, however, that leptin is not a medication that can solve the obesity problem for most people, who should instead explore exercise and moderation.

Quite by chance, another candidate factor for controlling weight emerged from a clinical trial of a potential treatment for motor neurone disease (CNTF, ciliary neurotrophic factor) when the subjects underwent unexpected massive weight reductions. Eventually, the effect was attributed to a similarity between the receptor for CNTF and the receptor for leptin.[368] Prospects for understanding the insidious problem of obesity are now looking better.

BETTER PROSPECTS FOR NEUROLOGICAL DISORDERS: Neurological disorders (i.e., those based on deficits of properly functioning brain cells) are an immense conceptual problem for those who wish to alleviate these most distressing of diseases. The most promising option seems to be the use of stem cells that can renew diseased tissues and certainly not the injection of protein medicines into the bloodstream. An important step in realising these ambitions is to identify the molecules that normally nurture the millions of different kinds of nerves in our bodies. Neurobiologists have advanced our understanding of nervous tissue by showing that neurones require special growth factors to ensure their survival, and gene technology has successfully identified several of these proteins. Although many factors of this kind are now known, the subject is still in its infancy, but judging by the ambitious plans of some American biotechnology companies, we can expect major advances in this area.

Challenges, Risks, and Rewards – a Summary

Gene technology now provides the means to make protein products generically in living factories that are identical to their counterparts in the human body. Proteins made in this way involve no animals, human bodily fluids, or human cadavers; moreover, they provoke no immunological reactions and are definitively free of viruses and prions. With one remarkable advance, all the risks previously taken when natural protein products were introduced into the human body have become avoidable. Virtually no evidence has been recorded in the years since the first Asilomar conference to suggest escaped genetic engineering vectors create an environmental hazard. Thirty years later, the merits of gene technology are evident in many kinds of commercial product. The facility to make variants of naturally occurring proteins with modified pharmacological properties to solve known clinical difficulties is amongst the most notable uses of this technology. Fast- and slow-acting insulin, for example, made by tiny changes in the amino acid sequence, provide diabetics with more flexible ways of managing their condition. Now that the human genome sequence is available, every protein product made by the human body can be made in living factories, and their potential for dealing with human ailments can be explored. The phrase "genetically modified," on the tongue of the Prince of Wales and others, may be a vacuous term of abuse, but in reality it signifies almost infinite possibilities for improving the human condition.

Whatever other scandals have discredited the industry, the safety of the fundamental idea is not an issue. Criticism is confined to the usual crimes of capitalism, unfair pricing and illegitimate use of products. Drugs such as erythropoietin, and growth hormone are widely used to enhance athletic performance, a development that was probably the first really profitable outcome of gene technology. Gene technology products are expensive, for understandable reasons. Once their efficacy is established and they become standard treatments, they are likely to strain the finances of national health services and to create difficulties in deciding who should receive treatment.

In insurance-funded health systems, the problem is simpler if less kind, because only the appropriately insured will qualify. Currently, the most expensive medicine known is Cerezyme®, an enzyme-replacement treatment for Gaucher's disease that costs between $100,000 and $400,000 a year. The disease afflicts a tiny group that is largely a captive market, although the manufacturer claims to support sufferers who cannot afford the treatment. A serious predicament arises for those who can just about afford this or any other very expensive treatment.[350] Unless a remarkable change occurs, the developing world will not benefit from new protein medicines for many years to come.

GENERAL SOURCES: 345, 348, 351, 354, 355.

13 Refurbishing the Body

Humane solutions to the problems of damaged bodies have always been the realm of surgeons. After the discoveries of antiseptic surgery and anaesthesia in the nineteenth century, surgical practice could make moderately safe and pain-free contributions to improved health care. However, with characteristic boldness, surgeons quickly extended their activities to other predicaments. Blood transfusion became increasingly routine in the 1920s and eventually had an enormous impact on most areas of medicine. More recently, open-heart surgery and organ transplantation have enabled many individuals to overcome congenital defects and chronic disease, extending their lives by many years. Bone marrow grafts have permitted cures of blood cancers and a variety of genetic diseases. Stem cell science and other aspects of cell biology have inaugurated the age of tissue engineering, which may have huge potential for the renovation of nervous and muscular tissues. Spectacular advances in techniques for refurbishing the body are not made without extraordinary audacity and risk.

The Surgical Tradition

The diary of Samuel Pepys describes how, in 1658, after several years of excruciating discomfort, he bravely decides that his bladder stone must be removed. Completely aware of the hideous unpleasantness of the operation, he sees the risks as preferable to the misery the condition will bring to the rest of his life. A few skilled surgeons existed in London at the time who evidently specialised in this operation with reasonable success.[369] Knowing it would be done without anaesthetic or any safeguard against infection, we are surprised at the audacity of the surgeon and at Pepys's confidence, but that is the remarkable tradition of surgery.

The removal of the appendix of Edward VII in 1902, a few months before his coronation, marked a time of rapidly improving fortunes for surgery. The craft was becoming moderately safe and pain-free because of advances in aseptic surgery and anaesthesia made during the previous century.[33] Amputations of severely damaged limbs, surgical removal of tumours and kidney stones, bone setting, and lancing of abscesses were conducted with some success, relying empirically on the phenomenal capacity of the human body to heal. In France, the first steps towards using grafts of skin to repair burn damage had already been taken.[112] Without antibiotics, mortality from surgery was high, but patients accepted the risk, like Pepys, when they reflected on the alternative. Even as late as 1940, the operation used to excise the royal appendix still claimed the lives of one in five patients.[112]

By the 1920s, surgeons in celebrated hospitals like the Mayo Clinic in Minnesota were carrying out many innovative operations to very high standards. Some, such as the radical mastectomy famously introduced by William Halsted for breast tumours, are now recognised as unnecessarily extensive. Elsewhere, a spirit of reckless adventure and showmanship uninformed by any scientific rationale guided some surgeons to perform operations that can only be described as lunatic.[112] Lengths of gut were removed to alleviate constipation; organs were rearranged according to incomprehensible theories; and the lungs of TB sufferers were deliberately collapsed in accordance with a flawed theory. Eighty children died each year in Britain in the 1930s following operations to remove tonsils, an operation now regarded as unnecessary.

The award of the Nobel Prize for Medicine and Physiology to the Portuguese neurologist Antonio Moniz in 1949 commemorated what the citation described as "one of the most important discoveries ever made in psychiatric therapy," the frontal leukotomy. Some would say it was the culmination of the most disturbing adventure in surgery ever recorded. Severely psychotic patients who successfully underwent this operation became more placid after the nerves connecting the brain to its frontal lobes were severed, but they usually also lost all imagination, foresight, and sensitivity in personal interactions. In spite of the profound hostility of psychiatrists to the operation, American neurologists developed it further into the "ice-pick lobotomy," in which a single slash through a hole above the eye orbits made it the work of a few minutes. In the 1940s, perhaps fifty thousand of these cost-efficient operations were performed in the United States and other countries, using consent procedures that would have no ethical validity today. The operation is now an historical curiosity, but the chilling horror of its impact on family dynamics is apparent in Tennessee Williams's play *Suddenly Last Summer*, which is based on memories of his schizophrenic sister, Rose, who was treated in 1943. The accidental discovery in 1952 of the calming effect of chlorpromazine on irrationally disturbed patients quickly eclipsed Moniz's achievement and marked the start of a kinder and more intelligent approach to the mentally ill (see Chapter 11).

In the 1950s, top-class surgeons turned to serious and difficult problems, such as congenital heart defects, and later to organ transplantation. With antibiotics available, surgery was safer than ever before, and with the arrival of immunosuppressives the stage was set for major advances in the way damaged human bodies could be refurbished.

The Biology of Renewal

If we are cut, burnt, or injured – not too severely – the most likely outcome is that the tissues will be restored to their original state by a natural process of repair and renewal. We lack the enviable capacity to restore lost limbs enjoyed by newts, but the mammalian liver has remarkable powers of regeneration that originate in an evolutionary adaptation that allowed animals to recover from liver poisoning brought on by food toxins. Two-thirds of the liver of an experimental animal can be removed and the remnant can regenerate to its original size and become fully functional, as every cell can divide once or twice.

Other tissues regenerate by an entirely different scheme involving a small population of cells known as stem cells. As we have seen, these cells renew themselves by cell division to an apparently unlimited extent, but they also generate cells that are characteristic of the mature tissue. Most stem cells specialise in making only one kind of differentiated daughter cell, such as red blood cells in the bone marrow and sperm cells in the testis. Cells of the skin or the lining of the small intestine that are sloughed from their respective surfaces are also replaced by stem cells. In a similar way, cartilage, bone, and lung can be regenerated. By contrast, muscles of the heart and skeleton, brain tissue, and spinal nerves have little or no capacity to regenerate and present the greatest challenge to those attempting to restore damaged human bodies.

Twentieth-Century Uses of Human Blood

Severe loss of blood following an injury was recognised as a bad sign for survival for centuries, and the idea of replacing blood by transfusion was explored by seventeenth-century French and British scientists. The idea never developed further until it was performed successfully for the first time in 1909 by Alexis Carrel, a French surgeon working at the newly founded Rockefeller Institute for Medical Research in New York. Carrel was already famous for the discovery of a technique for joining blood vessels, for which he was to receive the Nobel Prize in 1912. Carrel was summonsed to help save the life of a friend's newborn baby, who was bleeding internally. His response was to improvise a transfusion technique that became known as person-to-person transfusion. The radial artery, in the hand of the baby's father, was joined surgically to a minute vein on the back of the baby's leg. The daring experiment, carried out in desperation at the patient's home, was a complete success.[298] Blood clotting was not a problem, because the transfused blood was enclosed throughout and not exposed to air or other tissues.

BLOOD TRANSFUSION: For a few years, incredibly, person-to-person transfusion became a commonplace procedure in New York. About one-third of patients felt some kind of reaction to the transfusion originating in incompatibilities of blood type, and some subjects even died from the treatment. The importance of blood groups was already known from the experiments of Karl Landsteiner conducted in Vienna in 1900. This work indicated that blood could be classified into compatible classes or blood groups, which could cross-react with blood from different groups, except for blood group O (a potential universal donor). Later, when this work was more widely appreciated, blood would be typed by more refined criteria originating in more subtle features of the chemistry of blood cells.

Blood transfusion, as an integral part of emergency medicine, continued to develop throughout the twentieth century, often driven by the exigencies of war. A blood transfusion service was set up for the first time during the First World War by Captain Oswald Robertson of the American army in France.[370] A solution to the tricky problem of blood clotting had already been devised in New York. The problem was solved by adding sodium citrate to blood as soon as it was removed from the body to sequester calcium and completely prevent clotting. Citrate was also metabolised instantly within the patient's body and had no harmful effect on blood cells at the right concentration. Blood prepared by this method could be stored in a refrigerator for up to twenty-two days, typed using the Landsteiner system to identify universal donors, and tested to eliminate blood infected with syphilis. Transfusion was performed using needles and tubing, much as today. The value of the transfusion procedure in resuscitating battle casualties in order to permit surgery was widely recognised as an important advance. However, after the war, civilian hospitals were reluctant to adopt it out of fear of adverse reactions, including the danger of pyrogens – chemicals of bacterial origin that in the minutest quantities provoke a fever and other bad reactions. Only when better blood typing systems were devised, and when the citrate and equipment could be made free of pyrogens, were these fears finally dispelled.

The first civilian system, founded in London in 1921 by Percy Lane Oliver and Geoffrey Keynes (brother of Maynard, the economist), started as a small bureau that provided blood for hospitals. The two men set out to attract "respectable" donors, who would regard blood donating as a public-spirited but essentially unheroic act. Blood was typed and tested for syphilis, using the principles learned during the war. Similar systems emerged in most of Europe, but in the United States payments were made to donors, generating a substantial income that could be used to fund a much larger and more effective organisation.

In Britain in the late 1930s, with an imminent prospect of war, a robust system for supplying blood for civilian casualties in London was needed. The government was preoccupied with rearmament, but Janet Vaughan, a junior doctor at Hammersmith Hospital, with the enthusiastic support of other young doctors, began to collect and store blood – still from volunteers – initially in modified milk bottles. Financial support eventually came from the Medical Research Council on the first day of war. The system proved extremely effective. About 10 percent of civilian casualties from air raids needed transfusion, and by the end of hostilities half a million pints of blood had been dispensed without notable adverse consequences.[298]

BLOOD COMPONENTS: By the 1930s, interest in using the components of blood more effectively was growing, because stored whole blood was easily damaged and had a lifetime of only twenty-two days. The serum – the fluid part of blood, obtained after removing blood cells – was very beneficial in restoring lost fluid and preventing shock. No typing was needed, and when freeze-dried it could be reconstituted with water at a convenient moment.

When war came, the American blood authorities shipped vast amounts of serum overseas for the allied troops. However, with no oxygen-carrying capacity, serum was obviously less beneficial than whole blood, and if bacteria were not rigorously excluded from the reconstitution process, patients would almost certainly succumb to septicaemia.

In 1940s America a new industry emerged dedicated to preparing blood components and thereby ensuring a more efficient use of this valuable commodity. The key figure in this enterprise, Edwin Cohn, devised a system for separating blood components on an industrial scale. Albumin, a major constituent of serum, could be used as a serum substitute. The so-called gamma globulin fraction was a source of antibodies to many infections (such as measles and mumps) because it came from a large pool of donors' blood. This was given to American soldiers throughout the Second World War in the form of "shots" in order to confer passive immunity to these diseases, at a time when vaccines were not available. After the war, it remained in use to protect against polio, before the vaccine became available.

Cohn's imagination soared into increasingly sophisticated technology. Plasmapheresis was his scheme for continuously separating blood cells from the serum and returning them to the donor while the blood was being taken, with no harmful consequences. More serum was thereby obtained in a normal blood donation, because it was replaced more quickly than the blood cells. Cohn also devised methods to recover platelets, the strange tiny blood cells with a lifetime of only ten days that have the key role in blood clotting and wound healing. An infusion of platelets can avert a spontaneous haemorrhage in patients recovering from chemotherapy, whose normal level of platelets is greatly reduced.

Once the bleeding diseases haemophilia A and B were recognised as deficiencies of blood clotting factors VIII and IX, the logical next step was to overcome the deficiency using highly purified proteins to prevent bleeding. The pioneers of blood transfusion understood very clearly the potential danger of infection through blood, but they could only anticipate known risks such as syphilis. Viruses such as hepatitis B and later HIV and hepatitis C, whose existence was unknown, eventually presented serious challenges to blood transfusion services because they could exist in blood without causing an obvious disease. We have seen how "serum sickness" associated with transfusions was a severe nuisance to the American military from 1942 until the identification of the hepatitis B virus twenty-five years later.[298] In America, the "for-profit" blood banks that collected blood from all strata of society were particularly acutely affected.

NEMESIS: The triumph of protein chemistry that made haemophilia curable with factor VIII became a poignant tragedy once the blood supply was compromised. Because blood was bought and sold like a commodity in the Unitesd States, a surplus usually existed, from which factor VIII and other blood components could be made. Until a test for HIV became available (four years after the epidemic started), the virus was entering every kind of pooled blood resource from donors who were carriers. Ironically and tragically, the blood of male homosexuals was often eagerly accepted by blood banks because it was commonly rich in antihepatitis antibodies that were useful commercially. Indeed, factor VIII was isolated from this blood in the misguided belief that antihepatitis antibodies would improve its usefulness.[298] Meanwhile, the British authorities, dependent on voluntary donations of blood, were unable to make enough factor VIII to meet the demand and were driven to purchase these products from the United States, quite unaware of the risk.

Opportunities for containing the spread of the viruses were missed. Early in 1983, a proposal was made to use the hepatitis B antigen in the blood of donors as a surrogate marker for HIV, to identify blood donations at greatest risk of infection with HIV. The gay community vetoed this proposal out of fear of stigmatisation, and the danger persisted until HIV could be detected directly in blood in 1986. Pasteurisation was proposed as a method of inactivating HIV in factor VIII in 1983 (i.e., a heat treatment that would inactivate the virus), but delays in implementation of the process meant that contaminated stocks of plasma persisted longer than necessary. Outside of the United States, many countries – especially France, where the virus was first discovered – were unwilling to pay the vast royalties required for the first patented test kit, which was marketed by the American company Abbott Laboratories. In these confusing circumstances, more than half of American haemophiliacs became HIV positive; in Japan, Germany, France, and Britain, between 20 and 50 percent were infected by 1986.[298]

By the 1990s, the blood supply was considerably safer. Reliable tests for all contaminants, including newer diseases such as hepatitis C, were in use, and a campaign was initiated to make potential donors conscious of risky behaviour that might prejudice the blood supply. The risk to haemophiliacs is now very low, and blood factors made by gene technology are conclusively free of viruses. Nonetheless, the intrinsic danger of infection by blood transfusion may be hard to eliminate, as recent infections of patients with West Nile virus indicate.

Blue Babies and Coronary By-pass Operations

Fifty years ago, children born with congenital heart defects had little time to live and notoriously developed the characteristic bluish look that became known as the "blue baby syndrome" because they were unable to oxygenate their blood adequately. The number of babies born with congenital heart defects is surprisingly large (one percent), but the defects can often be corrected today, thanks to the efforts of daring and imaginative pioneer surgeons. In the decade after 1945, a succession of surgeons tackled the problem posed by the defect known as the Tetralogy of Fallot.[165] In this condition – the eponymous "hole in the heart" – oxygenated and deoxygenated blood mix through a hole in the septum separating the right and left sides of the heart. In addition, the artery that takes blood to the lungs is unusually narrow. Early efforts to correct such defects indicated that surgery would be impossible without opening the heart, and this would be possible only by using an alternative pump to take oxygen to the brain. Using a novel and daring variant of Alexis Carrel's idea of person-to-person transfusion, surgeons at the Mayo Clinic solved this problem in dramatic fashion in 1954. The blood circulation of an adult volunteer was connected to the circulation of a child with Fallot's Tetralogy, so that the volunteer's oxygenated blood pumped through the child's body. The surgeon then quickly opened the heart and enlarged the pulmonary valve. "Open-heart surgery" was born – within a few weeks the child was running about, and the volunteer was completely unharmed by the experience. The use of a volunteer's heart plainly put the life of a healthy person at risk and could not be countenanced as a routine procedure, but from that moment the future of open-heart surgery was rarely in doubt.

Over several years, a team at the Mayo Clinic perfected the design of a gadget that could replace the pumping action of the heart and the oxygenating action of lungs, without damaging blood cells. Setbacks occurred, in which patients died, but with more practice and continual improvements in equipment the operation eventually became routine and completely safe. By 1960, the heart-lung machine had become so dependable that any childhood heart defect could be corrected by heart surgeons. The time was ripe to tackle other problems of the heart.

In the 1960s, coronary heart disease was the most likely cause of death among middle-aged men in industrial countries. The disease originates in a build-up of atherosclerotic plaque in the coronary artery (the blood vessel in the heart muscle). Eventually, this culminates in a blockage of the artery that cuts off the surrounding tissue from oxygen, resulting in cell death. The precise position of such blockages or constrictions of the blood vessels could be revealed using a newly discovered technique called angiography. This involves outlining the blockage using a substance opaque to X-rays that is injected into the bloodstream. The apparently intractable problem of removing such blockages was elegantly solved by the Argentine surgeon Rene Favaloro at the Cleveland Clinic in 1967, when he replaced a blocked coronary artery with a vein from the patient's leg. Twenty-five years later, at a celebration to commemorate the anniversary of this momentous "by-pass" operation, the first patient was present, now aged seventy-nine.[371] The operation is still widely used, with more than ten thousand operations a year in Britain alone, often giving another twenty years of healthy life to the subject.

Tissue Engineering – Repairing Skin and Other Tissues

Damaged skin has a formidable capacity for regeneration, but occasionally damage can be so severe, particularly from burns, that the tissue cannot heal before infection sets in. Nineteenth-century French surgeons recognised this and conceived the idea of grafting the top layer of skin from an adjoining area to cover the wound. Realising that the grafted skin needed the nourishment normally provided by the bloodstream, they deliberately retained a connection between the graft and its blood supply. If the damage is to be repaired, cells from the deep layer of the graft must unite with the wounded area and deposit a gluelike protein called collagen in which more cells can proliferate. Eventually, new blood vessels sprout and connect the graft with the underlying tissue, and thereafter the wound quickly heals.

Stimulated by the pitiful sight of badly burnt young airmen, heroic efforts were made to improve skin-grafting techniques during the Second World War. By the 1970s, the old methods requiring many operations began to give way to a better technique for joining the capillaries of the skin graft to blood vessels in the wound. This enabled the surgeons to take larger grafts from more remote sites of the patient's body, so that a single operation was all that was necessary in some cases.

The surface cells of skin are renewed by constant cell division in the deep layers of skin by stem cells. In a truly radical application of this idea, Howard Green of Harvard Medical School and his colleagues have used skin stem cells, cultured outside of the body in large numbers, to repair skin.[372] Patients with more then 90 percent of their body surface burnt, who would otherwise have died, have been treated successfully using this remarkable scheme. Skin is taken from a small area (three square centimetres); the stem cells are then separated out and cultured over a four-week period in an area large enough to cover the burnt surface. This may be almost two square metres – an expansion in area of more than 5,000 times. The microscopically thin sheet of cells is then transferred onto the patient's burns. Within a week, the outer scaly layer of skin reappears; then the deep layer starts to develop, followed soon by other cell types normally found in skin. The deep layer is complete in four weeks and becomes progressively more tightly anchored to the underlying tissues, although two to five years are needed before the underlying connective tissue becomes completely normal. Bacterial infection is the major obstacle to the success of this process, but an intermediate graft that stimulates the host's defences against bacteria is used to overcome this.

The pioneering efforts to repair damage and disfigurement that started nearly a century ago have evolved into the new discipline of "tissue engineering," which now has an industrial base and uses all the resources of modern biology. A number of kinds of FDA-approved replacement skins (such as Apligraf) are now prepared by new biomedical companies such as Organogenesis, to be delivered to the surgeon in a frozen form ready for reconstructive surgery. These may contain cells of the superficial and deep layers of the skin that are assembled on a thin biodegradable scaffold in a bioreactor, using fibroblast cells cultured from the foreskin of newborn babies.[373] Once the cells start to grow, they secrete materials that stimulate repair of the patient's skin, and they will persist for about six months but produce no immunological reaction. As a commercial product, it incorporates many miraculous features, not least of which is the capacity to be stored in a frozen state ready for use when thawed. A number of related products with specialised uses, such as reestablishment of skin in diabetic ulcers, are also available.

Repair to severely damaged bone has moved a long way from the art of the nineteenth-century bone setter. Repair of shattered bones is possible using pieces of bone from a donor or from another part of the recipient's body to act as a scaffold for restoration of the bone, even if the transplanted bone dies. Once the initial disruption is mended, the correct shape is restored by filling the gaps in bone with a matrix composed of collagen, seeded with proteins known to stimulate bone growth.[374] In one remarkable instance, a new thumb was created in the laboratory to replace another lost in an accident, using a scaffold of coral that was seeded with the subject's own bone-making stem cells. This was then encased in a framework of flesh preserved from the remains of his thumb.

Cartilage, too, is the province of the tissue engineer, because the cells that make cartilage (the chondrocytes), when taken from a small piece of healthy cartilage removed from the patient's joint, can be grown in culture. These are then introduced to the injured region, where they restore normal cartilage formation.[373,375]

Perhaps the most remarkable indication of the promise of tissue engineering to date was the construction in Boston of an artificial bladder that, when implanted into a dog, conferred nearly normal function that persisted for almost a year. A tissue based on pig liver cells was grown into the shape of a bladder on a scaffold of biodegradable material, which was then seeded with canine bladder epithelium and smooth muscle cells.[376] This construction was critically dependent on a specific protein factor (keratinocyte growth factor) that helped the cells grow into the correct configuration. In a few years, human bladders created by essentially the same route will probably start clinical trials.[373] We can also probably anticipate other daring initiatives in the field of tissue engineering. A revolution in surgical possibilities is very likely in the next two decades.

Awareness of the physical principles involved in tissue formation is constantly increasing. Epithelial cells that line blood vessels characteristically sense hydrodynamic sheer – the force created by a moving liquid on the surface of a blood vessel – and can assume an appropriate shape in response. For example, cells cultured in the laboratory on a tubular scaffold knit together into artificial arteries with substantial mechanical strength. Using the same principle, tissues have already been created that can withstand the pressures to which normal healthy arteries are exposed.[377]

The tissue-engineering industry talks confidently of a time when they will be able to make replacement blood vessels, heart valves, and even a new heart. As more is learnt about the properties of stem cells and the growth factors that direct their development to particular roles, more startling developments are certain to occur.

Organ Transplantation

Almost a hundred years ago, Alexis Carrel successfully transplanted kidneys and blood vessels between animals, but within a month the recipient was emphatically rejecting the graft, and there seemed little possibility that this line of enquiry would develop any further. The deep-seated cause of rejection became clear only in the 1940s, when transplantation of mouse tumours between different strains of mice was investigated. Whether or not the tumour grafts were compatible (i.e., whether they were accepted or rejected) was controlled by a class of so-called histocompatibility genes.[142] So many alternative forms of these genes exist that apart from genetically identical individuals (i.e., twins), few people are sufficiently compatible to permit tissue matching.

But what causes rejection? The underlying events started to come into focus in wartime Britain, when desperate attempts, using skin grafts from donors, were being made to repair the severe burns suffered by airmen. The graft was usually rejected, and a second attempt was rejected even more fiercely. While seconded to a burns unit, the twenty-eight-year-old zoologist, Peter Medawar realised that skin grafts between genetically *unlike* fraternal twins, born to experimental animals, were not rejected, whereas grafts between similar animals born as singletons failed. The significant point that Medawar perceived was that although genetically unlike, the twins shared the mother's blood circulation. The immediate cause of tissue rejection was plain to see when he examined the material microscopically: white blood cells were attacking the cells of the graft in much the same way that they would attack microbes. This observation told him that during foetal life, the fraternal twins had learnt to tolerate the process that normally makes grafts fail. Medawar predicted that a transplanted tissue would be protected by antibodies directed at the recipient's otherwise hostile white blood cells. Experiments soon confirmed this reasoning, and the intellectual framework in which transplant rejection could be understood was established. The hunt was on for a practical method of suppressing the rejection process.

The kidney was the most promising candidate for transplantation, because only one artery, one vein, and the ureter needed to be joined up, and the kidney seemed to function perfectly well without a nerve supply. A live donor was also a possibility, as we all have two. Boston in 1954 was the scene of the first successful kidney transplant, which was made on a twenty-three year-old man whose kidneys were seriously diseased but who lived for another eight years. The kidney was donated by a live twin brother, ensuring the genetic identity necessary to prevent rejection and ensuring that the donated kidney would be in good condition compared to that of a cadaver. Such a fortunate occurrence is exceedingly rare, and little progress could be made until the rejection process could be suppressed effectively. Once azathioprine, the first immunosuppressive, became available in 1961, the way was open for using the kidneys of a recently deceased person for transplantation, rather than the ethically less satisfactory solution of using a live relative's kidney. Enthusiasm for trailblazing transplantation operations exploded immediately amongst surgeons. A number of unsuccessful attempts to transplant lungs and liver from recently dead donors into recipients who were close to death were made in 1963.

The prospect of being the first to transplant a human heart was a goal that tantalised many surgeons in spite of the seemingly insurmountable challenges it presented. Deterioration would start from the moment a heart was removed from the body of a potential donor (most probably a road accident victim) – limiting, among other things, the time for effective tissue typing. The electrifying news of the first successful heart transplant came in 1967, not from a great American medical school but from the Groote Schuur Hospital in Cape Town. Dr Christiaan Barnard had transplanted the heart of a young woman into Louis Washkansky. The operation was a technical success, but eighteen days later the patient was dead from pneumonia. Another extraordinary fact, invisible to the international press for almost four decades, was the crucial contribution to Barnard's success made by his black assistant, Hamilton Naki. Barnard openly admitted that Naki's surgical skills were greater than his own. Moreover, for the nine years before the famous operation Naki was intimately concerned with developing techniques on experimental animals, and he later trained hundreds of visitors who came to learn the technique. Apartheid forbade any recognition of his role – a wrong rectified only when the old regime crumbled.[378]

The floodgates opened in the following year. More than a hundred similar operations were performed, but just a few patients survived more than a year. The early transplants failed due to infection or rejection caused by lack of adequate tissue typing, and in some instances the failures were exacerbated by the collapse of other organs damaged by long-standing heart disease.[165] Heart grafts were a remarkable tour de force, but they were probably less demanding of surgeons than multiple heart valve operations and much simpler than liver grafts. The heart tissue is notably sensitive to deprival of oxygen, but a donor's heart could be removed and stitched into the recipient within forty minutes, before any harm was sustained. The part of the donor heart in the upper right atrium that controls heart beat was removed from the donor heart, in order to allow the vital nerve supply for the heart to function. Later, a complete heart-plus-lungs graft was introduced, because it required fewer joins than a just-heart graft, and there was no anxiety about the need to adapt a fresh heart to old lungs.

Cyclosporin, the first really effective immunosuppressive, initially found favour in Britain, where, starting in 1985, it was used for bone marrow grafts and eventually for organ transplants. Many of these transplants lasted for fifteen to twenty years, and recipients often returned to their old professions – even, in a few instances, to professional athletics. One British woman received a triple transplant (heart, lungs, and liver) in 1986 and survived twelve years to become the longest-lived survivor of this operation. Liver transplantation, although not requiring a nerve supply, is probably the most demanding of transplant operations because large numbers of blood vessels must be joined and the tissue is exceedingly susceptible to oxygen deprival, blood clots, and haemorrhages. Perhaps the most remarkable example of the transplant surgeon's skill and ambition was an operation performed in Florida in January 2004 on a six-month-old child, born with a rare severe genetic disease affecting smooth muscle that prevented her feeding. The operation, in which surgeons successfully replaced eight of the child's organs with organs from a dead donor, was a technical success, but the long-term prognosis is quite uncertain.

Transplant patients have every reason to be happy with their treatment compared to the alternative, although they are sometimes dogged by peculiar problems. These may include development of toxicity to cyclosporin, a significantly increased chance of a malignancy, and even the possibility of developing coronary heart disease in the transplanted heart. In the United States, transplants are so routine that they are covered by insurance, but a desperate shortage of organs exists wherever these treatments are available, which some attribute to the success of seat-belt laws.[373] Inevitably, this raises delicate ethical problems, as the supply of organs is unlikely to match the demand and people will inevitably die as they wait for a suitable donor. In the United States in 2001, just 21,000 people received transplants while 74,000 people were waiting. For similar reasons, the British National Health Service restricts heart transplant operations to people less than sixty years old. On the other hand, a social service department has felt obliged to request a High Court order imposing treatment on an unwilling child whose condition could only be cured by a heart transplant.[379] The exchange of organs for cash has been illegal in Britain since 1989, but the economics of organ transplantation create a commercial imperative that is hard to resist for the poor of the developing world. A thirty-three-year-old British property developer recently publicly admitted that he paid £25,000 for a replacement kidney inserted in Lahore, for which the donor, a twenty-two-year-old Pakistani woman, was paid £3,000.[380] More shameful still, a market exists in "organs for transplant" from executed Chinese prisoners.[381] Discussion of how the supply of organs for transplant could be increased usually centres on the willingness of the living to make a declaration that they are willing to be a donor. In eight European countries, the victims of accidents are usually considered potential donors unless they carry a document specifically withholding consent; consequently, demand for organs can almost be met in these countries.

In spite of obvious difficulties, the possibility of using organs from other animals has been actively considered as one way to meet the demand for organs. A primate, whatever its logical merits, is almost certainly completely unacceptable to the public, and the logistics of breeding colonies would be entirely impractical. The only credible candidates are pigs, which can be bred satisfactorily in just a few years and whose hearts are the right size. However, there is a major difficulty in that pig organs would provoke a massive so-called hyperacute reaction, because humans have antibodies that normally recognise a certain pig glycoprotein that covers their cells. A major breeding program to eliminate this gene could probably solve the problem, but there is also great anxiety that a hitherto unknown virus might transfer from pigs to the human population. Pigs carry retroviruses embedded in their genomes that cause no harm to pigs, but if their organs were transplanted to human bodies, the viruses might be activated with unpredictable consequences. Today, transplantation of pig organs into humans is unlikely to be undertaken until every possible hazard is eliminated.

Heart transplantation is only one of many treatments for heart conditions that the medical research community considers. An alternative to a living heart would be a very small electric heart designed to the exacting specifications of a human heart. This is a very tall order, but promising prototypes already exist. Devices that replace defective heart pacemakers have been available for forty years. The prototype, fastened to the outside of the chest, was designed to send an electric charge through the patient's body to the heart, but later versions were small enough to be implanted in the chest. Oddly, this life-enhancing advance was vilified by certain moral guardians, who believed the heart should not be tampered with because it is the seat of the soul.

TRANSPLANTATION OF THE ISLETS OF LANGERHANS: Although the quality of life for type 1 diabetics increased immeasurably with the availability of insulin, imperfect regulation of blood glucose levels continued to cause many undesirable health outcomes in later life. Transplantation of the glands that make insulin – the islets of Langerhans – or even the entire pancreas could solve this problem, but in the last thirty years, little progress was made until recently. The most elegant idea of recent times was to inject islet cells into a vein, which will transport them to the liver, where they will implant and start making insulin in a correctly regulated fashion. A Canadian team has found that immunosuppressives used in other transplant situations exacerbate diabetes but alternative immunosuppression regimes permit transplantation.[383] In the future, some kind of tissue-engineered organ made using stem cells,[384] or perhaps even pig organs, may be feasible.

BONE MARROW TRANSPLANTATION: The interior of the long bones and ribs of the mammalian body is a unique and special place that is the nursery for our blood cells from birth. Nature has packed these long spaces with a variety of cell types, usually known as bone marrow, that generate all of the many kinds of blood cell that exist. Blood cells have a complicated pedigree; they are constantly being renewed by the first step in this process, the stem cells (see Fig. 13.1). The bone marrow is one of the most active sites of cell division in the mammalian body. This feature makes the tissue particularly sensitive to X-ray irradiation, a vulnerability that became apparent with awful poignancy in the aftermath of the Japanese atomic bomb explosions of 1945. Experiments with mice then established the minimum lethal dose of X-rays that could destroy the bone marrow specifically. The bone marrow could then be repopulated by bone marrow cells from another animal, which would proliferate and prevent the mouse from dying.[142]

Throughout the 1950s, a time when the importance of tissue matching was not fully appreciated, physicians tried using whole-body irradiation or chemotherapy to destroy cancer cells, followed by bone marrow transplantation to restore the capacity to make blood. The first unambiguously successful treatment of a leukaemia patient using a bone marrow transplant following whole-body irradiation occurred in Paris in 1963.[142] After

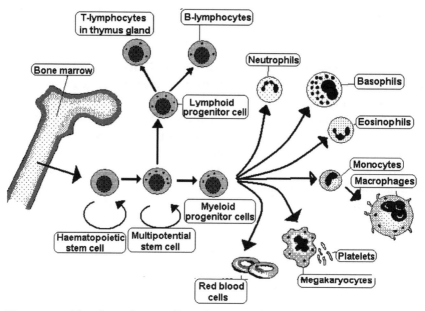

Figure 13.1. The place of stem cells in the origin of a variety of blood cell types.

eight months, the patient's blood was completely repopulated with donor red blood cells; the subject was in complete remission nine months later. The same principles are used today for treatment of acute leukaemia, lymphoma, and myeloma, but nearly thirty years of single-minded endeavour were required to reach the current high success rate. After intensive treatment to kill the diseased bone marrow, new bone marrow cells from an unrelated donor are introduced. These are aspirated from the donor's hip bone using a hypodermic needle and are then injected into the patient's bloodstream, whence they find their own way into the long bones. Quite soon, they are sufficiently numerous to restore the normal rate of blood cell formation.

Bone marrow transplantation has an obvious application in the treatment of human genetic diseases affecting blood cells. The first attempt was made with a condition known as SCID (Severe Combined Immune Deficiency). This syndrome has a number of causes but always makes its victims ultrasusceptible to infection because of their inability to make antibody-forming cells. The condition became familiar to the public through David Vetter, a young victim of the disease. His short life was passed in a plastic bubble to avoid infection, but in spite of this, he eventually succumbed to a latent infection of a virus that would not harm other children. In 1968, a five-month-old baby suffering from SCID caused by a deficiency of adenosine deaminase was cured by Robert Good using bone marrow transplantation. The bone marrow was obtained from one of the subject's four siblings, whose tissue type was almost an exact match. On the second attempt, a bone marrow transplant established blood formation with the character of the donor.[385] The long-term result was immensely gratifying, as at the time of Good's death in 2003, the first patient, a father of four children, was still alive. While this was a great conceptual advance in treating genetic disease, a fault line gaped between the intrepid physician and at least one scientist who feared the ethical implications of the treatment. Robert Good recalled how McFarlane Burnett, the eminent Australian immunologist, rebuked him in public for inflicting the experimental procedure on a child whose grim fate would be delayed only temporarily. Good's defiant response was to assert his moral obligation as a physician to attempt to cure any condition he had studied academically.[386] Happily, the subject of Good's work outlived both of the protagonists.

For bone marrow transplantation to be a success, the recipient's lymphocytes must not attack the donor cells, and in later versions of the treatment this was prevented by a radical and permanent destruction of the patient's bone marrow stem cells. This could be done using either total-body irradiation or a chemotherapeutic agent known as busulfan that kills bone marrow stem cells selectively. The donor's bone marrow stem cells then rapidly repopulate the patient's bone marrow and start making normal blood cells. Importantly, the donor must be well matched to the recipient; otherwise, donor lymphocytes may attack the recipient and cause what is known in the trade as graft-versus-host disease.

Robert Good's faith was certainly vindicated by the success of subsequent developments, in which treatments like the one he pioneered became commonplace. The list of diseases that have been successfully cured includes sickle-cell anaemia, thalassaemia, adenosine deaminase deficiency, Gaucher's disease, and fifty others.[387] Typically, patients with sickle-cell anaemia, who make a defective haemoglobin in otherwise normal red blood cells, are given total-body irradiation or busulfan to kill bone marrow stem cells. Donated bone marrow then permits the making of normal haemoglobin. Unfortunately, the chance of finding a donor with a matching tissue type is perhaps only one in four hundred, which makes the rarity of suitable donors a major problem. Happily, the practice of donating bone marrow has become, like blood donation, one of those humane altruistic gestures that enrich society.

The so-called cord blood of the umbilical cords of newborn babies contains sufficient stem cells to repopulate the bone marrow of a child. An American leukaemia patient was the first to receive an infusion of cord blood in 1972, although this was not widely known until much later. In 1989, a French child suffering from a potentially fatal genetic disease, Fanconi's anaemia, was cured by an infusion of the cord blood of his baby sister.[388] Although cord blood transfusions must be made to patients with similar tissue types, the requirements are less stringent than those for bone marrow transplantation. Potential donors of cord blood are very numerous compared to bone marrow donors, and umbilical cord blood can be tissue typed and screened for viruses immediately after birth and then stored while awaiting a request. There is a remote risk that the donor could have an undiagnosed genetic disease of the blood, but if the blood is stored until the donor is manifestly healthy there is no risk.

In the United States, stored cord blood of a child has become a kind of insurance against the unlikely eventuality – the odds are about one in a thousand – that the child may need chemotherapy for a childhood cancer. The cord blood, stored at birth, is available to repopulate the bone marrow and has already saved the lives of many children. Instead of the anxious search for a tissue-matched donor, the victim is assured of a perfect match. Unfortunately, the cord of one baby contains only sufficient stem cells to treat a single patient. However, as techniques for growing stem cells in the laboratory develop, cord blood could be used for adults.

Bone marrow transplantation was the focus of an ethical dilemma that became public knowledge in the United States in 1990. This concerned Anissa Ayala, a child with leukaemia that required chemotherapy and a bone marrow donor. Unable to locate a donor after a two-year search, the parents decided to conceive a second child, by the traditional method, for the express purpose of providing a bone marrow graft for the elder sister.[72] The chance of doing this successfully was very small: the mother was forty-two years old, and the father needed a vasectomy reversal. Most importantly, the baby would need to be of a tissue-type compatible with the elder sister, the chance of which was one in four. Nonetheless, they were successful, and a bone marrow graft taken from the baby at fourteen months cured Anissa. When, five years later, the full story was revealed on television, a chorus of outrage greeted the miracle, including death threats to the doctor involved and to the baby. Some of the shrillest critics believed it was better that Anissa should have died than a baby should be conceived to serve such an unethical manoeuvre. Voices of sanity such as Lee Silver eventually prevailed; they pointed out that a child conceived as a playmate for the firstborn or to cement a marriage was morally identical to a child conceived to save another's life. Silver added that in reality, most children were conceived unintentionally, but that irrespective of the circumstances of their conception, children generally received the unqualified love of their parents. Today, Anissa, now in her thirties, is alive and well, working for a Red Cross bone marrow program in California. Silver suspects that scores of similar, perfectly legitimate cases have since occurred without publicity.

The question of what is an acceptable purpose for a conception was news again in 1998 with the story of the "spare parts baby." This child, Adam Nash, was an early product of preimplantation genetic diagnosis (see Chapter 14); his umbilical cord blood was used to correct the usually fatal Fanconi's anaemia that afflicted his older sister, Molly.

Cell Therapy – Radical Uses of Stem Cells

When cells of the brain and heart die, these organs may sustain irreparable damage because the cells are not normally replaced by cell division. Similarly, neural tissues of the spine may be irreversibly damaged by trauma to the spine, resulting in paralysis of the victim that, at present, cannot be repaired. For some time, medical scientists have been wondering whether stem cells could be used to regenerate damaged tissues. There are indications that Parkinson's disease, other neurodegenerative diseases, and spinal injuries may be alleviated to some extent by cell therapy.

Parkinson's disease is caused by the death, for unknown reasons, of cells in a region of the brain called the *substantia nigra* that produces a neurotransmitter chemical called dopamine. When this capability is lost, patients lose control of their movements, their hands shake, and their muscles become rigid. For a period, the symptoms can be controlled by a drug called levodopa that is converted into dopamine, but the disease inevitably advances, because cells in the crucial region continue to die. The idea of introducing tissue that can make dopamine into the damaged region of the brain has been under consideration since the 1970s. The first tests were made in Sweden in 1985 by Anders Bjorklund, using grafts of foetal brain tissue, and the procedure has now become common practice in many countries.[389] This treatment is said to reduce the symptoms by 50 percent, although it is hard to know what an appropriate control would be or exactly how the technique works. The implanted neurones survive and grow because extensive connections with other neurones have not yet formed, and the glia, another class of brain cell present in the graft, may assist in repairing the lesion. The implant probably releases dopamine even if neural circuitry is not established. The technique, while promising, is of limited value; a treatment for a single individual would require tissue from six foetuses, and the preparations are not easily standardised. Experts hope to improve the situation by propagating immature neural progenitor cells obtained from brain tissue in culture before using them in cell replacement therapy. The same considerations apply to the use of foetal material for treatment of Huntington's disease.

The best prospect for repairing spinal injuries has emerged from the observation that severed spinal nerves of rats can regenerate if they are stimulated by growth factors made by the Schwann cells that ensheath them. The great hope for repair of spinal injuries is that these factors may be identified and delivered to the affected nerves, perhaps by cell therapy.[390]

The response of neurodegenerative disease to cell therapy is encouraging, but progress will probably be made only by using properly standardised stem cells instead of foetal material. The idea is to generate enough stem cells of an appropriate character in culture conditions where they acquire the "right" character. The conviction has been growing for twenty years that this will be possible, because cell lines derived from the embryos of mice, known as embryonic stem cells, can undergo exactly this kind of transformation. They can be cultured indefinitely, but under defined conditions they transform into neural, muscular, pancreatic, or other cell types. One interesting cell line can make the myelin sheath that surrounds neurones and can correct a demyelinating disease of mice. While that is encouraging, it would be a simpler option to introduce stem cells into a damaged tissue, such as a Parkinson's lesion, if biochemical cues could induce the implant to form the correct cell type.

Because of some recent developments, there is now a formal possibility that stem cell therapy for human diseases is possible. Embryonic stem cells have been obtained from human pre-embryos under the aegis of an American biotech company, the Geron Corporation. Unlike NIH grant holders, Geron is able to carry out this sort of research legitimately and has already found ways of generating heart muscle cells and three types of neurone.[391,392] These are important technical advances, but the goal of treating patients will be elusive until we can be certain that implants will not be rejected by the immune system. The cells must not acquire mutations that would make them malignant during the many cell cycles required before they are ready for use.

The cloning of Dolly the sheep suggests a way out of the immunological dilemma.[124] Dolly was created from a fully differentiated cell of the mammary gland of an adult animal by "reprogramming" a cell nucleus to resemble a newly fertilised egg. This developed into a sheep, genetically identical to the sheep from which the original nucleus came. The application to medicine that "reprogramming" suggests is that a cell nucleus from a patient – say, with Parkinson's disease – could be "reprogrammed" by nuclear transfer. Embryonic stem cells would then be obtained from a very early embryo cultured *in vitro* and could be used, in principle, to correct the brain lesion. These would be immunologically identical to the patient and should not be rejected. The scheme is at present entirely speculative, although experiments with mice suggest that it will eventually be possible.

The crucial step in this process is nuclear transplantation, a procedure that would raise two important ethical questions. The first, the destruction of a fertilised egg nucleus – to be replaced by the nucleus of a somatic cell – is the same moral dilemma that confronts any proposed manipulation of a fertilised egg. Has the zygote (the sperm-egg combination) achieved personhood at the time when it would be destroyed? These fertilised eggs are what Robert Edwards calls "spare embryos" (i.e., left over from IVF), so the ethical question is exactly the same as for IVF. Most biologists believe the answer is no and that the status of personhood is achieved only after the embryo is substantially formed. The second ethical checkpoint is the status of the reprogrammed nucleus, which if allowed to develop into a human child would be a clone of an existing person, just as Dolly the sheep is a clone of another, unnamed sheep. A child produced by reproductive cloning would almost certainly be physically harmed as a result. There is also wide agreement that such a child would have been wronged in a more general and obvious sense: it would be deprived of anything remotely like normal parentage. Moreover, what is usually called reproductive cloning would be utterly pointless except as a publicity stunt.

Scientists believe that there is a case for permitting a "transferred nucleus" to develop, for no more than two weeks, to create a tissue from which embryonic stem cells could be cultured. These could potentially be used in cell therapy for degenerative diseases of the type considered earlier, a scenario that is usually called "therapeutic cloning." Most scientists concerned believe that this is no different to the entirely legitimate use of "spare embryos," which has been sanctioned in British law since 1991.

Early in 2004, a team from South Korea reported in *Science* two remarkable landmark achievements: efficient "therapeutic cloning" and the successful culturing of human embryonic stem cells from the "clones." Nuclei from the normal diploid cells that surround developing eggs were transferred into egg cells and then cultured to the blastocyst stage, using novel procedures. The cells from the inner mass of some of these pre-embryos could be cultured as stem cells and could make bone, muscle, and nervous tissue. The authors attribute their success to the high-quality eggs they obtained from an unusually public-spirited group of donors along with their novel procedures. The wider biomedical community has the feeling that the prospect of treating Parkinson's disease and other intractable conditions is now significantly increased.

Recently enthusiasm for a challenge to this ethical checkpoint has diminished as the technical difficulties have come into focus. There is no lack of alternative strategies. One scheme, as yet undeveloped, is to find a totally novel way of reprogramming a nonembryonic cell type using the tools of molecular biology. Existing human stem cell lines could possibly be modified, uncontroversially, to make them resistant to any immunological assault before they are used for cell therapy.[393] How stem cells will be used to refurbish tissues and organs is still very vague, but probably the novel ideas of tissue engineering discussed earlier will be deployed.

Challenges, Risks, and Rewards – A Summary

Surgery as a solution to disorders of the human body has involved the severest of risks and the most gratifying rewards. The audacity and ambition of surgeons has steadily advanced the boundaries of what is technically possible in the interests of restoring damaged and diseased bodies. Just as remarkably, patients willingly put their lives in the hands of surgeons, from fear of the alternative. In the last century, blood transfusion, open-heart surgery, organ transplantation, and thousands of less spectacular developments have established surgery as a medical intervention that can decisively improve the quality of life, and indeed can extend life by many years. With hindsight, daring undertakings are always easy to criticise. The potential for disaster in blood transfusion was recognised by its earliest practitioners, but its potential value outweighed any hypothetical risks. Eventually, the system was severely tested by the hepatitis and HIV viruses, with tragic consequences, but armed with experience, blood transfusion services everywhere are finding better ways of eliminating potential risks. Similarly, after the pioneering struggles, organ transplantation began to provide salvation for a diverse group of people that would otherwise have died young. In such circumstances, the difficulty of meeting the demand for organs has created new problems in the social sphere, such as the "organs-for-cash" scandals. For a profession driven by audacious initiatives, the temptation to consider the organs of other animals as substitutes for human organs is perhaps inevitable, although this is clearly too dangerous a step to be undertaken at present. In skin grafts and bone marrow transplants, stem cells contribute importantly to an essentially surgical refurbishment of the human body. We are now on the brink of a revolution in which human stem cells will be used to grow replacement tissues. This novel technology will be used to overcome hitherto intractable problems, including genetic diseases, chronic neurological disorders, cancer, and accidental damage (such as spinal injuries that paralyse their victims). Compelling indications from studies of model animal systems suggest that these aspirations are plausible and may be applied to humans in several decades. While far-reaching benefits are a possibility, some of these objectives will not be reached without passing further ethical checkpoints.

GENERAL SOURCES: 142, 165, 298, 373.

14 Living with the Genetic Legacy

The recently completed draft of the human genome sequence is literally a vision of the genetic legacy of our species, revealing the relentless evolution of DNA as millions of differences in base sequence between different individuals. These differences create the diversity of appearance and character of individual humans that is an important part of the texture of the human condition, and also the origin of genetic disease. With perhaps one in thirty births in the industrial world adversely affected to some degree by genetic disorders, what should be done about this burden? Must we accept the cruelty of fate, or is it a humane and proper thing to find solutions for those who suffer from these distressing conditions? For those who know that they carry a genetic disease, the problems of parenting are especially poignant. Most geneticists believe it is a mark of our humanity to fight the dismal consequences of genetic disease rather than to accept them passively. Surgery or organ transplantation, appropriate diet, and replacement therapy can correct, to an extent, some of these disorders. However, avoiding the birth of genetically damaged children is becoming an increasingly realistic, if controversial, option.

The Burden of Genetic Disease

There can be few parental heartaches greater than the implacable advance of a genetic disease slowly crushing a child's unique personality. Still more awful is the burden that the child may carry as its mind shuttles between a brave effort to live like other children and a terrible wish that it had never been born. Genetic disease is a poignant burden, too, for prospective parents from families or communities afflicted by thalassaemia, Tay-Sachs disease, sickle-cell anaemia, and many other ancestral curses. Genetic disease is an intractable part of clinical practice; it is tackled with a few puny clinical tools and by advising prospective parents how to avoid the risk. Although the difficulties in curing these diseases are immense, significant progress in managing them has improved life expectancy and the quality of life of patients. In the 1950s, for example, cystic fibrosis patients usually died at four years of age; today, they live into their twenties if not beyond, albeit with a substantial burden of poor health, sustained by antibiotics and devoted care.

Substantial numbers of children are born with congenital abnormalities of varying seriousness that appear to be of genetic origin. Limb malformations afflict one in a thousand babies.[394] The Tetralogy of Fallot, the notorious hole-in-the-heart condition, is the most common form of congenital heart disease – one in three thousand live births – and is now confidently attributed to mutations.[395] Modern surgery can solve some of these problems, including heart defects, milder cases of spina bifida, cleft palate, and conditions such as Hirschsprung's disease in which the enervation of the gut is not completed. For these individuals, the story can have a happy ending as they often live a more or less normal life after surgery.

Rare and strange diseases that run in families have been recognised throughout recorded time. The Talmud, written more than two thousand years ago, reveals a clear understanding of the inheritability of haemophilia; it excuses sons from ritual circumcision if a brother or maternal cousin has already bled to death.[396] At the start of the last century, Archibald Garrod was the first to show how some diseases are inherited according to Mendel's rules and are caused by what he called "inborn errors of metabolism." Since Garrod's time, thousands of single-gene disorders that cause congenital birth defects have been recognised.

We are usually protected from the consequences of mutation because chromosomes exist in homologous pairs containing two complete sets of genes, so that, in general, a good copy of a gene can compensate for a mutant gene. To recapitulate, recessive genes present as one copy have relatively little effect, because a good member of the pair is present. However, if both parents have a mutation in the same gene, the offspring may develop the disease if they inherit both versions of the mutant gene (i.e., are homozygous). Dominant mutations cause disease when only one copy of the mutation is present that overcomes the effect of the good gene and usually appears in later life, after children, who may carry the fateful gene, have already been born. Huntington's disease, one of the most notorious, strikes in late middle age, first with involuntary dancing movements of the limbs and then with a slow progressive dementia.

Mutant genes carried on the sex chromosomes or on mitochondrial DNA are inherited in special ways. Mutant genes on the Y chromosome affect all male children. Male offspring are also victims if there is a recessive mutation on the X chromosome, because there will be no paired chromosome to provide a sound version. There are about fifty rare genetic diseases encoded in mitochondrial DNA, which are inherited only from the mother because mitochondrial DNA arriving in sperm is destroyed after fertilisation (see p. 53).

How frequent is genetic disease? One estimate suggests that in every thousand babies, roughly four to fourteen have mutations in single genes; about seven have chromosome abnormalities; and perhaps twenty have congenital malformations of unknown origin.[104] In addition, an unquantifiable component contributes to chronic age-related disorders. The average severity of these diseases is also unquantifiable and variable. Altogether, at least three percent of births in the industrial world are affected by genetic disturbances – an increasing contribution to health problems now that the threat of infectious disease is fading. The problem is largely invisible to the public and consequently not easily appreciated.

Genetic disorders causing an ambiguity of gender are distressing conditions that affect about one in 4,000 otherwise perfectly healthy children. They present doctors with unusual dilemmas requiring humane and decisive interventions. Such patients often have a strong perception that their true gender is different from the one their anatomy indicates. Indeed, there is now a suspicion that this psychological sense of male or female character may be established even before hormones are produced in the foetus. Sexual identity is normally established by the presence or absence of the Y chromosome, but mutations in an endocrinological function may make the sex at birth ambiguous. Very early in the development of a mammalian embryo, a specific region called the genital ridge is gradually transformed into genitalia. In an XY individual, this develops into male reproductive organs at seven weeks under the influence of male hormones. The same tissue develops into female organs by default in the absence of male hormones, but particularly in an XX individual. An XY foetus carrying a mutation that makes tissues insensitive to male hormones (AIS, androgen insensitivity syndrome) would develop testes but no penis or internal female sex organs, although breasts would develop in adolescence. An XX foetus with a congenitally abnormal adrenal gland may also develop genitalia of ambiguous appearance as a result of a surge in male hormones. Surgery at birth and management of the hormonal status of the child at puberty can, in principle, solve the immediate physical problem of assigning gender, but at birth it is impossible to anticipate how the child would feel about its gender. The hope must be that a test will be devised to provide this information.

Recessive mutations that cause genetic disease are puzzling because they do not die out during evolution. Rather, they persist in the population, hidden by a normal gene, only to reappear occasionally in an individual who inherits two mutant forms of the gene. Many recessive genetic diseases are extremely rare but may be more common in isolated social groups.

Certain mutations that cause serious diseases in the double-dose (i.e., homozygous) state, such as sickle-cell anaemia and thalassaemia, are very common in the single-dose or carrier state in some regions of the world. The sickle-cell trait originated in sub-Saharan Africa, where about one in thirteen people carry the gene, but it is now present in the United States and the Caribbean, introduced along with African slaves. Thalassaemia is found in most tropical and subtropical regions of the world. Individuals with the homozygous form (two mutant copies of the gene) of either disease die young, leaving no children; in contrast, in the heterozygous state, the gene persists because it protects the carrier against malaria. The WHO reckons that seven percent of the world's population is heterozygous for either of the two genes and that thalassaemia represents three-quarters of these cases.[104]

Sickle-cell anaemia results from a mutation in the globin gene that changes just one amino acid residue in the protein chain and renders haemoglobin insoluble when loaded with oxygen. Affected cells adopt a characteristic sickle shape that impedes their free circulation in blood vessels and causes anaemia, shortness of breath, localised bleeding, and acute pain throughout the body. Thalassaemia exists in many forms of varying severity, caused by different deletions of the coding regions of the four forms of globin.

Enhanced resistance to malaria arises because red blood cells (of the heterozygotes) infected with the malaria parasite (*Plasmodium*) are removed from the blood circulation more quickly than uninfected cells. Infections by the parasite are consequently significantly less serious than they would be for other people, but the heterozygotes are perceptibly anaemic.

A deficiency of the enzyme glucose-6-phosphate dehydrogenase also confers resistance to malaria because red blood cells burst when the parasites try to use oxygen,[397] and the life cycle is broken.

Although not as numerous as haemoglobin diseases, other genetic diseases, common in the heterozygotic form, afflict certain communities. The Ashkenazi Jews, whose ancestors lived in genetic isolation in Eastern Europe, suffer from a number of serious genetic diseases that are much less frequent elsewhere. Tay-Sachs disease, for example, is present in the single-dose form at the very high frequency of one in twenty-five and is usually fatal during childhood in the homozygous form. The disease is caused by a defect in one of the enzymes that recycle cell-surface components called glycolipids. These accumulate in the lysosomal compartment of cells, eventually insidiously compromising their ability to function as integral parts of muscle, the brain, or the eye. Why the mutant gene should be so prevalent is unknown. Some think it may protect against tuberculosis in the heterozygous state, but others believe it reflects conscious efforts by parents to have more children to compensate for their losses.[398]

About four percent of northern Europeans carry a mutation in the cystic fibrosis gene, a much higher frequency than exists elsewhere in the world. The disease is caused by a defect in an ion pump that catastrophically disrupts the proper functioning of the lungs and intestine when the mutant gene is in the homozygous state. In heterozygotes, the pump marginally underperforms, protecting those individuals with only one copy of the defective gene from dehydration during diarrhoeal diseases such as cholera or dysentery, possibly conferring a selective advantage to heterozygotes (see p. 74).[399]

The major chronic diseases that affect middle-aged people are not definitively genetic like the rare single-gene diseases, as their genetic component may be apparent only under particular circumstances. The health records of identical twins tell us that genetic factors are less important in determining longevity and incidence of cancer than life experience. However, in a number of chronic age-related diseases (such as insulin-independent or type 2 diabetes), the genetic component is evident when the disease is provoked by an environmental cue. Similarly, type 1 diabetes (insulin-dependent), manic depression, allergies, psoriasis, and other conditions are activated by an environmental cue in people with a genetic predisposition towards the disorder. Even in tuberculosis and leprosy, diseases that could not occur without infection by bacteria, susceptibility is a genetic factor.[104]

Avoiding Genetic Disease

Charles Darwin concluded – from his own experiments with plants and animals, from anecdotal evidence and from serious studies of humans – that in-breeding was potentially injurious to progeny. The strong preference for marriage to first cousins of the Darwin–Wedgewood family, including himself, could not have been far from his thoughts. The available data at that time, including that collected by Darwin's son George, suggests the practice was common only in geographically or socially isolated communities or amongst the aristocracy.[400] Mendel's contribution to genetics, once it became known, gave a tangible basis for the fear of in-breeding, as it was easy to see how inbred communities would have a greater prevalence of recessive genetic disorders. Indeed, Garrod (whom we met earlier) noticed a high frequency of alkaptonuria and other genetic diseases amongst the offspring of first cousins. The most recent estimates of the harm caused by cousin matings suggest that premature death and stillbirth is 40 percent more likely than in society generally.[401] The conclusion seems inescapable that the high frequency of the disease genes among Ashkenazi Jews is a consequence of in-breeding in the Eastern European schtetl. Marriage between cousins has probably been practised since prehistory as a device for maintaining family inheritances. The Old Testament patriarchs usually married cousins, and the custom continues in many parts of the world (particularly in the Middle East) at frequencies greater than 50 percent. The Council of Trent emphatically prohibited intercousin marriages in Catholic Europe in 1565, endorsing a trend that had been growing since medieval times. This almost uniquely Christian idea has no basis in the Old Testament but possibly reflects an appreciation of the possible dangers. The rate of cousin marriages in Europe and the United States has decreased in the twentieth century to less than 4 in 100,000. Moreover, cousin marriages were forbidden by law early in the last century in nineteen American states.[400,401]

The growing suspicion of cousin-cousin mating is quite distinct from the almost universal taboo against intersibling mating that has existed throughout human history – the royalty of ancient Egypt, Hawaii, and the Incas being the only exceptions. Anthropologists and evolutionary biologists suggest that the closeness of siblings during the first thirty months of life generates a profound psychological instinct that inhibits any interest in mating as adults. The laws of Moses, too, are quite unequivocally opposed to intersibling marriages (Leviticus 18:9-14), but oddly, Sarah, the wife of Abraham, who had such difficulty conceiving (see Fig. 2.1), was his half-sister, the child of his father's second wife.

Today, prospective parents who perceive a risk of genetic disease can be counselled to avoid this saddest outcome of pregnancy. In societies severely threatened by genetic diseases, community and religious leaders along with public health authorities actively encourage couples to avoid the possibility of such a birth. This is true in Cyprus and Sardinia (thalassaemia),[402] in Ashkenazi Jewish communities (Tay-Sachs disease),[404] in the Muslim world,[403] and in communist Cuba (sickle-cell anaemia).[402] Interestingly, some societies embrace genetic testing rather than abandoning intercousin marriages.[403] Tests that identify a carrier state were the important breakthrough in finding a route to avoiding genetic disease and are now available for thalassaemia, sickle-cell anaemia, Tay-Sachs disease, cystic fibrosis, and many others. These depend on blood tests by which heterozygotes are identified as having half the normal amount of gene product (the details need not detain us here).

Once carriers could be identified easily, the possibility of unfair discrimination against them became an immediate problem. Thirty years ago in the United States, insurance companies and employers sometimes confused the sickle-cell trait with the actual disease. Consequently, affected blacks frequently had difficulty in obtaining insurance and employment because of an absurd misinterpretation of their genetic status.

Members of the New York Hasidic Jewish community, like other Ashkenazi Jews, are frequently carriers of Tay-Sachs disease, which is fatal in the homozygous form. The community is totally committed to the exhortation of Genesis to "Go forth and multiply," and abortion is considered completely unacceptable. They have devised a remarkable and extremely effective system to guide their choice of marriage partners – the Dor Yeshorim program. The system rests on an anonymous, voluntary register of the young people of the community who have taken a Tay-Sachs disease test. Prospective marriage partners consult the register to see whether they could produce the double dose of the Tay-Sachs gene. To avoid any possibility of the birth of a baby with Tay-Sachs disease, a couple receiving this doleful warning regretfully abandon marriage plans.[405] The program has been remarkably effective, and Tay-Sachs disease has almost completely disappeared from the community.

Other communities notably afflicted by genetic diseases have come to believe that they have a moral right to defy this lottery, although few contemplate such a strong-minded scheme as the Dor Yeshorim program. Most other communities use prenatal diagnosis followed by therapeutic termination of the pregnancy if the foetus is likely to be affected. The practice is entirely legitimate almost everywhere, and even the religious leaders in Sardinia and Cyprus tacitly support their respective programs. Wherever sustained efforts to avoid the birth of babies with thalassaemia or Tay-Sachs disease have been undertaken, new cases of the disease have decreased notably.[402,405]

Scientists have been developing tests for genetic disease since the 1980s. The earliest was a test for thalassaemia that assessed the capacity of foetal blood cells to make haemoglobin. This was displaced by amniocentesis, a more versatile method in which cells from amniotic fluid – the liquid surrounding the foetus – are cultured until sufficient material is available for a test, late in the second trimester. This too has been abandoned in favour of chorion villus sampling, because late-stage termination is often distressing to the mother. Between nine and fourteen weeks after conception, cells removed from the outer layer of membranes that envelop the embryo are used to obtain foetal DNA for immediate analysis.

Older mothers are screened for the possibility of an abnormal chromosome configuration, such as Down's syndrome. Secretion of a substance called alpha-fetoprotein is now widely regarded as a warning sign of a potential neural tube defect originating from a chromosomal rearrangement. Direct examination of the foetus by ultrasound scans or X-rays to identify congenital malformations during the second trimester is the last safeguard available.

In vitro fertilisation technology makes genetic disease avoidable by another route. Eggs fertilised *in vitro* are cultured until the blastocyst stage (a ball of about one hundred cells) and then implanted into the mother. A peculiar feature of the eight-cell embryo is that one or two cells can be removed, and the remainder of the embryo still develops into a normal foetus. By culturing these cells, sufficient material is obtained to screen for harmful mutations or an abnormal chromosome complement. If there is a problem, the embryo is not implanted, and the procedure is repeated with a different egg until a conceptus is obtained that lacks the mutant gene or has a perfect chromosome configuration. This technique, usually known as preimplantation genetic diagnosis, was first performed in London in

1990.[406] Two children were born who had been "selected" from a number of embryos fertilised *in vitro* to carry both X chromosomes, because the paternal Y chromosomes carried genes for serious genetic diseases. The press was quick to sensationalise the event by labelling the products "designer babies," but this was a cheap canard, as the procedure bore no resemblance to the activities of the eugenicists whose spectre they wished to summon up.

Eugenics – the word was coined in 1883 by Francis Galton, Charles Darwin's cousin – was to be a scheme to eliminate "bad" genetic traits. Over three decades in America and Europe, the idea of curbing the reproductive powers of people perceived as genetically unfit gradually gripped a large section of middle-class opinion. Laws were enacted in America and Sweden that authorised the sterilisation of tens of thousands of individuals because of their feeblemindedness or criminality. The notion that "feeblemindedness" or "criminality" was inherited in a simple fashion, and that sterilisation of such individuals would remove the "bad" genes from the population, was recognised as nonsense by the genetics community. The pity is that the leaders of the discipline, such as Thomas Morgan and J. B. S. Haldane, never managed to challenge the political forces behind eugenics. The movement reached its dismal nadir in the depraved and lunatic Nazi regime and mercifully died with it.[407] Eugenics will be damned forever as gravely flawed science and an appalling affront to civil liberties. In the last decade, similar ideas have been surfacing again in China, where the government is convinced that in-breeding is the cause of congenital birth defects and genetic diseases. A law enacted in 1993 was intended to eliminate "inferior births" by forbidding those with hereditary diseases from marrying.[408]

We should be absolutely clear that the desire to avoid thalassaemia or Tay-Sachs disease differs in three important respects from the eugenics of the early twentieth century. Prospective parents participate of their own volition; nobody is sterilised; and the scientific basis of the traits in question is well understood. Moreover, geneticists believe that the genetic factors affecting personality are likely to be far too complex for a selective process to be remotely credible. An important special case is sex selection for its own sake, which, as we have seen, can be done after preimplantation genetic diagnosis. Sex selection is also possible using a sperm sample highly enriched for sperm carrying a Y chromosome, prepared by a novel "sorting" technique. Alternatively, natural pregnancies can be terminated once the sex of the baby is known. Private IVF clinics in Europe and the United States will assist in adjusting the "gender balance" of families, using either IVF or sperm sorting. In parts of Asia, the demand for sex selection is clearly very strong. The number of female births in some regions of the Punjab has already fallen to 750 per thousand boys, because of selective termination of pregnancies.[409] In November 2003, the British Human Fertilisation and Embryology Authority (HFEA) ruled that reproductive technology should not be used to adjust the "gender balance" of families. In making this judgement, they took into consideration an opinion poll indicating that 80 percent of the public disapproved of such selection. However, the HFEA has also ruled that an attempt to avoid the birth of a boy with a male-sex-linked genetic disease by prenatal genetic diagnosis is ethically acceptable.

Preimplantation genetic diagnosis has raised more subtle ethical problems, such as whether a child conceived by this intervention has been "wronged." This kind of case arises if an attempt is made to "select" embryos with a tissue type similar to that of an older sibling in order to provide what the press describes as "spare parts." Adam Nash, the so-called spare parts baby, was conceived by IVF at the Genetic Reproductive Institute in Chicago in 1999. He was one of fifteen fertilised eggs screened in order to find one with a tissue type identical to that of an older sister, Molly, who suffered from the rare and usually fatal Fanconi's anaemia.[410] His cord blood stem cells could then be used to correct her condition, using bone marrow transplantation. Molly was prepared for the graft by chemotherapy, to destroy her bone marrow, and was then transfused with her brother's cord blood. Her bone marrow became populated with stem cells that generated a population of normal healthy blood cells that grew and allowed Molly to live a normal life. Yuri Verlinsky, whose private clinic carried out the IVF, indicates that more babies of this kind are on the way, selected to have matching tissue types for a transfusion. In Britain, a similar proposal to use preimplantation genetic diagnosis to obtain a healthy child was fiercely contested. The aim, in this case, was to select a zygote with the same tissue type as an older sibling who suffered from thalassaemia. The child when born could then be a donor of umbilical cord blood or bone marrow cells. The media condemned the proposal as unethical, because a healthy child would be committed to be a donor for the older sibling, possibly forever. The HFEA initially approved the plan, but it was successfully challenged subsequently by a Catholic antiabortion group in the appeal court. The HFEA had no reason to comment on the Ayala case (see p. 284), in which a child was conceived by the traditional method to provide bone marrow for an elder sibling recovering from chemotherapy. Although that decision was initially unpopular in the United States, it was legal, and the outcome was a useful and happy life for the elder sibling.

The ethical debate about termination of pregnancies because of congenital imperfections has taken a sharper turn in recent years due to the activities of disability pressure groups. For example, the British Disability Rights Commission, under the influence of the Pro-Life antiabortion group, argued in 2002 that the British Abortion Act reinforces negative stereotypes of disability and is offensive and discriminatory. Similarly, spokespersons for the deaf community argue that attempts to avoid the birth of children with inherited deafness harm the interests of the deaf community.[411] At a personal level, a well-known bioethicist has revealed that he has *achondroplasia* (restricted growth of the limbs, caused by a deficiency of cartilage). The condition was inherited from his father and was passed on to his children, with no adverse consequences to the family's happiness.[412] Evidently, a delicate fault line separates two views of biomedical interventions that may allow a mother to avoid an unsatisfactory outcome of parenthood. On one side are individualists with aspirations for the most perfect birth possible; on the other are those who believe we should accept the machinations of fate, however disappointing or limiting that may seem. The discussion is frequently turbidified by sentimental reflection on the child who might have been born if the pregnancy had not been avoided. This seems to be a philosophical red herring, because, as we have discussed, the fate of most fertilised eggs is oblivion, and we do not reflect on their demise.

The situation of mentally handicapped persons who might conceive children presents more ethically challenging conundrums. Involuntary sterilisation, without consideration for the rights of the patient, was once commonplace in those countries where eugenics was practised, but these ideas are now widely condemned as unscientific and inhumane. Most countries insist that informed consent of the subject is necessary and that if this is not possible, such decisions must be made by courts, who usually conclude that sterilisation is not in the best interests of the patient.[413]

Treatment of Genetic Disease

Diabetes, though not strictly a genetic disease, was the first instance of a medical disorder that could be rectified by an injected protein. Other examples followed: administration of thyroxine from birth to correct a congenital underactive thyroid; growth hormone used to prevent brain damage and congenital insufficiency and severely stunted growth. The genetic diseases that affect blood clotting – haemophilia A and B – can be corrected by infusion with blood factors VIII and IX. Probably the most common genetic disease amongst Caucasians – and now one of the most manageable – is haemochromatosis, a painful and sometimes fatal condition originating in the iron deposited in many tissues. One in two hundred are homozygous and develop symptoms; almost one in ten Europeans are heterozygous, without symptoms but with elevated iron levels. Surprisingly, a simple and reasonably effective treatment predated the discovery of the gene (1996); this involved removal of a pint of blood once a month to reduce the iron load. The similarity of this practice to the bloodletting beloved of eighteenth-century physicians must make forward-looking biotechnologists weep. Sadly, of the thousands of genetic diseases known, very few can be successfully treated. In the case of sickle-cell disease, for example, we must admit that the discovery of the cause more than forty years ago has not greatly helped its victims.

Dietary management can overcome the worst effects of a few genetic diseases. Since the 1960s, newborn babies have been routinely tested for a rare disease called phenylketonuria. Those who are positive are provided with a diet deficient in phenylalanine, which prevents the neurological damage associated with the disease. An unusual tissue-type gene causes an immunological reaction to gluten, a protein present in wheat, that makes young patients feel ill and unable to thrive; the symptoms, usually called celiac disease, can be prevented by avoiding dietary gluten.[414]

The most radical treatment for genetic disease is the transplantation of bone marrow and other organs. Indeed, the principal use of organ transplantation is to overcome the organ-damaging consequences of genetic disease. In hyperoxaluria, for example, accumulated oxalic acid severely damages the liver and kidneys. Diabetics frequently require kidney transplants. Patients who have organ transplants live for many years free of their ancestral curse. Bone marrow transplants are used to treat sickle-cell anaemia, thalassaemia, adenosine deaminase deficiency, Gaucher's disease, and many others, if suitable donors with matching tissue types are available.[387]

As genetic diseases usually involve congenital absence of a protein, replacement therapy to restore the missing component is a logical idea. This is most feasible for proteins normally resident in the bloodstream (blood factor diseases) and is less plausible for proteins that function in cells. However, some important exceptions are the forty-strong family of storage diseases with defects in the ability to recycle the cell surface chemicals known as glycolipids, which cause complicated pathologies. As material destined for recycling is transported into the cell by vesicles that fuse with the cell surface, the same vesicles might import the missing enzymes. A treatment for Gaucher's disease of exactly this kind has been in routine use for more than ten years. The cause of this disease is the inability to break down gluco-cerebrosides (compounds of carbohydrate and lipid found in cell membranes); the condition ultimately causes anaemia and bone erosion in young adults. The remedy, a preparation of the enzyme glucosidase (Cerezyme®), is infused into the bloodstream every two weeks. The enzyme enters cells throughout the body from the bloodstream and then attacks the gluco-cerebroside, the immediate cause of the disease. Fabry's disease, a disorder with similarities to Gaucher's, responds to galactosidase delivered in the same manner, and the entire family of diseases may well be treatable in time using similar principles.[122]

The great lesson of human genetics is that a single amino acid substitution can cause a genetic disease; the affected protein is often destroyed by the proteasome because it cannot form a correct structure (see Chapter 4). Other systems recognise and destroy incorrectly folded proteins in the secretory pathway, thus obliterating a protein that would otherwise have had just reduced activity. Scientists see a glimmer of hope for treating genetic disease in this observation. They suggest that if a substance similar to the normal substrate of the protein were introduced into the body, it might help the protein to fold correctly and evade the disposal process. This stabilised protein would be sufficiently active to do its job in the correct location and prevent any serious biochemical deficiency. An exciting possibility is now opening up: many genetic diseases – including the lysosomal storage diseases, cystic fibrosis, emphysema, and many others – may be treatable by an innocuous small-molecule pill that helps the suspect protein fold correctly.[415]

Fixing Faults in the Genome – Gene Therapy

The idea of replacing a diseased gene with one that functions correctly is one of humankind's most radical notions for avoiding the fate dictated by the genes. In principle, it is similar to the genetic transformation that occurs naturally in bacteria. A DNA fragment encoding a protein-coding sequence and all its regulatory sequences should direct the making of a protein when introduced into an appropriate cell type and could compensate for any deficiency imposed by a mutation. In 1990, the pioneers of the process now called gene therapy thought it was going to be easy, and that it would be applicable not only to genetic disease but also to cancer and heart disease. Indeed, the hundreds of companies, mostly in the United States, preparing for this eventuality indicate that optimism is still riding high.

An important preliminary to approaching the problem has been the creation of lines of genetically engineered mice that develop handicaps exactly like those caused by human genetic diseases. Almost any genetic handicap that can be identified in humans – such as cystic fibrosis, Parkinson's disease, high blood cholesterol, and many others – can be reconstructed in a laboratory mouse. These animals are used for testing the efficacy of gene therapy protocols and other treatments. Many notable successes have been recorded with mice, but translating these experiences into a protocol suitable for human patients is a step that requires the utmost caution, because so many undesirable consequences are possible.

The strategy is to deliver therapeutic genes to a target tissue (e.g., the lungs for cystic fibrosis) in a harmless virus or as naked DNA. The viruses attach themselves to target cells and deliver their DNA into the cell, where the therapeutic gene becomes active. The long-standing favourites are the retroviruses, which insert their genes into the chromosome of the infected cell, ensuring that all the descendants of the infected cells will inherit the gene. A limitation is that these viruses infect only dividing cells (as in bone marrow, for example), thereby limiting their use to tissues where cell division occurs. The possibility that the virus might insert itself into an inappropriate site, where it could promote cancer, was once considered a remote possibility, although as we shall see shortly, this was a misjudgement. Another virus that infects both dividing and nondividing cells (adenovirus) is also a favourite, but unlike retroviruses it provokes an immune response and could be used only for a transient burst of therapeutic activity. A third candidate, which can insert itself into the chromosome – the adeno-associated virus – is gaining support. The virus was successfully employed in a trial to correct haemophilia B (lack of the blood-clotting factor IX). A new candidate, used recently in successful tests on mouse models, is the AIDS virus (HIV), stripped of all its dangerous attributes, which can integrate into the chromosomes of dividing and nondividing cells.

Blood diseases are the most accessible targets for gene therapy. Lymphocytes taken from the patient can be infected in the test tube and reintroduced into the body, where they can start to deliver a therapeutic protein. The first successful instance of gene therapy using this technique occurred in

1990 in the case of a four-year-old girl suffering from a serious immune deficiency caused by adenosine deaminase deficiency.[416] The child is now a healthy teenager with a fully functional immune system in which about a quarter of her T-lymphocytes are genetically modified, although she still receives a low dose of genetically engineered adenosine deaminase.

The technique has been improved substantially using better versions of the virus, and also by infecting the stem cells of the bone marrow to permit the virus to replicate indefinitely in the lymphocyte lineage. By 1999, a substantial number of children with X-SCID (the X-linked immunodeficiency that condemns its victims to live in a plastic bubble) and related conditions had been treated successfully. Immunity was restored, and they were able to live a normal life outside of the bubble. Gene therapy seemed to be working and was in use in several hospitals around the world. Then, in late 2002, two infant patients at the Necker Hospital in Paris developed leukaemia.[417] The cause was quickly traced to the unexpected insertion of retrovirus into one specific site in the genome at which an oncogene had been activated. After this very serious setback, trials of the procedure throughout the world were suspended while the matter was investigated, although many other approaches involving different vectors are still under consideration. This incident followed the tragic death from an unaccountable inflammatory reaction of an apparently healthy young subject, Jesse Gelsinger, in a trial therapy involving adenovirus in 1999. In the aftermath of the incident, industrial sponsors made strong petitions for secrecy to protect their commercial interests – a response guaranteed to arouse the distrust of the public.

These setbacks are a serious challenge to the whole idea of gene therapy, but with current interest in developing new methods it may still have a future. Typical of the wide-ranging enquiries still being made are investigations into the use of naked DNA as a means of gene therapy.[418] At first sight, the idea seems implausible, but as DNA causes no immunological reaction (unlike viruses) and is unlikely to integrate into chromosomes, it has attractions. Moreover, when DNA is encapsulated in a novel lipid, entry into cells is facilitated. In one notable trial, naked DNA was used, apparently successfully, to treat patients suffering from angina – the acute chest pain caused by a blockage of blood vessels in the heart muscle that often accompanies coronary heart disease. The idea was to stimulate new blood vessels to form around the blockage using naked DNA, encoding a protein called vascular epithelial growth factor (VEGF) that has a well-established role in making new blood vessels. The DNA was introduced into the heart muscle, where it stimulated blood vessel formation for a short period and allowed blood to by-pass the clot; this reduced angina and permitted the patient to exercise more strenuously. The same idea has potential for saving the limbs of diabetic patients whose circulation has been damaged by a blood clot, and for gene therapy of cystic fibrosis and haemophilia. From a scientific viewpoint reported successes must be considered conservatively, as there is rarely a realistic untreated control group to which the effects can be compared.

Germ-line Therapy

The possibility of deliberately changing the human inherited genetic constitution (i.e., the germ line) is a serious ethical checkpoint. The public is understandably apprehensive that well-established technology used for laboratory mice might be applied to humans. The fear is that if specific pieces of DNA can be introduced into a mouse, then it might be just a small step to doing the same thing with humans, with the intention of "enhancing" desirable human features. Most practising scientists and bioethicists believe that germ-line modification of humans is an unnecessary, pointless, and probably dangerous exercise. In most countries, genetic manipulations of this kind are illegal and likely to remain illegal for the foreseeable future. However, some scientists interested in exploring the possibilities believe it could be more effective than gene therapy of somatic cells.[419] How this will be done is not entirely clear. The former director of the Human Genome Project believes germ-line modification should be a choice left to prospective parents, and he apparently believes the idea of making children more attractive or more intelligent by genetical means is a worthy objective.[420] By contrast, most members of the profession believe the existing choice of techniques for avoiding genetic disease should be enough for any anxious couple without contemplating interventions that are ethically questionable or potentially hazardous to the child.[421,422]

The Human Genome Project – a Detailed Map of a New Frontier

During the year before February 2002, a veritable blizzard of human DNA sequence data emerged from more than a score of labs to be assembled into linear sequences. The entire genome contains three billion base pairs, compiled from millions of sections of sequence, read in units of less than one thousand bases. This was the culmination of many diverse developments of late twentieth-century biology, of which the mechanics of DNA sequence reading, the mapping of genes, and the availability of computers of astonishing power were only three of the crucial contributions.

About 30,000 or so genes are present in the genome – rather less than some people expected – but this is no reassuring indication that the problem is simpler than we thought. Many genes of completely unknown function are present, and importantly, most genes can be read in different ways to create a family of similar but not identical proteins. We already know about many genes that control metabolism, but now have the tantalising prospect of access to the sequences of every gene in the genome. Quite possibly, we will start to learn about genes that regulate human attributes such as personality, speech, behaviour, and creativity.

The idea of reading the entire human genome as a DNA sequence was initially floated in 1985 by Robert Sinsheimer, a veteran of the earliest days of molecular biology. The proposal was made at a time when sequencing costs were prohibitive, computer technology inadequate, and the entire conception fraught with serious obstacles. Nonetheless, some prominent scientific leaders needed no persuasion that the information was worth obtaining and could be obtained. The thrilling notion that gripped their minds was that the sum of the information encoded in the genome was a statement of what it was to be human, in every sense, that could come from no other source. Nobody supposed the result would be easy to comprehend, but the instigators believed that even if centuries elapsed before the genome could be understood in its entirety, the consequences for understanding human biology would be immense. If every gene was known, their contribution to the normal life process and to physical disease could eventually be assessed.

The U.S. Department of Energy was an early supporter of the project, envisaging the use of its resources in a project comparable to sending a man to the moon; by 1988, NIH was also keen. From that moment, the future of the project was secure, in spite of vociferous dissidents who viewed it as "a mindless big science project" that would drain funds from other science to no useful purpose.

Genes of known sequence were first mapped to particular chromosomes in the 1970s. Sites on the chromosomes were identified with alternative sequences in different individuals (so-called polymorphisms) that became landmarks for mapping genes for known diseases. By 1987, four hundred distinct landmarks of this kind had been mapped using the pedigrees of very large families from genetically isolated groups, such as the Mormons and the Pennsylvania Amish. The genes for Huntington's disease (1983), adult polycystic renal disease (1985), muscular dystrophy (1987), and cystic fibrosis (1989) were mapped relative to these landmarks during this period, and within a few years, the protein products of the genes were known. Once the genome project got going, geneticists found many more polymorphisms, making the entire process of mapping extremely robust.

Sequencing technology developed rapidly, in tandem with awesome computing power, so that when the time came to start reading the genome seriously, no difficulties were apparent. Sequencing organisations proliferated in the early 1990s; the Wellcome Trust and the MRC agreed to support a British centre for the project that would eventually contribute a third of the sequence. In 1993, it seemed the sequence of the entire genome would be complete by the year 2005, but further advances in sequencing technology hastened the process. A system for managing data collection, devised in 1995, enabled Craig Venter to sequence the entire genome of a bacterium (*Haemophilus* – 1.8 million base pairs) in less than a year. Three years later, in launching his own company (Celera) to sequence the human genome, Venter stunned everyone with his claim that the human genome could be finished in three years. Piqued by the prospect of many genes being patented, the Wellcome Trust doubled its support for the project and initiated a dramatic two-horse race that finished in a dead heat in February 2001, less than two years after the start.

The jubilation of the science community notwithstanding, bystanders generally were understandably mystified and bemused by an enterprise whose outcome had been discussed for ten years and whose benefits would not materialise for decades. Nonetheless, fascinating items emerged from the completed project.[423] Perhaps the single most interesting idea is that human DNA differs by a single base (known as a single-nucleotide polymorphism, or SNP) at literally millions of sites in different individuals. This is the most tangible evidence of why we all look so different and why the biological outcomes of our lives are so diverse. Together, they create the extraordinary diversity of physical and biological characteristics that form such an important part of the texture of human life, from eye and hair colour to much more complicated attributes, including those that affect our life expectancy. Some cause rare and terrible diseases that may terminate a life soon after it begins; others predispose individuals to hereditary diseases that are activated by an environmental cue.

Physicians have recognised important differences in the ways in which individuals assimilate and detoxify chemicals and medicines for many years.[424] The human genome sequence provided evidence of at least fifty genes that could metabolise and detoxify drugs but which were unexpectedly variable between individuals. Our bodies can modify most foreign chemicals they encounter – including medicines – into something harmless before they are excreted. Some individuals do this rapidly, others not at all, so that drugs can be ineffective in one individual and poisonous to another; others fail to respond to drugs because they are not absorbed. At a time when adverse drug reactions are reported to be a major health risk,[425] the genome reveals the scale of the problem and suggests how drug doses could be personalised in the future.

In September 2002, under the headline "Scientist to Sell Copies of Grim Reaper's Timetable," *The Times* of London reported an offer by Craig Venter's company to sequence the genome of anyone for £400,000. Customers would be provided with a CD containing their entire genome sequence, which could be further analysed in the future, and would be told whether they carry mutations that might predispose them to a "dread disease." Another reporter, somewhat less audaciously, paid £600 to Genovations, an American gene technology company, for a set of available genetic tests to be carried out on cells washed from the customer's mouth. No forecast of an early death from a notorious genetic disease emerged from this, only a sobering prediction that the reporter was vulnerable to Alzheimer's disease, atherosclerosis, and high blood pressure. The intrepid journalist was advised to improve his lifestyle and to take some vitamin supplements.

Whether genetic tests will become more incisive or more useful in the future is hard to know. The public, unlike brave reporters, are unlikely to take tests for diseases for which there is no cure. Even predicted conditions for which a change of lifestyle would be a benefit (such as coronary heart disease or stroke) would only be useful if individuals seriously attempted to reduce the risk. On the other hand, one can imagine that screening for treatable conditions in newborn babies could become routine. Families with a history of genetic disease could get better advice to inform choices about pregnancy.

Improved classification of diseases based on sequence information could reduce uncertainties about the course of a disease and indicate the most appropriate treatment. The news may not always be reassuring and would almost certainly involve agonising choices. A test that predicted the onset of a disease that could be averted only by unpleasant treatments would be hard to propose if the subject were a child. However, families at risk of hereditary cancers have lived with this situation for some time. Health authorities have been helping quietly and successfully with tests, advice, and follow-up treatments that may be the model for future genetic health care.

Ethical Dilemmas and the Genome Project

When James Watson became head of the Human Genome Project, one of his first acts was to insist that three percent of the resources of the project be used to examine its ethical consequences. In parallel, most developed countries have organisations to consider ethical and regulatory issues arising from genetics. In Britain, the Nuffield Council on Bioethics, an independent body composed of experts and lay people, has been active since 1991. Issues arising from reproductive biology and genetics are regulated by the Human Fertilisation and Embryology Authority and the Human Genetics Commission, which rules on specific proposals, formulates guidelines for policy, and responds to public opinion.

Early in the development of the genome project, two issues needed immediate attention – the consequences of genetic testing and the status of intellectual property rights originating from the project. Genetic tests were always likely to be contentious; they could be an obvious means of discrimination in insurance, employment, or in finding mates, and any wrongs perpetrated would be even greater if the information were inaccurate or disseminated irresponsibly. Life insurers have always had a right of access to information known to affect life expectancy, such as a family history of genetic disease, but the arrival of tests for these diseases has changed the character of the transaction. Insurers visualise a scenario in which a person with an incurable disease buys massive coverage that may be a serious and unfair drain on their business if a claim is made. On the other hand, to refuse insurance coverage or to make it prohibitively expensive would affect dependants and their homes and could be unfairly discriminatory. Belgium, followed by four other European countries and many American states, has prohibited genetic tests for insurance purposes. The Netherlands and now Britain require disclosure of genetic test results only if the coverage required is very large.[426] The British government has approved the use of tests for Huntington's disease and is considering a test for Alzheimer's disease. Germany requires applicants to reveal any genetic test results. The problem of how to administer insurance for people who take genetic tests is coming closer to a solution, but the danger remains that patients may forgo the medical treatment they really need because of anxiety about insurance.

The historic importance of patenting to the dynamics of the pharmaceutical industry meant that as soon as the genome project started, intellectual property rights became an issue. The founders of the genome project were against the patenting of raw human DNA sequences because they visualised protracted negotiation for access to data that would impede the use of sequence information. Their idealism was severely tested by Craig Venter's enthusiasm for patenting, first at NIH in 1991 and then through his own companies. We will know the consequences of these machinations only when products are developed that require patent protection. Meanwhile, the major biotech companies carefully draft their patents for protein medicines to try to establish unassailable rights to all royalties from substantially homologous versions.

The big ethical challenges arising from the genome project are likely to arise when the physical bases of emotional, behavioural, and mental traits become apparent in the still far-distant future. The profession formerly known as personnel management and now, with more menace, as "human resources" may see possibilities in genome research for devising tests of aptitude or personality. Genetic tests intended for discriminatory purposes are something we should all fear, primarily because they are very likely to remain offensively simplistic and inaccurate until the remote time when the physical basis of human personality is understood in great detail.

As proteins involved in personality are discovered, they will probably become targets in the search for new drugs to help treat disorders of the mind. It is a fair bet that a new generation of recreational drugs will emerge that will present our sensation-seeking species with entirely different kinds of challenges.

Challenges, Risks, and Rewards – a Summary

Our genetic legacy is an extraordinary diversity of human genomes that has created six billion people who, with the exception of twins, are completely individual. The forces that create this marvel – random genetic damage – also create genetic disease and malignancy. While damage of this kind follows inevitably from the evolutionary process, we have no moral obligation to accept its cruel consequences. It is a humane and proper activity to find ways of evading the dismal consequences of this process. Indeed, uncontroversial but only partially effective measures are available, through surgery, organ or tissue transplantation, dietary management, and enzyme replacement therapy. Most discussion of the future of clinical genetics focuses on the need for more effective methods for managing genetic disease and on the need to avoid the birth of children with serious genetic diseases. Apart from not breeding with cousins, the only preemptive procedure available is prenatal diagnosis followed by therapeutic termination of pregnancy or preimplantation genetic diagnosis. Such solutions are realistic only if prospective parents know, from their family history, whether their unborn children are at risk. Even supposing that couples want more genetic information, several decades will elapse before such information could be used routinely. Radical therapies that correct genetic lesions permanently, safely, and effectively may be possible in the future, but for the moment, progress is very slow and punctuated by serious setbacks. There is little justification, in the immediate future, for becoming excited about either good or bad prospects.

The belief of the eugenicists of the last century that they could improve the genetic quality of the species is scientifically flawed and lacking in appreciation of human rights. Their proposals were a prescription for the whole of society, whereas today's proposed genetic interventions reflect only the immediate interests of parents. However, at least five theoretically possible misuses of genetics still provoke more general apprehension, including germ-line therapy, reproductive cloning, attempts at genetic enhancement, preimplantation genetic diagnosis for social reasons, and discriminatory genetic testing. Many kinds of overseeing bodies exist not only to monitor such questionable activities but also to lead the public to an acceptable ethical consensus.

GENERAL SOURCES: 72, 104, 427, 428.

15 Epilogue: Signposts to "Wonderland"

In the developed world, human life expectancy has more than doubled in the last two centuries and is still increasing. The trend may not be sustainable, but life expectancy is evidently continually improving. Our bodies intrinsically lack the resources to provide foolproof maintenance long after our reproductive years, but our individual capacities are notably variable because of the heterogeneity of our genetic legacies and our life experiences. Even before birth, nutritional stress, toxic substances, and viral infections can have effects on the development of the foetus that will shorten an adult life span. After birth, accumulation of somatic mutations – sometimes originating in poor nutrition or chronic inflammation – creates a risk of cancer and other diseases of late middle age and may advance the ageing process. However, although our bodies may be unable to repair all biochemical damage, we can, in principle, reduce some of the risk to health and life expectancy it imposes.

New health challenges originating in life experience still emerge. Breast cancer, allergic reactions, and obesity are three very different kinds of health threat that are becoming increasingly common in the developed world. Breast cancer was well known before the twentieth century, but the lifetime risk of developing such a malignancy in Britain and the United States (and probably in many other countries) has now reached one in eight. Steady improvements in detection and treatment mean that the disease is not fatal to the majority of its victims, but the reasons for its advance remain poorly understood. Changes in the historic pattern of reproduction seem to be significant, but they are not the only possibility. Allergies have also become more common over the last five decades, for unknown reasons. A widely held suspicion is that children born recently have not been challenged by microbes as babies in the same way as children born in less hygienic times, which may predispose them to allergy. Without conclusive support for the "hygiene theory" of allergy, the finger of suspicion points at the immense accumulation of novel chemicals on the planet. Obesity is the third emerging health problem, with serious consequences for health in middle age. This has arisen from a fatal combination of sedentary lifestyle and excessive food consumption, to which the "fast food" industry contributes importantly. However, the alarming growth of obesity in very young children may originate in a predisposition to obesity acquired *in utero* through the mother's lifestyle.

In the last century, the human condition has improved dramatically because of a vast cooperative effort involving public health authorities, clinical expertise, interventions by civil administrations, and the growth of citizenly feelings of obligation to improve hygienic practices. Critical contributions from pharmaceutical science came relatively late but quickly met the tangible and historic needs of people from all strata of society, in a competitive mass market at affordable prices. The stream of novel and efficacious pharmaceuticals – antibiotics, vaccines, steroids, the contraceptive pill – that started in the 1950s continues to flow in ever-greater profusion. Similarly, the birth of a gene technology industry able to make human proteins – such as insulin and growth hormone – in living factories has met well-established clinical needs. Indeed, we have now entered an era of momentous possibilities arising from deep knowledge of biological systems and access to the entire human genome sequence. Many hitherto intractable medical conditions – such as chronic neurological disorders, cancers, genetic diseases, and accidental damage – may be overcome, perhaps in two decades, using appropriately designed drugs or other treatments.

Steadily improving medical care means that unprecedented numbers of people are now living into their ninth and tenth decades. As birth rates fall – in some countries below the replacement rate – and the populations of the developed world become older, economists wring their hands at the oppressive burden that will fall on a younger generation. However, those who acknowledge the overcrowded state of the planet must surely regard the news of declining birth rates as the greater good, even if our social arrangements must change to match the new circumstances. This increased longevity appears to be accompanied by continued good health for many people, and we must hope that any further extension of life span is not a time of extreme dependence.

Since the 1950s, the health care industry has become an important determinant of our biological future because of its ability to provide life-saving medication and a host of other life support systems. People in late middle age throughout the industrial world are being steadily drawn into a critical dependence on the industry. People whose health is unimpaired are already being encouraged to take aspirin, blood pressure medication, and statins to reduce the risk of chronic ill health. However, the first duty of the companies is to enrich their shareholders, so we should not be surprised that most see their best business opportunities among the affluent elderly of North America. Several prominent American biotechnology companies see their "mission" as the extension of the human life span using medication and are, apparently, already looking for compounds that will fulfil this breathtaking ambition. Americans already lead the world in their appetite for medication and are likely to be highly receptive to any kind of anti-ageing pill, no matter how weak its benefits. In 2000, the average American was spending four times as much as the average Briton on medications,[429] a feat attributed more to the power of advertising than to any perceived need for medication. Indeed, the statistics show that the companies, on average, spend more on drug promotion than on research[342] and frequently make misleading claims in order to attract customers.[430] If the appetite of Americans for prescription drugs is surprising, the strength of alternative medicine in the United States and elsewhere is astonishing. Americans spend two dollars on alternative medicines for every ten dollars spent on prescription drugs – and the situation is probably similar in Europe. Many of them are taken because they supposedly delay the ageing process.[25]

A growing disquiet is palpable in medical journals that clinical trials may be losing objectivity. Until about fifteen years ago, clinical trials of new drugs were made in great medical centres under the control of physicians who had access to patients. Increasingly, the entire process of drug development and clinical trials is now being done in organisations directly controlled by the drug companies.[337] Many clinicians involved in trials receive financial support from the drug companies; others involved in studying drugs claim their academic freedom is being curtailed by pressure from companies that fund research.[338] Nobody supposes that safety is seriously compromised, but there is significant evidence that the efficacy of some new drugs may be misrepresented and the case for using them consequently oversold. The chief bulwark against any diminution of objective standards is the movement towards evidence-based medicine based on the international network of Cochrane centres, which aim to foster this approach. These organisations systematically review current medical practice to identify the best procedures to solve health problems.

The biomedical industry now addresses much more than plain clinical needs. The industry has diversified into products that publicists and the press describe as "lifestyle" or "designer" drugs. To earn this sobriquet, drugs need a reputation – usually earned through public relations ploys – for extending personal capacities in a variety of ways; they usually have no life-or-death consequences in the clinical arena. Such drugs include antiobesity drugs for those who cannot face the obvious alternative and oestrogen for hormone replacement therapy. Viagra, a drug that has made "erectile dysfunction" a routine topic for family newspapers, is now said to be a lifestyle drug for men. Many drugs are used for bodybuilding and improved athletic performance, either to satisfy personal vanity or to cheat in competition. A variety of mood-affecting drugs such as Prozac – and before that, Valium and Librium – can rescue people unaccountably blighted by depression or anxiety but have been shamelessly overprescribed to individuals whose problems are only marginal. Most remarkably, in the late 1990s ten percent of American children were said to be hyperactive and were given Ritalin as an aid to greater concentration, with the full support of their parents and teachers. This marks a watershed in our evolution, as the public choose the convenience of a pharmacological remedy, with uncertain long-term consequences, in preference to traditional methods of child rearing. Regardless of its merits for especially troublesome children, nobody doubts that Ritalin was grossly overused in the United States in the 1990s.

One likely consequence of the gene technology revolution will be the discovery of more compounds with potent effects on mood, pain, perceptions, libido, and even memory that can be taken as pills. In a climate in which substance abuse is rife, such innovations will easily find markets, clandestine or otherwise. The old temperamental difference between "Puritans" and "Cavaliers" persists in all matters relating to drugs. Puritans instinctively avoid unnatural pharmacological supports and see unnecessary drugs as a slippery slope of medical and psychological risk – and will come to no harm. Cavaliers, on the other hand, have no preconceived objection to exploring new frontiers of sensation and performance enhancement, and live dangerously. While the ethical problems posed by genetic ideas attract serious intellectual attention, small-molecule lifestyle drugs are likely to have more far-reaching effects on society and will be intrinsically more difficult to regulate.

Novel gene-technology-based medicines with particular value in treating serious diseases could become luxuries affordable only by the wealthy or by those prepared to make major sacrifices. These drugs are expensive; competition does little to reduce prices, as in other economic sectors; and signs of cost inflation in health care are very apparent in the United States. Drugs that are useful for large numbers of patients are likely to strain the finances of national health services that aim to offer treatments to all. Indeed, expensive treatments of this kind (interferon for victims of multiple sclerosis; monoclonal antibodies for certain cancers) have already provoked controversy in Britain after episodes of inequitable rationing. The overwhelming dominance of the American industry in the development of new medicines means that sales, for the foreseeable future, will be a significant foreign exchange cost even for relatively rich countries. Gene and stem cell therapy will almost certainly be affordable only by the affluent.

The expense of high-technology medicine makes these products irrelevant to poorer countries. Low life expectancy in the developing world originates in the lack of resources needed to control infectious disease, infant mortality and malnutrition. The facilities that were so decisive in the improvement in public health in late nineteenth-century Europe (clean piped water and closed sewers) are still uncommon in huge areas of the world. Continuing high birth rates also means that any benefit from increasing family income is likely to be dissipated as family size increases. The existing treatments for Third World infectious diseases are too expensive for the slim resources available, and major research efforts are still needed. On the other hand, nutritional deficiencies that could be treated at relatively little cost still occur. With the AIDS and tuberculosis emergency becoming steadily more serious in sub-Saharan Africa, the people wait and wonder if anything can relieve their plight. International pharmaceutical companies have found no way to help, but a price war between generic drug producers and the major companies has at least reduced the annual cost of the AIDS triple-drug therapy for an individual to $1,330.[431] Indian companies hope to provide similar treatments at just $350–600, but even this is too much for 95 percent of the infected population of those regions. Hope is once again invested in philanthropists and international organisations who might pledge substantial donations to pay for drugs to fight AIDS, malaria, and TB. Even so, these diseases are unlikely to come under control until vaccines become available.[432] Some Third World countries risk legal retribution and manufacture their own antibiotics at very low prices by ignoring patents. Colombia and Guatemala manufacture the antibiotic Ciprofloxacin for five cents a tablet, compared to $3.40 in the United States.[431]

On the long metaphorical journey from Eden to the present day, our biological limitations have often been challenged by new lifestyles, but with persistence and ingenuity, our will has prevailed. Today, our understanding of heredity and our knowledge of the human genome equips us to extend our biological capabilities more than ever. The intractable problems of the ageing process and the quirks of our individual genetic legacies seem increasingly amenable to investigation; it is likely that an extension of our healthy life span will be achieved through uncontroversial schemes. As we have seen, other ideas originating in modern genetics arouse controversy at several levels. They are wholly incompatible with certain traditional ethical principles, but there is also a more widespread feeling that society should not be bamboozled into adopting any radical schemes. Most developed countries now have regulatory bodies, incorporating all shades of opinion, that are charged with overseeing human genetic technology and developing an appropriate legal framework. These same countries, with a few notable exceptions, explicitly prohibit human cloning, germ-line gene therapy, and genetic enhancement of human embryos. Nonetheless, the idea that ethically reprehensible procedures may be performed under an undemanding legal jurisdiction provokes justifiable anxiety. The existing legal framework regulating genetic and reproductive technology does little to limit speculation about genetic dystopias of the kind envisaged more than seventy years ago in *Brave New World*. Some excitable communicators take these scenarios seriously, although how or when these fantasy societies could arise is rarely apparent. This does not stop them from calling for governments to rein in the biomedical industry (for examples, see references 72, 407, 428, and 433). Their worries are a quaintly abstract trinity: the decline in the quality of "human nature," "commodification" of human life, and the loss of "authenticity" of human experience. Generally, this existential angst is confined to affluent enclaves; the meek, who famously "will inherit the earth," remain deeply indifferent and will continue to respond constructively and humanely to new situations and opportunities. Anguish over a hypothetical "loss of humanity" from biomedical innovation in some far distant time seems oddly priggish in a world in which authentic crimes against humanity – genocide, terrorism, and retributive war – are all pervasive.

The idea that forthcoming biomedical advances will be detrimental to human perceptions of dignity, self-esteem, and autonomy is wholly unconvincing, too. Previous generations – when they considered the alternative – accepted gladly the benefits conferred by insulin, hearing aids, spectacles,

artificial hips, blood transfusions, anaesthetics, and hundreds of other in-
ventions; our descendants will rejoice in different innovations. We can be
certain that if or when effective treatments for Parkinson's or Alzheimer's
disease become available, the beneficiaries will be utterly mystified by current doubts about the ethical use of stem cells. The bogeyman of "commodification" of reproduction already exists legitimately, if rather dubi-
ously, in the United States. Commercial surrogate motherhood and a brisk
trade in Third World babies are well established, and they are not clearly
more reprehensible than business relationships with the "oldest profession." Nostalgia for authentic human experience is faintly comical and
patronising in a world still groaning with the bracing realities of life in less
developed times.

This book catalogues, sketchily, examples, ancient and modern, of the
negative biological fallout that often accompanies the novel undertakings
of humans. We are peculiarly vulnerable to the very human innovation of
psychological persuasion that we use to inveigle our fellows into biological adventures with unknowable risks. Using techniques from seductive
whispers between friends to the satanic art of advertising, legitimate or
otherwise, the aim is to engineer sentiment in favour of immediate grati-
fication and to dispel any suspicion of longer-term negative consequences
for ourselves or our progeny. Only dreary puritans consistently reject the
charms of convenient novelties or pharmacological satisfaction. Cigarettes,
narcotics, alcoholic drinks, fast food, artificial baby food, medicines for
hyperactive children, alternative medicines, HRT, and mood-affecting or
performance-enhancing drugs are the story of human caprice in the twentieth century.

Paradoxically, while individuals happily embrace personal risk, mass
public opinion and institutions are increasingly and perhaps hysterically
risk-averse. For this odd situation we usually thank the media, – for whom
the discovery of victims, disquieting revelations, and the assignment of
blame are key business strategies. We should not be surprised that members of the public view biomedical science with a somewhat jaundiced eye,
even though remarkable numbers of people born in the last six decades have
not yet needed a doctor's expertise. The biomedical industry, driven by the
same economic imperatives as any other business, seeks special situations
where they can enrich their shareholders. This results in what is probably
excessive medicalisation of problems (e.g., obesity and childhood hyper-
activity) that may have other solutions or no real solution (e.g., ageing).
Cures for executive gastric ulcers have secured some of the most notable

financial triumphs of recent times (see Chapter 11), while drugs and vaccines against Third World diseases are neglected for want of an assured profit. The search for desperately needed new antibiotics is being curtailed because they are unlikely to be profitable from now on. In the estimation of some, the search for effective anticancer therapies proceeds at a snail's pace because the industry will not abandon a strategy whose chief merit is profitability.[273] At a time when the drug industry is hailed as an important engine for future economic growth, it is worth remembering that many of our ills could, in principle, be prevented rather than cured. At least 80 percent of the risk of cancer, for example, originates in our life experience, and the risk of allergy and obesity even more so. Prevention rather than cure is a logical and worthy objective, but sustained progress in this direction seems elusive.

The "signposts" in the title of this concluding chapter point to paths meandering into unexplored territories. Unhelpfully, perhaps – but appropriately for such a Wonderland – they can only carry a question mark. Indeed, the entire discourse raises profound unanswerable questions. Will biomedicine deliver smooth and welcome progress indefinitely? Will all the peoples of the world benefit from it? Will future human societies avoid serious self-inflicted injuries? Will biomedicine have anything to offer if our social fabric deteriorates dramatically because our oil reserves are exhausted, or because of global warming or an outbreak of war? Our prehistoric ancestors survived – by the narrowest margins – mass culls from famine or disease originating in climate change and geological disaster and bounced back to populate complex and dynamic societies. In the tale of the Tree of Knowledge of Good and Evil, these same people fashioned a prescient metaphor for that most original of human traits, the desire to challenge vicissitude in all its forms with exploration and innovation. The fruit of the Tree was a visa that allowed them to leave the benign and eternal refuge of Eden for a land with stern challenges and consequences that could be hugely beneficial or catastrophically ugly.

References

1 Harpending, H., and Rogers, A. (2000) Genetic perspectives on human origins and differentiation. *Ann. Rev. Genomics Hum. Genet.* 1, 361–85.
2 Haviland, W. (2000) *Human evolution and prehistory.* Harcourt.
3 Neiburger, E. J. (2000) Dentistry in ancient Mesopotamia. *J. Mass. Dent. Soc.* 49, 16–19.
4 Keynes, R. (2002) *Annie's Box.* Fourth Estate.
5 Coale, A. J. (1974) The history of the human population. *Sci. Am.* 231, 41.
6 Eaton, S. B., and Konner, M. (1985) Palaeolithic nutrition: A consideration of its nature and current implications. *N. Engl. J. Med.* 312, 283–9.
7 McClaren, A. (1978) *Birth control in nineteenth century England.* Croom Helm.
8 Hardy, A. (2001) *Health and medicine in Britain since 1860.* Palgrave.
9 Riley, J. (2001) *Rising life expectancy: a global history.* Cambridge University Press.
10 Fisher, D. (1978) The Rockefeller Foundation and the development of scientific medicine in Great Britain. *Minerva* 16, 20–41.
11 Tomes, N. (1998) *The gospel of germs: Men, women, and the microbe in American life.* Harvard University Press.
12 Miles, A. (1966) The Lister Institute of Preventive Medicine 1891–1966. *Nature* 212, 559.
13 Medawar, P. B. (1963) The MRC fifty years ago and now. *Nature* 200, 1039–43.
14 Tuchman, B. (1962) *The proud tower.* MacMillan.
15 Austoker, J. (1988) *A history of the Imperial Cancer Research Fund, 1902–1986.* Oxford University Press.
16 Hare, R. (1970) *The birth of penicillin.* Allen and Unwin.
17 Fosdick, R. (1952) *The history of the Rockefeller Foundation.* Harper.
18 Corner, G. (1964) *A history of the Rockefeller Institute.* Rockefeller University Press.
19 Edwards, R., and Steptoe, P. (1980) *A matter of life.* Hutchinson.
20 Perutz, M. F. (1999) Will medicine outgrow support? *Nature* 399, 299.
21 Abelson, P. (1968) The advancement of the nation's health. *Science* 158, 51.

22 Strickland, S. (1971) Integration of medical research and health policies. *Science* **173**, 1093.

23 Editorial. (1965) Does the United States have a science policy? *Nature* **217**, 225–31.

24 Kornberg, A. (1987) The NIH did it! *Science* **278**, 1863.

25 Strauss, S. (2000) www:nccam.nih.gov/ne/testimony/houseoflords.html.

26 Doll, R. (1992) Sir Austin Bradford Hill and the progress of medical science. *BMJ* **305**, 1521.

27 Obituary. (2003) Louis Lasagna. *BMJ* **327**, 565.

28 Lewin, B. (1970) Second golden age of molecular biology. *Nature* **227**, 1009–13.

29 Smith, F. B. (1979) *The people's health 1830–1910*. Holmes and Meier.

30 Diamond, J. (1998) *Guns, Germs, and Steel*. Vintage.

31 Shaw, G. B. (1911) Preface to *The Doctor's Dilemma*. Penguin.

32 Rosenberg, D., and Bloom, H. (1990) *The Story of J*. Faber and Faber.

33 Youngson, A. J. (1979) *The scientific revolution in Victorian medicine*. Holmes and Meier.

34 Loudon, I. (1992) The transformation of maternal mortality. *BMJ* **305**, 1557.

35 Rosenberg, K. R., and Trevathan, W. R. (2001) The evolution of human birth. *Sci. Am.* **285**, 72–7.

36 Cohen, J. (1996) Doctor James Young Simpson, Rabbi Abraham de Sola and Genesis Chapter 3, Verse 16. *Obstetrics and Gynaecology* **88**, 895.

37 Caton, D. (1996) Who said childbirth is natural? The medical mission of Grantly Dick Read. *Anaesthesiology* **84**, 955–64.

38 Behague, D. P., Victora, C. G., and Barros, F. C. (2002) Consumer demand for caesarean sections in Brazil. *BMJ* **324**, 942–5.

39 Short, R. V. (1976) The evolution of human reproduction. *Proc. Roy. Soc. B.* **195**, 3.

40 May, R. (1978) Human reproduction reconsidered. *Nature* **272**, 491–5.

41 Tanner, J. (1989) *Foetus into man: Physical growth from conception to maturity*. Castlemead.

42 Short, R. (1984) Breast feeding. *Sci. Am.* **250**, 23.

43 Mepham, T. B. (1993) "Humanizing" milk: The formulation of artificial feeds for infants (1850–1910). *Med. Hist.* **37**, 225–49.

44 Smith, E. (1988) *The retreat of tuberculosis, 1850–1950*. Croom Helm.

45 Cuthbertson, W. (2000) Evolution of infant nutrition. *Brit. J. Nut.* **81**, 359–371.

46 Forsyth, J., et al. (2003) Long chain polyunsaturated fatty acid supplementation in infant formula and blood pressure in later childhood: Follow up of a randomised controlled trial. *BMJ* **326**, 958–62.

47 McNeill, W. (1976) *Plagues and peoples*. Basil Blackwell.

48 CIA. (2002) *World Fact Book*. www.cia.gov.

49 Langer, W. (1974) Infanticide: A historical survey. *Hist. Child. Q.* (Winter), 353–66.

50 Parliament. (1999) House of Commons research paper 99/111. House of Commons.

51 McKeown, T. (1988) *The origins of human disease*. Basil Blackwell.
52 Halsey, A. (1972) *Trends in British society since 1900*. Macmillan.
53 Stevenson, J. (1984) *British society 1914–45*. Penguin.
54 Djerassi, C. (1962) *The Pill, pygmy chimps, and Degas' horse*. Basic Books.
55 Wardell, D. (1980) Margaret Sanger: Birth control's successful revolutionary. *Am. J. Public Health* 70, 736–42.
56 Black, I. (2002) EU replaces cash denied to UN family planning by US. *The Guardian*, London, July 24.
57 Harris, J. (1992) *Clones, genes, and immortality*. Oxford University Press.
58 Editorial. (1968) Ethics and abortion. *BMJ* 2, 3.
59 Paintin, D. (1985) Legal abortion in England and Wales. *CIBA Foundation Symposium* 115, 4–20.
60 Ho, P. C. (2000) Emergency contraception: Methods and efficacy. *Curr. Opin. Obstet. Gynecol.* 12, 175–9.
61 Ullman, A. (2000) The development of Mifepristone: A pharmaceutical drama in three acts. *Am. Med. Women's Ass.* 55 (suppl.), 117–20.
62 Kolata, G. (1978) In vitro fertilization: Is it safe and repeatable? *Science* 201, 698.
63 Trounson, A. L. J., Wood, C., Webb, J., and Wood, J. (1981) Pregnancies in humans by fertilization in vitro and embryo transfer in the controlled ovulatory cycle. *Science* 212, 681–2.
64 Edwards, R., and Steptoe, P. (1983) Current status of *in-vitro* fertilisation and implantation of human embryos. *The Lancet* 2, 1265–9.
65 Tesarik, J., and Mendoza, C. (1999) *In vitro* fertilization by intracytoplasmic sperm injection. *Bioessays* 21, 791–801.
66 Powell, K. (2003) Seeds of doubt. *Nature* 422, 656–8.
67 Gosden, R. (1999) *Designer babies*. Phoenix.
68 Porcu, E., et al. (2000) Clinical experience and applications of oocyte cryopreservation. *Mol. Cell. Endocrinol.* 169, 33–7.
69 Harris, R. (2002) Human babies in germ-line juggernaut. *Current Biology* 11, R539–40.
70 Editorial. (1985) Powell bites dust: The British government needs to think hard about surrogacy. *Nature* 314, 568.
71 Stromberg, B., et al. (2002) Neurological sequelae in children born after *in-vitro* fertilisation: A population-based study. *The Lancet* 359, 461–5.
72 Silver, L. (1998) *Remaking Eden: Cloning, genetic engineering and the future of humankind?* Phoenix.
73 McCarty, M. (1985) *The transforming principle*. Norton.
74 Judson, T. (1979) *The eighth day of creation*. Jonathan Cape.
75 Olby, R. (2003) Quiet debut of the double helix. *Nature* 421, 402–5.
76 Crick, F. (1993) Looking backwards: A birthday card for the double helix. *Gene* 135, 15–18.
77 Griffith, J. S. (1967) Self-replication and scrapie. *Nature* 215, 1043–4.
78 Prusiner, S. B. (1998) Prions. *Proc. Natl. Acad. Sci. USA* 95, 13363–83.
79 Melov, S., et al. (1999) Mitochondrial DNA rearrangements in ageing human brain and in situ PCR of mtDNA. *Neurobiol. Ageing* 20, 565–71.

80   Wallace, D. C. (1997) Mitochondrial DNA in ageing and disease. *Sci. Am.* **277**, 40–7.

81   Gemmell, N. J., and Sin, F. Y. (2002) Mitochondrial mutations may drive Y chromosome evolution. *Bioessays* **24**, 275–9.

82   Ponamarev, M. V., et al. (2002) Active site mutation in DNA polymerase gamma associated with progressive external ophthalmoplegia causes error-prone DNA synthesis. *J. Biol. Chem.* **277**, 15225–8.

83   Melton, L. (2000) Womb wars. *Sci. Am.* **283**, 24, 26.

84   Yu, N., et al. (2002) Disputed maternity leading to identification of tetragametic chimerism *N. Engl. J. Med.* **346**, 1545–52.

85   Monod, J. (1972) *Chance and necessity.* Collins.

86   Crow, J. F. (1997) The high spontaneous mutation rate: Is it a health risk? *Proc. Natl. Acad. Sci. USA* **94**, 8380–6.

87   Friedberg, E. C. (2001) How nucleotide excision repair protects against cancer. *Nature Rev. Cancer* **1**, 22–33.

88   Bartek, J., and Lukas, J. (2003) Damage alert. *Nature* **421**, 486–7.

89   Crow, J. F. (2000) The origins, patterns and implications of human spontaneous mutation. *Nat. Rev. Genet.* **1**, 40–7.

90   Deininger, P. L., and Batzer, M. A. (1999) *Alu* repeats and human disease. *Mol. Genet. Metab.* **67**, 183–93.

91   Li, X., et al. (2001) Frequency of recent retrotransposition events in the human factor IX gene. *Hum. Mutat.* **17**, 511–19.

92   International Human Genome Sequencing Consortium. (2001) Initial sequencing and analysis of the human genome. *Nature* **409**, 860–921.

93   Greaves, M. (2002) Childhood leukaemia. *BMJ* **324**, 283–7.

94   Nekrutenko, A., and Li, W. H. (2001) Transposable elements are found in a large number of human protein-coding genes. *Trends Genet.* **17**, 619–21.

95   Smit, A. F. (1999) Interspersed repeats and other mementoes of transposable elements in mammalian genomes. *Curr. Opin. Genet. Dev.* **9**, 657–63.

96   Symer, D. (2002) Human L1 retrotransposition is associated with genetic instability in vivo. *Cell* **110**, 327–8.

97   Hughes, V. M., and Datta, N. (1983) Conjugative plasmids in bacteria of the 'pre-antibiotic' era. *Nature* **302**, 725–6.

98   Karaolis, D. K., Lan, R., Kaper, J. B., and Reeves, P. R. (2001) Comparison of *Vibrio cholerae* pathogenicity islands in sixth and seventh pandemic strains. *Infect. Immun.* **69**, 1947–52.

99   Stephens, C., and Murray, W. (2001) Pathogen evolution: How good bacteria go bad. *Curr. Biol.* **11**, R53–6.

100 Groisman, E. A., and Ochman, H. (1997) How *Salmonella* became a pathogen. *Trends Microbiol.* **5**, 343–9.

101 Herskind, A. M., et al. (1996) The heritability of human longevity: A population-based study of 2872 Danish twin pairs born 1870–1900. *Hum. Genet.* **97**, 319–23.

102 Iachine, I. A., et al. (1998) How heritable is individual susceptibility to death? The results of an analysis of survival data on Danish, Swedish and Finnish twins. *Twin Res.* **1**, 196–205.

103 Lichtenstein, P., et al. (2000) Environmental and heritable factors in the causation of cancer – analyses of cohorts of twins from Sweden, Denmark, and Finland. *N. Engl. J. Med.* **343**, 78–85.

104 Weatherall, D. (1991) *The new genetics and clinical practice.* Oxford University Press.

105 Marx, J. (2002) Unravelling the cause of diabetes. *Science* **296**, 686–9.

106 Cole, T. J. (2000) Secular trends in growth. *Proc. Nutr. Soc.* **59**, 317–24.

107 Brundtland, G. H., Liestol, K., and Walloe, L. (1980) Height, weight and menarcheal age of Oslo schoolchildren during the last 60 years. *Ann. Hum. Biol.* **7**, 307–22.

108 Floud, R., Wachter, K., and Gregory, A. (1990) *Height, health and history.* Cambridge University Press.

109 Wilmoth, J. R., Deegan, L. J., Lundstrom, H., and Horiuchi, S. (2000) Increase of maximum life-span in Sweden, 1861–1999. *Science* **289**, 2366–8.

110 Alberts, B., et al. (2002) *The molecular biology of the cell.* Garland.

111 Ridley, M. (2003) *Nature via nurture.* Fourth Estate.

112 Porter, R. (1997) *The greatest benefit to mankind.* Oxford University Press.

113 Masutomi, K., et al. (2003) Telomerase maintains telomere structure in normal human cells. *Cell* **114**, 241–53.

114 Wong, J. M., and Collins, K. (2003) Telomere maintenance and disease. *Lancet* **362**, 983–8.

115 Cohen, M., Lee, K. K., Wilson, K. L., and Gruenbaum, Y. (2001) Transcriptional repression, apoptosis, human disease and the functional evolution of the nuclear lamina. *Trends Biochem. Sci.* **26**, 41–7.

116 Yaffe, M. (1999) The machinery of mitochondrial inheritance and behaviour. *Science* **283**, 1493–7.

117 Wallace, D. C. (1999) Mitochondrial diseases in man and mouse. *Science* **283**, 1482–8.

118 Marchington, D. R., et al. (1998) Evidence from human oocytes for a genetic bottleneck in an mtDNA disease. *Am. J. Hum. Genet.* **63**, 769–75.

119 Larson, J. E., and Cohen, J. C. (2000) Cystic fibrosis revisited. *Mol. Genet. Metab.* **71**, 470–7.

120 McNeil, P., Miyake, K., and Vogel, S. (2003) The endomembrane requirement for cell surface repair. *PNAS* **100**, 4592–7.

121 Brady, S. T. (1993) Motor neurons and neurofilaments in sickness and in health. *Cell* **73**, 1–3.

122 Winchester, B., Vellodi, A., and Young, E. (2000) The molecular basis of lysosomal storage diseases and their treatment. *Biochem. Soc. Trans.* **28**, 150–4.

123 Keller, J., Gee, J., and Ding, Q. (2002) The proteasome in brain ageing. *Ageing Res. Rev.* **1**, 279–93.

124 Wilmut, I., et al. (1997) Viable offspring derived from fetal and adult mammalian cells. *Nature* **385**, 810–13.

125 Levi-Montalcini, R. (1987) The nerve growth factor 35 years later. *Science* **237**, 1154–62.

126 Jacobson, M. D., Weil, M., and Raff, M. C. (1997) Programmed cell death in animal development. *Cell* 88, 347–54.
127 Raff, M. C. (1992) Social controls on cell survival and cell death. *Nature* 356, 397–400.
128 Krammer, P. H. (2000) CD95's deadly mission in the immune system. *Nature* 407, 789–95.
129 Hetts, S. W. (1998) To die or not to die: An overview of apoptosis and its role in disease. *JAMA* 279, 300–7.
130 Fleisher, T. A., Straus, S. E., and Bleesing, J. J. (2001) A genetic disorder of lymphocyte apoptosis involving the *Fas* pathway: The autoimmune lymphoproliferative syndrome. *Curr. Allergy Asthma Rep.* 1, 534–40.
131 Niederkorn, J. Y. (1999) The immune privilege of corneal allografts. *Transplantation* 67, 1503–8.
132 Uckan, D., et al. (1997) Trophoblasts express *Fas* ligand: A proposed mechanism for immune privilege in placenta and maternal invasion. *Mol. Hum. Reprod.* 3, 655–62.
133 Brash, D. (1996) Cellular proofreading. *Nature Medicine* 2, 525–6.
134 Fadeel, B., Orrenius, S., and Zhivotovsky, B. (1999) Apoptosis in human disease: A new skin for the old ceremony? *Biochem. Biophys. Res. Commun.* 266, 699–717.
135 Levine, A. (1997) p53 the cellular gatekeeper for growth and division. *Cell* 88, 323–31.
136 Hayflick, L. (1994) *How and why we age.* Ballantine Books.
137 Ferri, K. F., and Kroemer, G. (2001) Mitochondria – the suicide organelles. *Bioessays* 23, 111–15.
138 Fuchs, E. (1997) Of mice and men: Genetic disorders of the cytoskeleton (Keith R. Porter Lecture, 1996). *Mol. Biol. Cell* 8, 189–203.
139 Nicklas, R. B. (1997) How cells get the right chromosomes. *Science* 275, 632–7.
140 Dynes, M. (2003) Miracle birth of the elusive baby found growing on mother's liver. *The Times*, May 24.
141 Nilsson, L. (1990) *A child is born.* Dell.
142 Brent, L., and Owen, R. (1996) *A history of transplantation immunology.* Academic Press.
143 Barinaga, M. (2002) Cells exchanged during pregnancy live on. *Science* 296, 2169–72.
144 Barker, D. (1998) *Mothers, babies and health in later life.* Churchill Livingston.
145 Wolpert, L. (1991) *The triumph of the embryo.* Oxford University Press.
146 Qumsiyeh, M. B., Kim, K. R., Ahmed, M. N., and Bradford, W. (2000) Cytogenetics and mechanisms of spontaneous abortions: Increased apoptosis and decreased cell proliferation in chromosomally abnormal villi. *Cytogenet. Cell Genet.* 88, 230–5.
147 Griffin, D. K. (1996) The incidence, origin, and etiology of aneuploidy. *Int. Rev. Cytol.* 167, 263–96.

148 Norimura, T., et al. (1996) p53-dependent apoptosis suppresses radiation-induced teratogenesis. *Nat. Med.* **2**, 577–80.

149 Heyer, B. S., MacAuley, A., Behrendtsen, O., and Werb, Z. (2000) Hypersensitivity to DNA damage leads to increased apoptosis during early mouse development. *Genes Dev.* **14**, 2072–84.

150 Stewart, R. J., Sheppard, H., Preece, R., and Waterlow, J. C. (1980) The effect of rehabilitation at different stages of development of rats marginally malnourished for ten to twelve generations. *Br. J. Nutr.* **43**, 403–12.

151 Gluckman, P. D. (2001) Editorial: Nutrition, glucocorticoids, birth size, and adult disease. *Endocrinology* **142**, 1689–91.

152 Lumbers, E. R., Yu, Z. Y., and Gibson, K. J. (2001) The selfish brain and the Barker hypothesis. *Clin. Exp. Pharmacol. Physiol.* **28**, 942–7.

153 Garofano, A., Czernichow, P., and Breant, B. (1997) In utero undernutrition impairs rat beta-cell development. *Diabetologia* **40**, 1231–4.

154 Kwong, W. Y., et al. (2000) Maternal undernutrition during the preimplantation period of rat development causes blastocyst abnormalities and programming of postnatal hypertension. *Development* **127**, 4195–202.

155 Butterworth, C. E., Jr., and Bendich, A. (1996) Folic acid and the prevention of birth defects. *Ann. Rev. Nutr.* **16**, 73–97.

156 Biever, C. (2003) Bring me sunshine. *The New Scientist* **179**, 30–1.

157 van der Meulen, J., et al. (1999) Fetal programming associated with the Dutch famine 1944–5. In *Fetal Programming: Influences on Development and Disease in Later Life*, ed. P. O'Brien, T. Wheeler, and D. J. P. Barker. Royal College of Obstetricians and Gynaecologists, pp. 45–56.

158 Stein, Z., Susser, M., Saenger, G., and Marolla, F. (1975) *Famine and human development: The Dutch hunger winter of 1944–1945.* Oxford University Press.

159 Boyd-Orr, J. (1936) *Food, health and income.* Macmillan.

160 Eriksson, J., et al. (1999) Catch-up growth in childhood and death from coronary heart disease: Longitudinal study. *BMJ* **318**, 427–31.

161 Trowell, H., and Burkitt, D. (1981) *Western diseases, their emergence and prevention.* Harvard University Press.

162 Diamond, J. (2003) The double puzzle of diabetes. *Nature* **423**, 599–602.

163 Prentice, A. (2003) Intrauterine factors, adiposity and hyperinsulinaemia. *BMJ* **327**, 880–1.

164 Porter, D. (1999) *Health, civilisation and the state.* Routledge.

165 Le Fanu, J. (1999) *The rise and fall of modern medicine.* Little, Brown.

166 Osmond, C., and Barker, D. J. (2000) Fetal, infant, and childhood growth are predictors of coronary heart disease, diabetes, and hypertension in adult men and women. *Environ. Health. Perspect.* **108** (suppl. 3), 545–53.

167 Chandra, R. K. (1991) Interactions between early nutrition and the immune system. *Ciba. Found. Symp.* **156**, 77–89.

168 Moore, S. E., et al. (1997) Season of birth predicts mortality in rural Gambia. *Nature* **388**, 434.

169   Stephens, T., and Brynner, R. (2001) *Dark remedy: The impact of thalidomide and its revival as a vital medicine.* Perseus.

170   Colborn, T., Dumanoski, D., and Myers, J. (1997) *Our stolen future.* Abacus.

171   Ibarreta, D., and Swan, S. (2002) The DES story: Long term consequences of prenatal exposure. In *The Precautionary Principle in the 20th Century: Late Lessons from Early Warnings,* ed. Poul Harremoës et al. European Environment Agency, pp. 90–9.

172   Kaiser, J. (2000) Endocrine disrupters: Panel cautiously confirms low-dose effects. *Science* 290, 695–7.

173   Hunt, P. A., et al. (2003) Bisphenol exposure causes meiotic aneuploidy in the female mouse. *Curr. Biol.* 13, 546–53.

174   Rennie, J. (1996) Perinatal management of the lower margin of viability. *Arch. Dis. Child* 74, F214-F218.

175   Christie, B. (2000) Premature babies have high death and disability rate. *BMJ* 321, 467.

176   Wickelgren, I. (2004) Resetting pregnancy's clock. *Science* 304, 666–8.

177   Tin, W., Wariyar, U., and Hey, E. (1997) Changing prognosis for babies of less than 28 weeks' gestation in the north of England between 1983 and 1994. *BMJ* 314, 107–11.

178   Avery, M. E. (2000) Surfactant deficiency in hyaline membrane disease: The story of discovery. *Am. J. Respir. Crit. Care Med.* 161, 1074–5.

179   Uauy, R., and Hoffman, D. (2003) Essential fat requirements of preterm infants. *Am. J. Clin. Nutr.* 71, 245–50.

180   Field, T. (1988) Stimulation of preterm infants. *Pediatr. Rev.* 10, 149–53.

181   Schanberg, S. M., and Field, T. M. (1987) Sensory deprivation stress and supplemental stimulation in the rat pup and preterm human neonate. *Child Dev.* 58, 1431–47.

182   Hack, M., and Merkatz, I. R. (1995) Preterm delivery and low birth weight – a dire legacy. *N. Engl. J. Med.* 333, 1772–4.

183   Liu, Y., Albertsson-Wikland, K., and Karlberg, J. (2000) Long-term consequences of early linear growth retardation (stunting) in Swedish children. *Pediatr. Res.* 47, 475–80.

184   Karlberg, J., and Luo, Z. C. (2000) Foetal size to final height. *Acta Paediatr.* 89, 632–6.

185   Bogin, B. (1999) *Patterns of human growth.* Cambridge University Press.

186   Janeway, C. A., Jr. (2001) How the immune system works to protect the host from infection: A personal view. *Proc. Natl. Acad. Sci. USA* 98, 7461–8.

187   Zasloff, M. (2002) Antimicrobial peptides of multicellular organisms. *Nature* 415, 389–95.

188   Carswell, E. A., et al. (1975) An endotoxin-induced serum factor that causes necrosis of tumors. *Proc. Natl. Acad. Sci. USA* 72, 3666–70.

189   McCormick, J. K., Yarwood, J. M., and Schlievert, P. M. (2001) Toxic shock syndrome and bacterial superantigens: An update. *Ann. Rev. Microbiol.* 55, 77–104.

190   Libby, P. (2002) Atherosclerosis: The new view. *Sci. Am.* 286, 46–55.

191 Yazdanbakhsh, M., Kremsner, P. G., and van Ree, R. (2002) Allergy, parasites, and the hygiene hypothesis. *Science* **296**, 490–4.

192 Martinez, F. D., and Holt, P. G. (1999) Role of microbial burden in aetiology of allergy and asthma. *Lancet* **354** (suppl. 2). SII12–15.

193 Mackay, I. R. (2000) Tolerance and autoimmunity. *BMJ* **321**, 93–6.

194 Goldbach-Mansky, R., and Lipsky, P. E. (2003) New concepts in the treatment of rheumatoid arthritis. *Ann. Rev. Med.* **54**, 197–216.

195 Todd, J. A. (2001) Human genetics: Tackling common disease. *Nature* **411**, 537, 539.

196 Ye, X., et al. (2000) Engineering the provitamin A (beta-carotene) biosynthetic pathway into (carotenoid-free) rice endosperm. *Science* **287**, 303–5.

197 Rajakumar, K. (2000) Pellagra in the United States: A historical perspective. *South. Med. J.* **93**, 272–7.

198 Chan, A. C. (1998) Vitamin E and atherosclerosis. *J. Nutr.* **128**, 1593–6.

199 Bang, H. O., and Dyerberg, J. (1981) Personal reflections on the incidence of ischaemic heart disease in Oslo during the Second World War. *Acta. Med. Scand.* **210**, 245–8.

200 Connor, W. E. (2000) Importance of n-3 fatty acids in health and disease. *Am. J. Clin. Nutr.* **71**, 171S–5S.

201 Lamartiniere, C. A. (2000) Protection against breast cancer with genistein: A component of soy. *Am. J. Clin. Nutr.* **71**, 1705S–7S; discussion 1708S–9S.

202 Taubes, G. (2001) Nutrition: The soft science of dietary fat. *Science* **291**, 2536–45.

203 Fahey, J. W., Zhang, Y., and Talalay, P. (1997) Broccoli sprouts: An exceptionally rich source of inducers of enzymes that protect against chemical carcinogens. *Proc. Natl. Acad. Sci. USA* **94**, 10367–72.

204 Brosens, J., and Parker, M. (2003) Oestrogen receptor hijacked. *Nature* **423**, 487–8.

205 Kogevinas, M. (2001) Human health effects of dioxins: Cancer, reproductive and endocrine system effects. *Hum. Reprod. Update* **7**, 331–9.

206 Ames, B., Magaw, R., and Gold, L. (1987) Ranking possible carcinogenic hazards. *Science* **236**, 271–280.

207 Janeway, C., Travers, P., Walport, M., and Shlomchik, M. (2001) *Immunobiology*. Garland Publishing.

208 Tuljapurkar, S., Li, N., and Boe, C. (2000) A universal pattern of mortality decline in the G7 countries. *Nature* **405**, 789–92.

209 Kirkwood, T. B., and Austad, S. N. (2000) Why do we age? *Nature* **408**, 233–8.

210 Holliday, R. (1995) *Understanding ageing*. Cambridge University Press.

211 Hayflick, L. (2000) The future of ageing. *Nature* **408**, 267–9.

212 Davey, B., Halliday, T., and Hirst, M. (2001) *Human biology and health: An evolutionary approach*. Open University.

213 Harman, D. (1956) Aging: A theory based on free radical and radiation chemistry. *J. Gerontol.* *1956* **11**, 298–300.

214   Beckman, K. B., and Ames, B. N. (1998) The free radical theory of aging matures. *Physiol. Rev.* 78, 547–81.
215   Melov, S., et al. (2000) Extension of life-span with superoxide dismutase/catalase mimetics. *Science* 289, 1567–9.
216   McKenzie, D., et al. (2002) Mitochondrial DNA deletion mutations: A causal role in sarcopenia. *Eur. J. Biochem.* 269, 2010–15.
217   Brunk, U. T., and Terman, A. (2002) The mitochondrial-lysosomal axis theory of aging: Accumulation of damaged mitochondria as a result of imperfect autophagocytosis. *Eur. J. Biochem.* 269, 1996–2002.
218   Suter, M., et al. (2000) Age-related macular degeneration: The lipofusion component N-retinyl-N-retinylidene ethanolamine detaches proapoptotic proteins from mitochondria and induces apoptosis in mammalian retinal pigment epithelial cells. *J. Biol. Chem.* 275, 39625–30.
219   Dirks, A., and Leeuwenburgh, C. (2002) Apoptosis in skeletal muscle with aging. *Am. J. Physiol. Regul. Integr. Comp. Physiol.* 282, R519–27.
220   Lamberts, S. W., van den Beld, A. W., and van der Lely, A. J. (1997) The endocrinology of aging. *Science* 278, 419–24.
221   Rudman, D., et al. (1990) Effects of human growth hormone in men over 60 years old. *N. Engl. J. Med.* 323, 1–6.
222   Laron, Z. (2002) Effects of growth hormone and insulin-like growth factor 1 deficiency on ageing and longevity. *Novartis Found. Symp.* 242, 125–37; discussion 137–42.
223   Yen, S. S. (2001) Dehydroepiandrosterone sulfate and longevity: New clues for an old friend. *Proc. Natl. Acad. Sci. USA* 98, 8167–9.
224   Lane, M. A., Ingram, D. K., Ball, S. S., and Roth, G. S. (1997) Dehydroepiandrosterone sulfate: A biomarker of primate aging slowed by calorie restriction. *J. Clin. Endocrinol. Metab.* 82, 2093–6.
225   Lord, J. M., et al. (2001) Neutrophil ageing and immunesenescence. *Mech. Ageing Dev.* 122, 1521–35.
226   Finkel, T., and Holbrook, N. J. (2000) Oxidants, oxidative stress and the biology of ageing. *Nature* 408, 239–47.
227   Finch, C. E., and Tanzi, R. E. (1997) Genetics of aging. *Science* 278, 407–11.
228   Perls, T. T. (1995) The oldest old. *Sci. Am.* 272, 70–5.
229   Martin, G. M., and Oshima, J. (2000) Lessons from human progeroid syndromes. *Nature* 408, 263–6.
230   Hsin, H., and Kenyon, C. (1999) Signals from the reproductive system regulate the lifespan of C. elegans. *Nature* 399, 362–6.
231   Hamilton, M. L., et al. (2001) Does oxidative damage to DNA increase with age? *Proc. Natl. Acad. Sci. USA* 98, 10469–74.
232   Kayo, T., Allison, D. B., Weindruch, R., and Prolla, T. A. (2001) Influences of aging and caloric restriction on the transcriptional profile of skeletal muscle from rhesus monkeys. *Proc. Natl. Acad. Sci. USA* 98, 5093–8.
233   Cristofalo, V. J., et al. (1998) Relationship between donor age and the replicative lifespan of human cells in culture: A reevaluation. *Proc. Natl. Acad. Sci. USA* 95, 10614–9.

234 Faragher, R. G., and Kipling, D. (1998) How might replicative senescence contribute to human ageing? *Bioessays* 20, 985–91.
235 Kassem, M., et al. (1997) Demonstration of cellular aging and senescence in serially passaged long-term cultures of human trabecular osteoblasts. *Osteoporosis Int.* 7, 514–24.
236 Hayflick, L. (2002) DNA replication and traintracks. *Science* 296, 1611–12.
237 von Zglinicki, T. (2001) Telomeres and replicative senescence: Is it only length that counts? *Cancer Lett.* 168, 111–16.
238 Shay, J. W., and Wright, W. E. (2001) Ageing and cancer: The telomere and telomerase connection. *Novartis Found. Symp.* 235, 116–25; discussion 125–9, 146–9.
239 Johnson, F. B., Marciniak, R. A., and Guarente, L. (1998) Telomeres, the nucleolus and aging. *Curr. Opin. Cell Biol.* 10, 332–8.
240 Schwarcz, J. (2001) *The genie in the bottle.* Freeman.
241 Lock, S. (1983) "O that I were young again": Yeats and the Steinach operation. *Br. Med. J. (Clin. Res. Ed.)* 287, 1964–8.
242 Alcor Life Extension Foundation. (2000) www.alcor.org.
243 deGrey, A., et al. (2003) Antiageing technology and pseudoscience. *Science* 296, 5568–9.
244 Hall, S. (2003) In vivo vitalis? Compounds activate life-extending genes. *Science* 301, 1165.
245 Kirkwood, T. *The time of our lives.* Weidenfeld and Nicholson.
246 Rettig, R. (1977) *Cancer crusade.* Princeton University Press.
247 Weisburger, E. K. (1989) Current carcinogen perspectives: De minimis, Delaney and decisions. *Sci. Total Environ.* 86, 5–13.
248 Stehelin, D., Varmus, H. E., Bishop, J. M., and Vogt, P. K. (1976) DNA related to the transforming gene(s) of avian sarcoma viruses is present in normal avian DNA. *Nature* 260, 170–3.
249 Culliton, B. (1972) Cancer virus theories: Focus of research debate. *Science* 177, 44–7.
250 Peto, J. (2001) Cancer epidemiology in the last century and the next decade. *Nature* 411, 390–5.
251 Hanahan, D., and Weinberg, R. A. (2000) The hallmarks of cancer. *Cell* 100, 57–70.
252 Look, A. (1997) Oncogenic transcription factors in the human acute leukaemias. *Science* 278, 1059–64.
253 Marx, J. (2003) Mutant stem cells may seed cancer. *Science* 301, 1308–10.
254 Campisi, J. (2000) Cancer, aging and cellular senescence. *In Vivo* 14, 183–8.
255 Artandi, S. E., and DePinho, R. A. (2000) A critical role for telomeres in suppressing and facilitating carcinogenesis. *Curr. Opin. Genet. Dev.* 10, 39–46.
256 Nordstrom, D. K. (2002) Public health. Worldwide occurrences of arsenic in ground water. *Science* 296, 2143–5.
257 Ames, B. N. (1979) Identifying environmental chemicals causing mutations and cancer. *Science* 204, 587–93.

258 Ohshima, H., and Bartsch, H. (1994) Chronic infections and inflammatory processes as cancer risk factors: Possible role of nitric oxide in carcinogenesis. *Mutat. Res.* **305**, 253–64.

259 Iversen, L. (2001) *Drugs: A very short introduction.* Oxford University Press.

260 Weinberg, R. (1997) *Racing to the start of the road: The search for the origin of cancer.* Bantam.

261 Peto, R., et al. (1992) Mortality from tobacco in developed countries: Indirect estimation from national vital statistics. *Lancet* **339**, 1268–78.

262 Kmietowicz, Z. (2001) Tobacco company claims that smokers help the economy. *BMJ* **323**, 126.

263 Scotto, J., and Bailar, J. (1969) Rigoni-Stern and medical statistics: A nineteenth-century approach to cancer research. *J. Hist. Med. Allied Sci.* **24**, 65–75.

264 Pierce, D. A., et al. (1996) Studies of the mortality of atomic bomb survivors. Report 12, Part I. Cancer: 1950–1990. *Radiat. Res.* **146**, 1–27.

265 Obituary. (2002) Dr. Alice Stewart. *New York Times.*

266 Goldman, M. (1996) Cancer risk of low-level exposure. *Science* **271**, 1821–2.

267 Coussens, L. M., and Werb, Z. (2002) Inflammation and cancer. *Nature* **420**, 860–7.

268 Gee, D., and Greenberg, M. (2002) Asbestos: From 'magic' to malevolent mineral. In *The Precautionary Principle in the 20th Century: Late Lessons from Early Warnings*, ed. Poul Harremoës et al. European Environment Agency, pp. 49–63.

269 Ferber, D. (2002) Virology: Monkey virus link to cancer grows stronger. *Science* **296**, 1012–15.

270 Ames, B. N. (1999) Cancer prevention and diet: Help from single nucleotide polymorphisms. *Proc. Natl. Acad. Sci. USA* **96**, 12216–8.

271 Blacklock, C. J., et al. (2001) Salicylic acid in the serum of subjects not taking aspirin: Comparison of salicylic acid concentrations in the serum of vegetarians, non-vegetarians, and patients taking low dose aspirin. *J. Clin. Pathol.* **54**, 553–5.

272 Doll, R., and Armstrong, B. (1980) Cancer. In *Western Diseases, Their Emergence and Prevention*, ed. H. Trowell and D. Burkitt. Harvard University Press, pp. 93–112.

273 Leaf, C. (2004) Why we're losing the war on cancer and how to win. *Fortune*, March 22, 42–9.

274 Drancourt, M., and Raoult, D. (2002) Molecular insights into the history of plague. *Microbes Infect.* **4**, 105–9.

275 Halleyday, S. (1999) *The great stink of London: Sir Joseph Bazalgette and the cleansing of the Victorian metropolis.* Sutton Publishing.

276 Daniel, T. M. (2000) The origins and precolonial epidemiology of tuberculosis in the Americas: Can we figure them out? *Int. J. Tuberc. Lung Dis.* **4**, 395–400.

277 Baxby, D. (1996) Two hundred years of vaccination. *Curr. Biol.* **6**, 769–72.

278 Baker, J. P. (2000) Immunization and the American way: Four childhood vaccines. *Am. J. Public Health* **90**, 199–207.

279 Vitek, C. R., and Wharton, M. (1998) Diphtheria in the former Soviet Union: Reemergence of a pandemic disease. *Emerg. Infect. Dis.* **4**, 539–50.

280 Robertson, S. E., et al. (1996) Yellow fever: A decade of re-emergence. *JAMA* **276**, 1157–62.

281 Meldrum, M. (1998) "A calculated risk": The Salk polio vaccine field trials of 1954. *BMJ* **317**, 1233–6.

282 Blume, S., and Geesink, I. (2000) A brief history of polio vaccines. *Science* **288**, 1593–4.

283 Nathanson, N., and Fine, P. (2002) Virology: Poliomyelitis eradication – a dangerous endgame. *Science* **296**, 269–70.

284 Weiss, R. A. (2001) Animal origins of human infectious disease (Leeuwenhoek Lecture 2001). *Philos. Trans. R. Soc. Lond. B. Biol. Sci.* **356**, 957–77.

285 Ferber, D. (2002) Public health: Creeping consensus on SV40 and polio vaccine. *Science* **298**, 725–7.

286 Behr, M. A. (2001) Comparative genomics of BCG vaccines. *Tuberculosis (Edinb.)* **81**, 165–8.

287 Weiner, D. B., and Kennedy, R. C. (1999) Genetic vaccines. *Sci. Am.* **281**, 50–7.

288 Madsen, K. M., et al. (2002) A population-based study of measles, mumps, and rubella vaccination and autism. *N. Engl. J. Med.* **347**, 1477–82.

289 Loudon, I. (1987) Puerperal fever, the streptococcus, and the sulphonamides, 1911–1945. *BMJ* **295**, 485–490.

290 Levy, S. B. (1998) The challenge of antibiotic resistance. *Sci. Am.* **278**, 46–53.

291 Ferber, D. (2003) WHO advises kicking the livestock antibiotic habit. *Science* **301**, 1027.

292 Crisostomo, M. I., et al. (2001) The evolution of methicillin resistance in *Staphylococcus aureus*: Similarity of genetic backgrounds in historically early methicillin-susceptible and -resistant isolates and contemporary epidemic clones. *Proc. Natl. Acad. Sci. USA* **98**, 9865–70.

293 Ferber, D. (2002) Livestock feed ban preserves drug's power. *Science* **295**, 27–8.

294 Clarke, T. (2003) Drug companies snub antibiotics as pipeline threatens to run dry. *Nature* **425**, 225.

295 Reid, S. D., et al. (2000) Parallel evolution of virulence in pathogenic Escherichia coli. *Nature* **406**, 64–7.

296 Eisen, J. A. (2001) Gastrogenomics. *Nature* **409**, 463, 465–6.

297 Pupo, G. M., Lan, R., and Reeves, P. R. (2000) Multiple independent origins of *Shigella* clones of *Escherichia coli* and convergent evolution of many of their characteristics. *Proc. Natl. Acad. Sci. USA* **97**, 10567–72.

298 Starr, D. (1999) *Blood*. Warner.

299 Defrancesco, L. (2002) US in grip of West Nile Virus. *Nature Medicine* **8**, 1051.

300 Cyranoski, D., and Abbott, A. (2003) Virus detectives seek source of SARS in China's wild animals. *Nature* **423**, 467.

301 Balter, M. (2000) Emerging diseases: On the trail of Ebola and Marburg viruses. *Science* **290**, 923–5.

302 Berche, P. (2001) The threat of smallpox and bioterrorism. *Trends Microbiol.* 9, 15–18.

303 Cosby, A. (1989) *America's forgotten pandemic.* Cambridge University Press.

304 Taubenberger, J. K., Reid, A. H., and Fanning, T. G. (2000) The 1918 influenza virus: A killer comes into view. *Virology* 274, 241–5.

305 Webster, R. (2001) A molecular whodunit. *Science* 293, 1773–5.

306 Shortridge, K. F. (1999) The 1918 'Spanish' flu: Pearls from swine? *Nat. Med.* 5, 384–5.

307 Kilbourne, E. D. (1997) In pursuit of influenza: Fort Monmouth to Valhalla (and back). *Bioessays* 19, 641–50.

308 Prusiner, S., Montagnier, L., and Gallo, R. (2002) Discovering the cause of AIDS. *Science* 298, 1726–31.

309 Hillis, D. (2000) Origins of HIV. *Science* 288, 1757–8.

310 Marx, P., Alcabes, P., and Drucker, E. (2001) Serial human passage of simian immunodeficiency virus by unsterile injections and the emergence of epidemic HIV in Africa. *Phil. Trans. Roy. Soc.* 356, 911–20.

311 Hooper, E. (2000) *The river: A journey back to the source of HIV and AIDS.* Penguin.

312 Weiss, R. A. (2001) Polio vaccines exonerated. *Nature* 410, 1035–6.

313 Richman, D. D. (2001) HIV chemotherapy. *Nature* 410, 995–1001.

314 Weller, I. V., and Williams, I. G. (2001) ABC of AIDS: Antiretroviral drugs. *BMJ* 322, 1410–12.

315 Collinge, J. (1999) Variant Creutzfeldt-Jakob disease. *Lancet* 354, 317–23.

316 Zobeley, E., et al. (1999) Infectivity of scrapie prions bound to a stainless steel surface. *Mol. Med.* 5, 240–3.

317 Sun, M. (1985) Gene-spliced hormone for growth approved. *Science* 230, 523.

318 Ghani, A. C., Ferguson, N. M., Donnelly, C. A., and Anderson, R. M. (2003) Factors determining the pattern of the variant Creutzfeldt-Jakob disease (vCJD) epidemic in the UK. *Proc. R. Soc. Lond. B. Biol. Sci.* 270, 689–98.

319 Mead, S., Stumpf, M., and Whitfield, J. (2003) Balancing selection at the prion protein gene consistent with prehistoric kurulike epidemics. *Science* 300, 640–3.

320 Carter, R., and Mendis, K. N. (2002) Evolutionary and historical aspects of the burden of malaria. *Clin. Microbiol. Rev.* 15, 564–94.

321 Spielman, A., and D'Antonio, M. (2001) *Mosquito: The story of man's deadliest foe.* Faber and Faber.

322 Gubler, D. J. (2004) Cities spawn epidemic dengue viruses. *Nature, Medicine* 10, 129–30.

323 Packard, R. M., and Gadehla, P. (1997) A land filled with mosquitoes: Fred L. Soper, the Rockefeller Foundation, and the anopheles gambiae invasion of Brazil. *Med. Anthropol.* 17, 215–38.

324 Zajtchuk, R. (1997) *Textbook of military medicine: Medical aspects of chemical and biological warfare.* http://www.vnh.org/MedAspChemBioWar/chapters.

325   Meselson, M., et al. (1994) The Sverdlovsk anthrax outbreak of 1979. *Science*
      266, 1202–8.
326   Henderson, D. A. (1999) The looming threat of bioterrorism. *Science* **283**,
      1279–82.
327   Check, E. (2002) Poliovirus advance sparks fears of data curbs. *Nature* **418**,
      265.
328   Sneador, W. (1985) *Drug discovery: The evolution of modern medicines.*
      Wiley.
329   Tulp, M., and Bohlin, L. (2002) Functional versus chemical diversity: Is
      biodiversity important for drug discovery? *Trends Pharmacol. Sci.* **23**,
      225–31.
330   Plunkett, M. J., and Ellman, J. A. (1997) Combinatorial chemistry and new
      drugs. *Sci. Am.* **276**, 68–73.
331   Moellering, R. C., Jr. (1999) A novel antimicrobial agent joins the battle
      against resistant bacteria. *Ann. Intern. Med.* **130**, 155–7.
332   Haseltine, W. A. (2001) Beyond chicken soup. *Sci. Am.* **285**, 56–63.
333   Fabbro, D., et al. (2002) Protein kinases as targets for anticancer agents: From
      inhibitors to useful drugs. *Pharmacol. Ther.* **93**, 79–98.
334   Pfeffer, N. (1993) *The stork and the syringe: A political history of reproductive
      medicine.* Polity Press.
335   Pilcher, H. (2003) The ups and downs of lithium. *Nature* **425**, 118–20.
336   Healy, D. (2003) *The creation of psychopharmacology.* Harvard University
      Press.
337   Bodenheimer, T. (2000) Uneasy alliance – clinical investigators and the phar-
      maceutical industry. *N. Engl. J. Med.* **342**, 1539–44.
338   Dyer, O. (2001) University accused of violating academic freedom to safe-
      guard funding from drug companies. *BMJ* **323**, 591.
339   Safer, D. J. (2000) Are stimulants overprescribed for youths with ADHD?
      Ann. Clin. Psychiatry **12**, 55–62.
340   Vane, J. (1994) Towards a better aspirin. *Nature* **367**, 215–16.
341   Editorial. (2003) How a statin might destroy a drug company. *The Lancet*
      361, 793.
342   Henry, D., and Lexchin, J. (2002) The pharmaceutical industry as a medicines
      provider. *The Lancet* **16**, 1590–5.
343   Alper, J. (2003) Biotech thinking comes to academic medical centres. *Science*
      299, 1303–5.
344   Goeddel, D. V., and Levinson, A. D. (2000) Robert A. Swanson (1947–99).
      Nature **403**, 264.
345   Hall, S. (1988) *Invisible frontiers.* Atlantic.
346   Vajo, Z., Fawcett, J., and Duckworth, W. C. (2001) Recombinant DNA
      technology in the treatment of diabetes: Insulin analogs. *Endocr. Rev.* **22**,
      706–17.
347   Barinaga, M. (2000) Asilomar revisited: Lessons for today? *Science* **287**,
      1584–5.
348   Kornberg, A. (1995) *The golden helix: Inside biotech ventures.* University
      Science Books.

349 Dalton, R., and Schiermeier, Q. (1999) Genentech pays $200m over growth hormone 'theft'. *Nature* 402, 335.
350 Maeder, T. (2003) The orphan drug backlash. *Sci. Am.* 288, 70–7.
351 Lahteenmaki, R., and Fletcher, L. (2002) Public biotechnology 2001 – the numbers. *Nat. Biotechnol.* 20, 551–5.
352 Agres, T. (2003) Licenses worth a billion. *The Scientist* 17, 55.
353 Ziomek, C. (1998) Commercialisation of proteins produced in the mammary gland. *Theriogeniology* 49, 139.
354 Steinberg, F. M., and Raso, J. (1998) Biotech pharmaceuticals and biotherapy: An overview. *J. Pharm. Pharm. Sci.* 1, 48–59.
355 Ezzell, C. (2001) Magic bullets fly again. *Sci. Am.* 285, 34–41.
356 Stevens, D. L., et al. (1996) Group A streptococcal bacteremia: The role of tumor necrosis factor in shock and organ failure. *J. Infect. Dis.* 173, 619–26.
357 Inoue, N., Takeuchi, M., Ohashi, H., and Suzuki, T. (1995) The production of recombinant human erythropoietin. *Biotechnol. Annu. Rev.* 1, 297–313.
358 Pfeffer, L., et al. (1998) Biological properties of recombinant alpha-interferons: 40th anniversary of the discovery of interferons. *Cancer Res.* 58, 2489–99.
359 Gutterman, J. (1994) Cytokine therapeutics: Lessons from interferon alpha. *Proc. Natl. Acad. Sci.* 91, 1198–1205.
360 Brown, J. B. (1986) Gonadotrophins. In *Infertility: Male and Female*, ed. V. Insler and B. Lunenfield. Churchill Livingston, pp. 359–96.
361 Laffan, M., and Tuddenham, E. (1998) Science, medicine, and the future: Assessing thrombotic risk. *BMJ* 317, 520–3.
362 DeFrancesco, L. (2001) First sepsis drug nears market. *Nature* 7, 516–17.
363 Lu, Y., et al. (2001) Recombinant vascular endothelial growth factor secreted from tissue-engineered bioartificial muscles promotes localized angiogenesis. *Circulation* 104, 594–9.
364 Griffioen, A. W., and Molema, G. (2000) Angiogenesis: Potentials for pharmacologic intervention in the treatment of cancer, cardiovascular diseases, and chronic inflammation. *Pharmacol. Rev.* 52, 237–68.
365 Li, R. H., and Wozney, J. M. (2001) Delivering on the promise of bone morphogenetic proteins. *Trends Biotechnol.* 19, 255–65.
366 Teitelbaum, S. L., and Ross, F. P. (2003) Genetic regulation of osteoclast development and function. *Nat. Rev. Genet.* 4, 638–49.
367 Chiurel, M. (2000) Whatever happened to leptin? *Nature* 404, 538–40.
368 Mattson, M. P. (2001) Lose weight STAT: CNTF tops leptin. *TINS* 24, 313–14.
369 Tomalin, C. (2002) *Samuel Pepys: The unequalled self.* Penguin Viking.
370 Hess, J., and Schmidt, P. (1984) The first blood banker: Oswald Hope Robertson. *Transfusion* 24, 404–7.
371 Obituary. (2000) Rene Favaloro. *The Times* (London), July 31.
372 Green, H. (1991) Cultured cells for the treatment of disease. *Sci. Am.* 265, 96–102.

373 Petit-Zeman, S. (2001) Regenerative medicine. *Nature Biotechnology* **19**, 201–6.

374 Rodan, G. M., TJ. (2000) Therapeutic approaches to bone diseases. *Science* **289**, 1508.

375 Lindahl, A., Brittberg, M., and Peterson, L. (2003) Cartilage repair with chondrocytes: Clinical and cellular aspects. *Novartis Found. Symp.* **249**, 175–86.

376 Oberpenning, F., Meng, J., Yoo, J. J., and Atala, A. (1999) *De novo* reconstitution of a functional mammalian urinary bladder by tissue engineering. *Nat. Biotechnol.* **17**, 149–55.

377 Griffith, L. G., and Naughton, G. (2002) Tissue engineering – current challenges and expanding opportunities. *Science* **295**, 1009–14.

378 Dynes, M. (2003) Heart transplant secret of the hospital gardiner. *The Times* (London), April 26.

379 Dyer, C. (1999) English teenager given heart transplant against her will. *BMJ* **319**, 209.

380 Mayes, T. (2003) Ailing tycoon buys Pakistan girl's kidney. *The Sunday Times* (London), March 23.

381 Mufson, S. (2001) Chinese doctor tells of organ removals after executions. *Washington Post*, June 27, p. A1.

382 Platt, J. L. (2000) Xenotransplantation: New risks, new gains. *Nature* **407**, 27, 29–30.

383 Buhler, L., et al. (2001) Transplantation of islets of Langerhans: New developments. *Swiss Med. Wkly.* **131**, 671–80.

384 Lumelsky, N., et al. (2001) Differentiation of embryonic stem cells to insulin-secreting structures similar to pancreatic islets. *Science* **292**, 1389–94.

385 Good, R. A., Gatti, R. A., Hong, R., and Meuwissen, H. J. (1969) Graft treatment of immunological deficiency. *Lancet* **1**, 1162.

386 Good, R. (1971) Immunodeficiency in developmental perspective. *The Harvey Lectures*, series 67, 1–107.

387 Parkman, R. (1986) The application of bone marrow transplantation to the treatment of genetic diseases. *Science* **232**, 1373–8.

388 Gluckman, E., et al. (1989) Hematopoietic reconstitution in a patient with Fanconi's anemia by means of umbilical-cord blood from an HLA-identical sibling. *N. Engl. J. Med.* **321**, 1174–8.

389 Barinaga, M. (2000) Fetal neuron grafts pave the way for stem cell therapies. *Science* **287**, 1421–2.

390 Li, Y., Field, P., and Raisman, G. (1997) Repair of adult rat corticospinal tract by transplants of olfactory ensheathing cells. *Science* **277**, 2000–2.

391 Thomson, J. (1998) Embryonic stem cell lines derived from human blastocysts. *Science* **282**, 1145–7.

392 Marshall, E. (2000) The business of stem cells. *Science* **287**, 1419–21.

393 Aldous, P. (2001) Can they rebuild this? *Nature* **410**, 622–5.

394 Manouvrier-Hanu, S., Holder-Espinasse, M., and Lyonnet, S. (1999) Genetics of limb anomalies in humans. *TIGS* **15**, 409–17.

395 Eldadah, Z. A., et al. (2001) Familial tetralogy of fallot caused by mutation in the *jagged* gene. *Hum. Mol. Genet.* 10, 163–9.

396 Rosner, F. (1969) Hemophilia in the Talmud and rabbinic writings. *Ann. Intern. Med.* 70, 833–7.

397 Roth, E. F., Raventos-Suarez, C., Rinaldi, A., and Nagel, R. L. (1983) Glucose-6-phosphate dehydrogenase deficiency inhibits in vitro growth of *Plasmodium falciparum*. *Proc. Natl. Acad. Sci. USA* 80, 298–9.

398 Chakravarti, A., and Chakraborty, R. (1978) Elevated frequency of Tay-Sachs disease among Ashkenazic Jews unlikely by genetic drift alone. *Am. J. Hum. Genet.* 30, 256–61.

399 Hogenauer, C., et al. (2000) Active intestinal chloride secretion in human carriers of cystic fibrosis mutations: An evaluation of the hypothesis that heterozygotes have subnormal active intestinal chloride secretion. *Am. J. Hum. Genet.* 67, 1422–7.

400 Kuper, A. (2003) Incest, cousin marriage, and the origin of the human sciences in the nineteenth century. *Past and Present Society* 174, 155–83.

401 Gibbons, A. (1993) The risks of inbreeding. *Science* 259, 1252.

402 Cao, A., and Rosatelli, M. C. (1993) Screening and prenatal diagnosis of the haemoglobinopathies. *Baillieres Clin. Haematol.* 6, 263–86.

403 Modell, B., and Darr, A. (2002) Genetic counselling and customary consanguineous marriage. *Nature, Genetics* 3, 225–9.

404 Kaback, M. M. (2001) Screening and prevention in Tay-Sachs disease: Origins, update, and impact. *Adv. Genet.* 44, 253–65.

405 Ekstein, J., and Katzenstein, H. (2001) The Dor Yeshorim story: Community-based carrier screening for Tay-Sachs disease. *Adv. Genet.* 44, 297–310.

406 Handyside, A. H., Kontogianni, E. H., Hardy, K., and Winston, R. M. (1990) Pregnancies from biopsied human preimplantation embryos sexed by Y-specific DNA amplification. *Nature* 344, 768–70.

407 Rifkin, J. (1998) *The biotech century.* Victor Gollanz.

408 Editorial. (1994) China's misconception of eugenics. *Nature* 367, 3.

409 Kumar, S. (2003) Ratio of girls to boys in India continues to decline. *BMJ* 327, 1007.

410 Crockin, S. (2001) Adam Nash: Legally speaking, a happy ending or slippery slope? *Reprod. Biomed. Online* 2, 6–7.

411 Levy, N. (2002) Deafness, culture and choice. *J. Med. Ethics* 28, 284–5.

412 Shakespeare, T. (2003) How perfect do we want our babies to be? *The Sunday Times* (London), November 23.

413 Dyer, C. (2000) Appeal court bars sterilisation of woman with learning disabilities. *The Guardian*, May 19.

414 Papadopoulos, G. K., Wijmenga, C., and Koning, F. (2001) Interplay between genetics and the environment in the development of celiac disease: Perspectives for a healthy life. *J. Clin. Invest.* 108, 1261–6.

415 Cohen, F. E., and Kelly, J. W. (2003) Therapeutic approaches to protein misfolding diseases. *Nature* 426, 405–9.

416 Anderson, W. F. (1992) Human gene therapy. *Science* 256, 808–13.

417 Williams, D., and Baum, C. (2003) Medicine: Gene therapy – new challenges ahead. *Science* 302, 400–1.

418 Ferber, D. (2001) Gene therapy: Safer and virus-free? *Science* 294, 1638–42.

419 Stock, G., and Campbell, J. (2000) *Engineering the human germline: An exploration of the science and ethics of altering the genes we pass to our children.* Oxford University Press.

420 Lindee, M. (2003) Watson's world. *Science* 300, 432–4.

421 McLaren, A. (1998) Correspondence. *Nature* 392, 645.

422 Nielsen, T. (1997) Human germline therapy. *McGill Journal of Medicine* 3, 126–32.

423 Peltonen, L., and McKusick, V. A. (2001) Genomics and medicine: Dissecting human disease in the postgenomic era. *Science* 291, 1224–9.

424 Evans, W. E., and Relling, M. V. (1999) Pharmacogenomics: Translating functional genomics into rational therapeutics. *Science* 286, 487–91.

425 Wolf, C., Smith, G., and Smith, R. L. (2000) Science, medicine, and the future: Pharmacogenetics. *BMJ* 320, 987.

426 Mossialos, E., and Dixon, A. (2001) Genetic testing and insurance: Opportunities and challenges for society. *Trends Mol. Med.* 7, 323–4.

427 Ridley, M. (1999) *Genome: The autobiography of a species in 23 chapters.* Fourth Estate.

428 Appleyard, B. (1999) *Brave new worlds.* HarperCollins.

429 Charatan, F. (2001) US spending on prescription drugs rose by 19% in 2000. *BMJ* 322, 1198.

430 Gottlieb, S. (2002) Congress criticises drugs industry for misleading advertising. *BMJ* 325, 1379.

431 Binswanger, H. (2003) HIV/AIDS treatment for millions. *Science* 292, 221.

432 Cohen, J. (2003) Companies, donor's pledge to close gap in AIDS treatment. *Science* 289, 368.

433 Caplan, A. (2004) Is biomedical research too dangerous to pursue? *Science* 303, 1142.

Index

Definitions of technical words are on the page corresponding to the first entry in the index.